AKADEMIE DER WISSENSCHAFTEN DER DDR

Veröffentlichungen des Forschungsbereichs Geo- und Kosmoswissenschaften

Herausgegeben von Heinz Kautzleben

HEFT 18

Middle Atmosphere of Venus

Edited by Klaus Schäfer and Dietrich Spänkuch

With 163 Figures and 19 Tables

AKADEMIE-VERLAG BERLIN

1990

Editors:

Prof. Dr. sc. nat. Klaus Schäfer
Akademie der Wissenschaften der DDR
Heinrich-Hertz-Institut für Atmosphärenforschung
und Geomagnetismus
DDR-1199 Berlin
German Democratic Republic

Dr. sc. nat. Dietrich Spänkuch
Meteorologischer Dienst der DDR
Meteorologisches Hauptobservatorium
DDR-1561 Potsdam
German Democratic Republic

ISBN 3-05-500591-0
ISSN 0138-4600

Erschienen im Akademie-Verlag Berlin,
Leipziger Str. 3—4, Berlin, DDR-1086

Lizenznummer: 202 · 100/90
Lektor: Dipl.-Met. Heide Deutscher
Printed in the German Democratic Republic
Gesamtherstellung: VEB Druckhaus „Maxim Gorki", Altenburg, 7400
LSV 1495
Bestellnummer: 7638896 (2156/18)

Preface

During more than 20 years of planetary research by means of space techniques have led to a rapid growth of knowledge about our solar system. The research of Earth-like planets is of high interest and has direct relevance to the geophysics and meteorology of our own planet. Comparative studies of planets are a suitable tool for investigations of the development of the Earth and the understanding of its present state. As with all planets of our solar system considerable knowledge about Venus has been gathered. The Venus International Reference Atmosphere documents the present knowledge of its mean atmospheric state. More advanced models as well as a deeper understanding of the physical and chemical processes that take place within the Venusian atmosphere need more reliable data of its atmospheric phenomena, their detailed structure and variability. The infrared Fourier spectrometer measurements aboard Venera-15 added some new results after the Pioneer Venus mission. Based on the experiences gained by the development of space-borne Fourier spectrometers on Soviet weather satellites in the seventies, the first deep space experiment of the GDR, jointly made with the USSR, was aimed at measuring radiance spectra of Venus' outgoing longwave radiation from 250 to 1600 cm^{-1} (6.25 to 40 μm) with a spectral resolution of 5 to 10 cm^{-1}. For a two-month period, i.e. from October 12 to December 14, 1983, more than 1500 spectra of high quality were measured. These spectra have been interpreted as part of an Intercosmos activity in close cooperation between the Academy of Sciences of the USSR, the Academy of Sciences of the GDR, and the Meteorological Service of the GDR. The first part of this book summarizes the methodology and basic results of the infrared Fourier spectrometer experiment of Venera-15. These results were published in Russian in the Soviet journal "Kosmicheskie Issledovanija" Volume 23, Number 2, pp. 191–258 already in 1985. The second part contains the papers of the International Symposium "Middle Atmosphere and Clouds of Venus" organized by the Heinrich-Hertz-Institute of Atmospheric Research and Geomagnetism of the Academy of Sciences of the GDR from June 22 to 26, 1987 in Potsdam, GDR. Both editors, actively, engaged in the Venera-15 Fourier spectrometer mission, were members of the scientific programm committee of this symposium.

We wish to thank Mrs. D. Kopsch for her great support in translation from Russian in English, Mrs. I. Winkler and Mrs. M. Wack for drawing figures, as well Mrs. E. Pöschel and Mrs. Schlunke for typing manuscripts.

K. Schäfer D. Spänkuch

Contents

1. Infrared experiment aboard the automatic interplanetary stations Venera 15 and Venera 16

1. Methodics and first results

D. Oertel, V. I. Moroz, J. Nopirakowski, V. M. Linkin, H. Jahn, R. S. Kremnjov, H. Becker-Ross, W. Stadthaus, V. V. Kerzhanovich, I. A. Macygorin, A. V. D'jachkov, L. I. Khljustova, W. Bärwald, M. Uhlig, H. Driescher, W. Skrbek, H. Studemund, R. Schuster, G. Kaiser, S. P. Ignatova, I. A. Zelenov, I. D. Cerenin, D. Spänkuch, W. Döhler, K. Schäfer, L. V. Zasova, E. A. Ustinov, G. Fellberg, A. N. Lipatov, A. A. Shurupov, N. G. Khavenson

Abstract

Spectra of the planet Venus were measured in the 250—1600 cm^{-1} (6—40 μm) range with a resolution of approximately 5 cm^{-1} by means of infrared Fourier spectrometers aboard Venera 15 and Venera 16. The measurements were made mainly at nothern hemispheric latitudes with $\varphi \geqq 20°$. The field of view of the instrument made it possible to discern near the pericentre ($\varphi \simeq 62°$) on the planet regions with an extension up to 60 km and spacing between two registrations of about 200 km. The construction of the instrument, its characteristics and its orbital program are described. There are characteristic patterns of spectra and profiles of brightness temperature along the route of measurements. Some of their peculiarities are mentioned: different character of spectra in polar and middle latitudes, existence of hot spots, absence of the so-called thermal asymmetry.

The Venusian infrared spectrum is an important source of information on the chemical composition and the physical characteristics of its atmosphere. The radiation spectrum with wave lengths $\lambda > 4\ \mu m$ is of special interest because it contains the strongest absorption bands of atmospheric gases to which belong, first of all, the ν_2-CO_2-fundamental band at 667 cm^{-1} ($\lambda = 15\ \mu m$). In the centre of this band the level with the optical thickness $\tau = 1$ is at about 90—95 km altitude. In the "continuous" spectrum (i.e. outside the absorption bands) the level $\tau = 1$ is nearly at 60 km altitude (in the region of wave lengths 20—25 μm). Therefore, knowing the radiation spectrum in the fundamental CO_2 band, one can construct the profile of atmospheric temperature in a relatively great altitude range.

Further, other components — H_2O, SO_2, H_2SO_4 bands — are expected in the Venusian spectrum. The measurements of planetary infrared spectra aboard artificial satellites are of special interest since high spectral and spatial resolution can be gained simultaneously and thus latitudinal and local-temporal variations can be investigated. For the planet Venus such an experiment was at first carried out aboard the Soviet automatic interplanetary stations Venera 15 and Venera 16 launched in 1983.

By means of a Fourier spectrometer aboard these satellites the spectrum of the planet was recorded in the range of 250 to 1600 cm^{-1} (6—40 μm) with a spectral resolution of 5 to 7.5 cm^{-1}. The field of view of the instrument made it possible to identify areas of the dimension up to 60 km.

Previous investigations of the Venusian infrared spectrum can be divided into two groups: Earth based and satellite based investigations. A resolution of 0.25 cm^{-1} [1] was reached in the Earth based spectroscopic measurements, but radiation was taken from the whole disk of the planet. Satellite infrared measurements of Venus had previously been carried out only by means of radiometers with one (Mariner 5 [2], Mariner 10 [3], Venera 9 and Venera 10 [4]) or several filters (Pioneer-Venus [5]). The combination of a sufficiently high spectral and spatial resolution has firstly been gained in the Venera 15 and Venera 16 experiment. It offers — as it will be shown in the following — basically new possibilities.

1. Fourier spectrometer FS 1/4

The most essential restrictions which determined the choice of the instrument can be defined in the following manner:

1) Venusian brightness temperatures (in the range to be investigated) are in the interval 160—250 K; 2) the spectrum must be determined in a few seconds with a signal to noise ratio in the order of 100; 3) under real experimental conditions only a non-cooled receiver can be used. In modern infrared spectrometry only one instrument type can guarantee the solution of the task considering these restrictions — the Fourier spectrometer; it has a great throughout spectral radiance and guarantees a multiplexity of spectral measurements [6, 7].

The main part of the Fourier spectrometer is the optical interferometer — an element which is very sensitive to mechanical and temperature influences. A sufficiently reliable construction of a satellite Fourier spectrometer is a complicated technical task, and therefore these instruments are only sporadically used in cosmic experiments.

For the first time, Fourier spectroscopy aboard a spacecraft has been acquired by the group of R. Hanel (spectra of the Earth [8], Mars [9], and in the past time Jupiter [10], Saturn [11], Uranian system [12]). 3 satellite Fourier spectrometers were developed and built in the GDR in 1971—1979 under the direction of V. Kempe. These instruments were used aboard Meteor satellites [13] within the Intercosmos program for investigations of the spectrum of the outgoing radiation of the Earth.

The Venera 15 and Venera 16 experiment (1983) was carried out according to the Intercosmos program, too. The two instruments (Fourier spectrometers FS 1/4) were developed and implemented by several scientific institutes and facilities of the Academy of Sciences of the GDR. Soviet specialists elaborated the requirements on the Fourier spectrometers FS 1/4 from the viewpoint of scientific tasks and coupling with the spacecraft, tested it within the satellite configuration and carried out information reception and processing. Interpretation is made by scientists of both countries. A first presentation of experimental results is given in [14]. In the following we give a more detailed description of methodics, instrument and observation data. In the next 4 papers we give their preliminary interpretation.

The blockdiagram of the instrument FS 1/4 is shown in Fig. 1, and its most essential parameters are the following:

spectral range, cm^{-1}	280–1600[1])
	400–1600[2])
spectral resolution: nonapodized cm^{-1}	2.5
apodized fast Fourier transformation mode, cm^{-1}	5.0
rectangular field of view	$4° \times 4°$
interferogram recording time, s	5.5
full length of one working cycle, s	20
noise-equivalent radiance, $W/cm^2 \times sr \times cm^{-1}$	2×10^{-8}

[1]) The Venera 15 instrument has a CsJ beam splitter.
[2]) The Venera 16 instrument has a KBr beam splitter.

The instrument consists of 4 modules — one optical and three electronic ones. All modules are installed outside the hermetical part of the spacecraft on its shadow side. The optical module with the interferometer, the entrance rotating mirror system and the autonomous temperature control system weights 15 kg and is installed on a special console which guarantees a certain orientation of the viewing direction (along the z axis of the automatic interplanetary station). The fixing elements of the optical module to the control ensure the thermal decoupling. To minimize vibrational effects the optical module is installed additionaly on attenuators. The elctronic modules (with a total weight of 15 kg) are installed outside the hermetical part of the automatic interplanetary station but have in difference to the optical module a good thermal contact with the spacecraft.

Two important improvements of the FS 1/4 instruments were made compared with the Meteor-version: 1) a non-hermetical construction of the instrument, 2) installation of a board computer with one of its task the Fourier transformation in real time operation (with following transmission of complete spectra to the Earth).

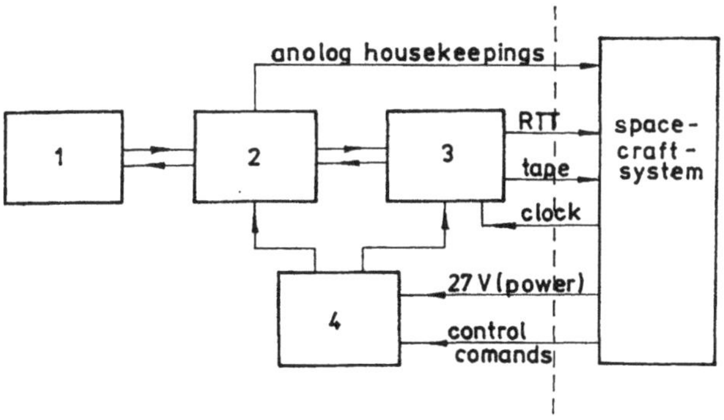

Fig. 1. Blockdiagram of Fourier spectrometer FS 1/4
1 — optical module, 2 — analog-digital-electronics module, 3 — module for microprocessor-control and operative information processing, 4— power unit

2. Interferometer and rotating mirror system

The optical module consists of two basic elements — the interferometer and the rotating mirror system. The diagram of the interferometer is presented in Fig. 2. It consists of the entrance IR-filter 1 which cuts off the short-wave part of the spectrum in the wavelength range $\lambda < 6\ \mu m$, the beam splitter with compensating plate 2, the fixed mirror 3 and the moved mirror 4.

Radiation is divided by the semi-transparent beam-splitter into two beams which are reflected on the plane mirrors 3, 4. After reflection, radiation is directed to the beam-splitter where the beams are joined again (with a loss of half of the intensity). In the united beam a radiation interference takes place. By means of the collimating mirror 5 the radiation interference is focussed on the receiver (bolometer) 6. The beam-splitter of the instrument installed on Venera 15 is made of CsJ, and in the Fourier spectrometer of Venera 16 a KBr-beam-splitter was used. Special multi-layered interference coatings were used to ensure a sufficiently wide spectral range. The bolometer records the interferogram

$$i(\delta) = \int_{-\nu_0}^{\nu_0} a(\nu) \times [I(\nu) - B_R(\nu)] \times \exp\{-j \times [\Phi(\nu) - 2 \times \pi \times \nu \times \delta]\} \times d\nu, \qquad (1)$$

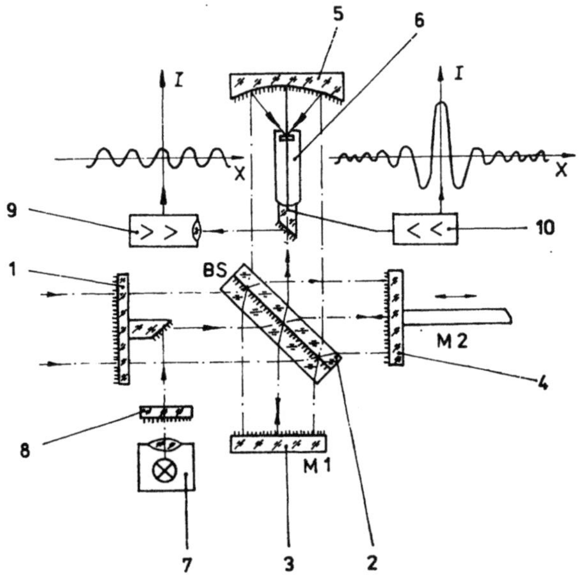

The Michelson-Interferometer (with the central reference channel)

Fig. 2. Interferometer (with reference channel in central part)
1 — entrance IR filter (separating the short wave part), 2 — beam-splitter with compensator, 3 — fixed mirror, 4 — moved mirror, 5 — collimator mirror, 6 — bolometer, 7 — neon source of reference channel, 8 — interference filter for $\lambda = 703$ nm, 9 — receiver and amplifier of reference channel, 10 — amplifier of IR channel signal

where δ — the optical path difference of the beams in the interferometer, ν — the wave number, ν_0 — maximum wave number in the spectrum, $a(\nu)$ — the responsivity of the interferometer, $\Phi(\nu)$ — the phase characteristics of the interferometer, $I(\nu)$ — the radiance of the source, i.e. of input beam, $B_R(\nu)$ — self-radiance of the receiver. In case of a non-cooled receiver, the radiance $B_R(\nu)$ can be greater than the incoming radiance $I(\nu)$. In the FS 1/4 instruments, the receiver temperature T_R was $T_{OM} + 20\,K = 303\,K$, where $T_{OM} = 283\,K$ — is the temperature of the optical module at the fixing point of the bolometer housing maintained constant during the whole experiment by means of an autonomous thermoregulation system.

The moved mirror 4 moves during the working cycle with constant speed, and the samples of the interferogram are taken discretely in identical time intervals. The path difference δ is

$$\delta = 2 \times v_0 \times t, \tag{2}$$

where v_0 — the speed of forward movement of the mirror, t — the time measured from the moment where $\delta = 0$ (i.e. when mirrors 3 and 4 are at the same distance from the beamsplitter).

The signal frequency f of the interferogram, the wave number ν of the spectral component which participates in the signal generation at frequency f, and the speed of the moved mirror v_0 are connected by

$$f = 2 \times \nu \times v_0. \tag{3}$$

The interferogram at the bolometer output enters the IR-channel signal amplifier 10. The interferogram is there upon transformed from analogous into digital form by means of an analog to digital converter (ADC). The following transformation of the interferograms into spectra by means of the discrete fast Fourier transformation requires a spatial uniformity of the interferogram samples. The control pulses for the sampling circuit are separated from the signal of the interferometer reference channel by a trigger. The optical elements of the reference channel are the Ne-discharge lamp 7, the interference filter 8, and the receiver with amplifier 9 shown also in Fig. 2. The reference channel signal is also used in the regulation chain of the forward movement of mirror 4. In the case of a monochromatic source and a linear uniform movement of mirror 4 a sinusoidal signal occurs at the interferometer output, with a frequency determined by relation (3). In the FS 1/4 instruments a reference line with a wave length $\lambda_{Ref} = 703$ nm is used. The spectral resolution of the interferometer with apodization [6, pp. 20—29] is

$$\Delta\nu_{ap} \approx \frac{1}{\delta_{max}}, \tag{4}$$

resulting in a maximum path difference δ_{max} of 2 mm for a required spectral resolution $\Delta\nu_{ap} = 5\ cm^{-1}$. Recording an interferogram with identical number of samples left and right of the maximum, mirror 4 must be moved by $l = 2$ mm on the time $t_c = 5.5$ s. The speed of the mirror 4 is $v_0 = l/t_c = 0.36$ mm/s, the frequency f_R of the sinusoidal signal of the reference channel is, according to equation (3), $f_R = 1024$ Hz. Without going into details of the electronic regulation of the moved mirror drive, we state that basically an enlarged phase-look-system ensures a high stability of the speed v_0 by comparing the reference signal with the signal of quartz generator.

With the maximum wave number of the measured spectral radiance at $\nu_{max} = 1660$ cm^{-1}, the upper limiting frequency of the interferogram signal is $f_{max} = 120$ Hz according to (3). For sampling such a signal, according to the Kotelnikov theorem, a frequency of $f_s \geqq 2 \times f_{max} = 240$ Hz is necessary. Considering the possibility of interpolation after information transmission to Earth [19], the interferogram sampling frequency was chosen at $f_s = 512$ Hz $> 4 \times f_{max}$. The number of samples in one interferogram amounts

$$N_{in} = \frac{2 \times \delta_{max}}{2 \times \lambda_{Ref}} = \frac{l}{\lambda_{Ref}} = 2845. \tag{5}$$

The density of the noise equivalent power of the infrared Fourier spectrometer is determined by the equation [18, 20, 21]:

$$\varrho_{min} = [\eta \times D^* \times \Delta\nu_{ap} \times (F \times \Omega_1 \times \Omega_2 \times t_c)^{1/2}]^{-1} \tag{6}$$

where (in brackets the FS 1/4 parameters are given): η — the optical efficiency (0.1), D^* — the specific detectivity of the bolometer (5×10^8 cm $\times$ Hz$^{1/2} \times W^{-1}$), $\Delta\nu_{ap}$ — the spectral resolution with apodization (5 cm^{-1}), F — the entry aperture (8 cm^2), Ω_1 — the field of view of the spectrometer (5×10^{-3} sr), Ω_2 — the field of view of the receiver (1.0 sr), t_c — the recording time (5.5 s).

With the FS 1/4 parameters we get from formula (6):

$$\varrho_{min} \approx 1 \times 10^{-8} \text{ W/cm}^2 \times \text{cm}^{-1} \times \text{sr}.$$

The transformation of the interferogram into the spectrum is based on the approximation of the complex spectrum (see (1)):

$$S(\nu) = a(\nu) \times [I(\nu) - B_R(\nu)] \times \exp[j \times \Phi(\nu)] = |S(\nu)| \times \exp[j \times \Phi(\nu)] \tag{7}$$

by means of the Fourier series with a finite term number [6, p. 23]:

$$S_N(\nu) = \Delta \times \sum_{n=-N}^{N} \alpha_n \times i_n \times \exp(2 \times \pi \times j \times \nu \times \Delta), \tag{8}$$

where $\Delta = 2 \times \lambda_R$, $\alpha_n = 0.54 + 0.46 \times \cos(\pi \times n/N)$, $N = (N_{in} \times 1/2) = 1422$, n — the actual interferogram sample (with $n = 0$ in the point of the mean maximum of the interferogram at $\delta \approx 0$). One interferogram $i_n(-N \leqq n \leqq N)$ is not sufficient for obtaining information about the absolute spectrum $I(\nu)$. The procedure of obtaining absolute spectra is described in papers [6, 12, 20, 22]. To obtain the absolute spectrum of the planet $I_{pl}(\nu)$, auxiliary radiation measurements of two absolute black bodies with known temperatures are necessary. We call spectra obtained from interferograms of the planet (pl), the reference black body (bb) and the cold cosmos (c) by fast Fourier transformation difference spectra (S_{pl}, S_{bb}, and S_c). These three difference spectra are sufficient for the determination of

a) the responsivity of the instrument

$$a(\nu) = \frac{S_{bb}(\nu) - S_c(\nu)}{B_{bb}(\nu)}, \tag{9}$$

b) the self-radiation of the receiver

$$B_R(\nu) = \frac{S_c(\nu)}{a(\nu)}, \tag{10}$$

c) the absolute radiance spectrum of the planet

$$I_{\mathrm{pl}}(\nu) = \frac{S_{\mathrm{c}}(\nu) - S_{\mathrm{pl}}(\nu)}{a(\nu)} \tag{11}$$

where B_{bb} is the spectral radiance of the black body, calculated according to the Planck function for the measured black body temperature.

A rotating mirror system shown in Fig. 3 turns the field of view of the interferometer into the positions 'planet', 'black body' and 'cosmos'.

Without going into details of this element of FS 1/4 (which is described in detail in [23]), we state that solar sensors in the tubes (18) 'cosmos' and (16) 'planet' avoid the damage of parts of the interferometer due to intense solar radiation by producing a blocking signal: If the sun gets into the cone with a half angle 15° relating to the optical axis of the tubes, the inherent automatic of the FS 1/4 prevents a turn of the field of view of the interferometer in direction to 'planet' or 'cosmos'. In that case the field of view of the interferometer remains in the initial position 'black body', where it is always in the switch-off-mode of the FS-1/4. The design rules of the one board black body is dealt with according to the procedure described in [22]. With the required photometric measuring accuracy of the absolute planetary spectrum $\Delta I \lesssim (2-3) \times 10^{-8}$ W/cm² $\times$ cm^{-1} $\times$ sr and a working temperature of the black body and the interferometer of approximately 283 K it is necessary to guarantee a temperature stability of the bolometer within $T = 303 \pm 0.05$ K even for the time of measurement of the three difference spectra S_{pl}, S_{bb} and S_{c}. Additionally, the required measuring accuracy of the

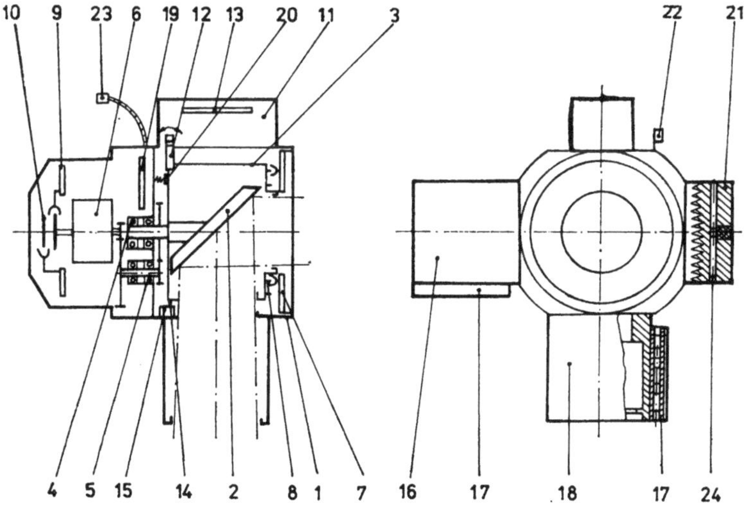

Fig. 3. Rotating mirror system

1 — casing, 2 — rotating mirror, 3 — closing cylinder, 4 — bearing of rotating mirror, 5 — reducer, 6 — step motor, 7 — sensor 'position', 8 — key disk 'position', 9 — position sensor 'nonius', 10 — key disk 'nonius', 11 — blocking unit, 12 — latch, 13 — blocking electronics, 14 — bolt, 15 — 90° support, 16 — tube 'planet', 17 — casing of Sun sensor, 18 — tube 'cosmos', 19 — electronics of Sun sensor, 20— friction body, 21 — black body, 22, 23 — plug in connection, 24 — temperature sensor

black body temperature is $\Delta T_{bb} \leqq 0.03$ K (within $\varepsilon_{bb} \geqq 0.99$). The evaluation of the dynamic range which is necessary for a quantitative representation of interferograms and spectra not distorted by quantization noise allows the following conclusions:

1. Maximum amplitudes in the interferograms occur when radiation emitted by the instrument into the cosmos is recorded, i.e. when $I(\nu) = 0$.
2. For the processing system of the interferograms a dynamic range $\pm 16 \times 10^3$ is necessary. With an automatic decrease of amplifications for 8 times in the maximum range of the interferogram after ad-conversation, digital words of 13 bits for the samples are necessary which 12 bits for the ad-conversation (considering the interferogram sign) and 1 bit for the factor of amplification in the analogous part of the system, which can differ by 8.
3. A dynamic range of 2.5×10^3 is necessary for presentation digital words consisting of 12 bits ($2^{12} = 4096 > 2500$).

The optical module 1 (see Fig. 1), contains besides the interferometer, the moved mirror system, the black body and autonomous thermoregulation system the following electronic elements: amplifiers of the interferometer signals and of the reference channel signals, the fine measurement system of black body and interferometer temperature, the electronic of the rotating mirror system and the drive of the moved mirror of the interferometer, but also the blocking and de-blocking mechanisms of the rotating mirror system and of the drive of the interferometer moved mirror.

The following components are contained in the electronic module 2: A power transverter unit of stand-by or service mode of operation, control circuits of the rotating mirror system and of the drive of the moved mirror of the interferometer, circuit of formation of analogous housekeeping signals entering the telemetric system of the automatic interplanetary station, and a system of analog-digital telemetering. The control and operative processing unit 3 has two identical microprocessor systems (reserving each other) which control the instrument during measurements, perform digital processing of interferograms and housekeeping information. Other tasks are fast Fourier transformation and coding, but also information output to telemetry of the automatic interplanetary station in direct transfer operation with a speed of 8 kbit/s or in recording mode to the memory of the automatic interplanetary station with a speed of 3 kbit/s.

The main power transverter unit contains the input and storage circuits of control, function and numerical commands of the service system of the automatic interplanetary station and the transverter circuits of voltage of the on board supply (27 V) into supply voltage of electronic circuits of FS 1/4, which are working only in measuring mode.

The measuring mode of operation is that state of the instrument when it is completely ready to record planetary radiation. In this stage all units are working, except the so-called 'cold reserves', and the power consumption in that mode is maximum (30—40 W).

During the overfly to Venus and during that part of the orbit where readiness for measurements was not intended the instrument FS 1/4 operated in a service mode. The main task of this mode was to keep the temperature of the optical module in certain limits. The following elements work in the service mode: the autonomous system of thermoregulation, the power transverter unit of the service mode of operation, the command reception and storage circuit, the fine measurement system of black body and interferometer temperature circuits for the formation of analog system house keeping signals. The instrument FS 1/4 consumes 5 to 20 W in the service mode in dependence on the on/off switching ratio of the heating elements of the autonomous system of thermoregulation.

The FS 1/4 instrument is the first spaceborn Fourierspectrometer which performs the transformation of the interferograms $I(\nu)$ into spectra $S(\nu)$ digitally on board in real time increasing, thus, significantly the reliability of information transfer on the Venus-Earth route. The true reconstruction of the spectra from interferograms after their transmission to the Earth is impossible, if random errors occur as a result of insertion or falling out of bits or digital words in form of packets (i.e. several errors take place successively in the transmitted interferogram samples) [7]. From the other side, if such errors occur in spectra transformed on board and than transmitted to the Earth, the loss is much smaller: only some spectral elements fall out, and all other elements of the spectrum can be used for scientific interpretation.

In the on board fast Fourier transformation, 2048 out of 2845 samples of the interferograms are used, where two successive samples regarding redundance ($f_s > 4 \times f_{max}$, see above) are combined into one. Thus, one can confine in fast Fourier transformation to an amount of 1024 points.

Each of the two microprocessors can do this operation with a 16-bit processing width in 4 s. In such a way a small symmetric "cutting" of the interferograms leads to a spectral resolution in the 'on board spectrum' of about 7 cm^{-1} instead of 5 cm^{-1}, in case of apodization of interferograms are transmitted to Earth and the Fourier transformation is made by Earth based electronic data processing. The following different modes of information output of each FS 1/4 instrument can be selected by means of control commands:

1. Mode of direct transfer 1: A data frame is generated in the FS 1/4 during 20 s for direct transfer to the Earth which consists of a sequence of two identical interferograms and two identical spectra ('on board') with a bitrate of 8 kBit/s.
2. Mode of direct transfer 2: A data frame is formed during 20 s consisting of a sequence of three identical interferograms with a bitrate of 8 kBit/s.
3. Memory mode 1: A data sequence from FS 1/4 is during 20 s put out into the memory of the automatic interplanetary station which consists of two identical 'on board' spectra (bit rate $\approx$ 3 kBit/s).
4. Memory mode 2: A data sequence consisting of one interferogram is put out (bit rate $\approx$3 kBit/s) during 20 s.

Thus, the maximum redundance which can be used in primary ground processing is contained in the data frames of mode direct transfer 1, and the minimum one — in memory mode 2. In dependence on the special conditions of the so called 'radiovisibility' all modes had been used in the experiment.

3. Measurements on the orbit

The orbits of the artificial satellites had been corrected several times. They are similar for both satellites. For example the orbital elements of Venera 15 before the first correction are given. The pericenter lies above the northern hemisphere (latitude $\varphi = 62°$). The optical axis of the 'planet' tube has a fixed space orientation. This orientation is such that the tube 'planet' is nearly north-south directed. Measurements begin approximately 10 min before intersection of the limb and consist of 20-second cycles. During one measurement cycle spectra of the planet, black body and cosmos are recorded in the one or the other sequence. For obtaining a maximum quantity of planetary spectra a special mode can be used in which the rotating mirror is directed by turns to the 'planet' tube and to the black body in the first 10 minutes, and then 25 min only to the 'planet' tube. In the last 10 min after intersection of the 2nd limb again radiation of the 'planet' and black body tube is recorded by turns. During the first and the last 10 min the 'planet' tube is directed into the open cosmos. The geometry of sounding is explained by the scheme on Fig. 4.

The first cycle of measurements was carried out soon after the turn of Venera 15 in the orbit — on October 12, 1983. 20 high-quality spectra were obtained by means of the onboard fast Fourier transformation. On October 14 already the special mode for obtaining maximum number of planetary spectra was switched on, and 57 spectra were recorded at the coordinates indicated on Fig. 5. They comprise latitudes from approximately 20 to 87° both on the night and on the day side of the planet. 49 measuring cycles had been carried out aboard Venera 15 in the period from October 12 till December 14, 1983, 47 of them with standard orientation of the instrument and 2 with a turn of the viewing direction by 90° to north-south.

No measurements were made after December 14, 1983. Venera 16 was mainly used for radiolocation survey of the planet's surface restricting, thus, the possibilities of its

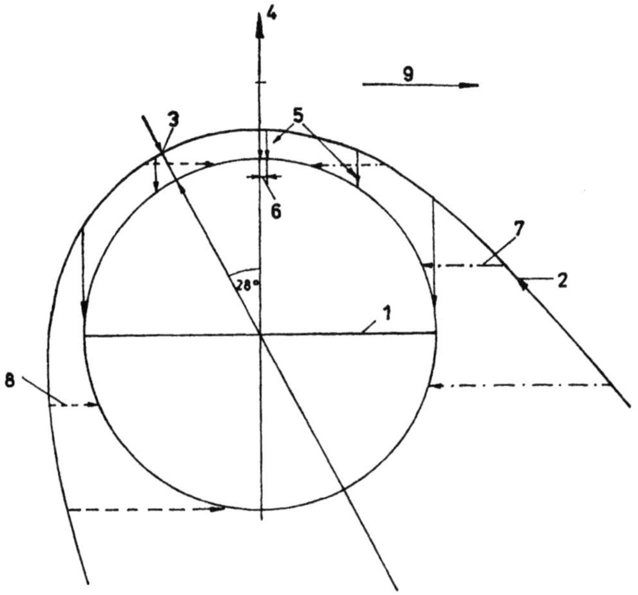

Fig. 4. Diagram of geometry of the Fourier spectrometer sounding aboard spacecrafts Venera 15 and Venera 16 in projection on the orbit plane

1 — Venus equator, 2 — direction of movement of space-crafts along the orbit, 3 — pericenter point, 4 — direction to north pole of ecliptic, 5 — viewing direction with standard orientation of spacecrafts, 6 — minimum distance between measurements: ~200 km, 7 and 8 — viewing direction (other orientation of spacecraft) with 90° turn, 9 — direction to the Earth at beginning of experiment

application for the infrared experiment with only 10 cycles altogether. Altogether more than 1500 spectra of high quality were obtained between latitudes of −66° to 87° and local solar time from 4 to 10 h (morning) and from 16 to 22 h (evening). 9 cycles were carried out in direct transfer mode with a spectral resolution of 5 cm^{-1} with Earth based processing of interferograms, the other in memory mode with onboard processing and a spectral resolution of 7.5 cm^{-1}.

4. Primary Earth based processing of information

The simplified block diagram of primary processing of FS 1/4 information on the Earth is represented in Fig. 6. The primary processing contains all calculation operations necessary for obtaining abolute spectra of Venusian radiation with spatial-temporal co-ordination. From this diagram it can be seen that after correction of 'raw' information at the level of the bit data stream 7 and separation of the house keeping information 8 of the scheme of Fig. 6 information processing is clearly subdivided into two branches:

- achieving absolute spectra of the interferograms,
- achieving absolute spectra of the 'onboard' difference spectra.

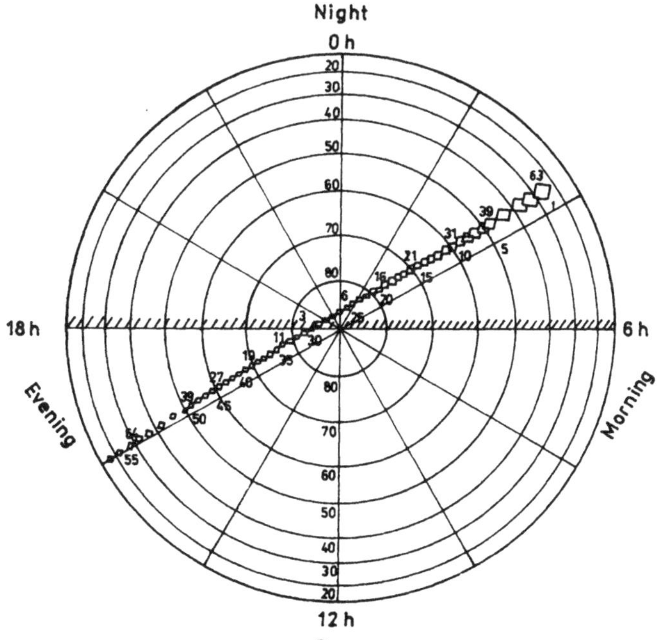

Fig. 5. Example of a measurement route. The spots give the local positions of the radiation spectra obtained with a Fourier spectrometer of Venera 15 in the measuring series on October 14, 1983. The dimensions of the rectangles approximately correspond to the dimensions of the real field of view. The figure is given in normal projection of the northern hemisphere, the position of the terminator is shown. Along the outer circle local solar time is given. Numbers along the route are the following: from below: number of spectra, from above: zenith distances of the viewing direction

Correction of raw data and house keeping information are necessary for the qualitative calibration (e.g., temperature parameter of black body, position of the moved mirror system in the given measurement etc.) and for spatial-temporal coordination of the Venusian spectra.

The spectra obtained from interferograms on Earth have a wave number step of $\sigma_{\nu_1} = 1/4096 \times \Delta = 1.736\ \text{cm}^{-1}$ between the spectral elements. The onboard processed spectra have $\sigma_{\nu_2} = 2 \times \sigma_{\nu_1} = 3.472\ \text{cm}^{-1}$. Their comparison is allowed, but only after the calibration procedure since the amplitude of the difference spectra in the short-wave part of the spectrum drops by 20% due to the combination of two interrogation points into one before the onboard fast Fourier transformation.

5. Examples of infrared spectra

Fig. 7 presents three of the 57 spectra obtained by Venera 15 on October 14. The most prominent feature is the CO_2 absorption band at 667 cm^{-1} (15 μm). In fact, in this region there are many CO_2 bands, but the greatest contribution is supplied by absorption at the basic frequency ν_2, which corresponds to the bending vibration of CO_2 molecules. The optical thickness $\tau = 1$ in the center of this band is in quasi nadir geometry at the pressure level of about 0.5 hPa, i.e. in an altitude of approximately 90 km. Along the 15 μm band profile the levels with optical thickness $\tau = 1$ are from about 60 to 90 km altitude.

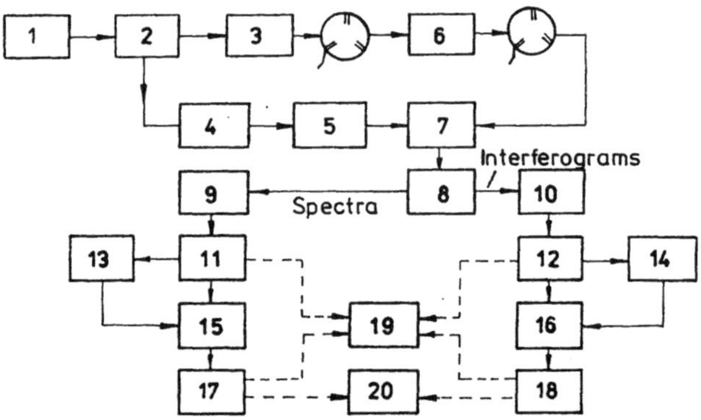

Fig. 6. Simplified block diagram of primary processing of FS 1/4 information on the Earth

1 — receiver and demodulation (in the Deep Space Communication Link Center), 2 — separation of FS 1/4 information out of the general data stream and intermediate recording, 3 — recording of information of several cycles on magnetic tape, 4 — transmission of express information per phone channel to the Institute of Space Research of the Academy of Sciences of the USSR, 5 — recording of express information, quick-look, 6 — transport of the magnetic tapes from the Deep Space Communication Link Center to the Institute of Space Research, 7 — decoding, detecting and marking of errors, correction on bit level, 8 — separation of house keeping information, 9 — interpolation of onboard spectra, testing, 10 — interpolation of interferograms, testing, fast Fourier transformation, 11, 12 — obtaining of spectra (module and phase), recording and output to the plotter 19, 13, 14 — records of difference spectra of black body and cosmos, 15, 16 — calibration, i.e. obtaining absolute spectra of Venus, output to plotter, 17, 18 — records of Venusian absolute spectra, 19 — plotter, 20 — recording on magnetic tape

The curves in Fig. 7 show a series of other, weaker CO_2 bands, namely of 'hot' bands (544, 791, 961 and 1064 cm^{-1}) and of 'isotopic' bands, which belong to $^{12}C\,^{16}O\,^{18}O$ (1259 and 1366 cm^{-1}).

For the first time, lines of the H_2O rotational band — they occur in the region 280 to 475 cm^{-1} — are recorded in the Venusian thermal radiation spectrum. Possibly, the fundamental $\nu_3 SO_2$ band contributes to the absorption near 1360 cm^{-1}. In addition to the gas absorption bands one can identify bands of sulphuric acid (diffuse features near 450 and 900 cm^{-1}) in the obtained spectra.

Fig. 7 contains also radiance densities measured in the infrared experiment aboard the satellite Pioneer-Venus. They coincide well with our spectra obtained in the same latitudes. However, the number of spectral channels on Pioneer-Venus is limited, and it is clear that our spectra contain significantly more information.

The analysis of the spectra obtained on October 14 gives at least three characteristic groups which correspond to different regions of the planet demonstrated in Fig. 7 by one individual example for each group.

They are completed by latitudinal profiles of brightness temperature, represented in Fig. 8 for some spectral channels. In the following we will denote the brightness temperatures for wave number ν (cm^{-1}) with the symbol $T_B(\nu)$.

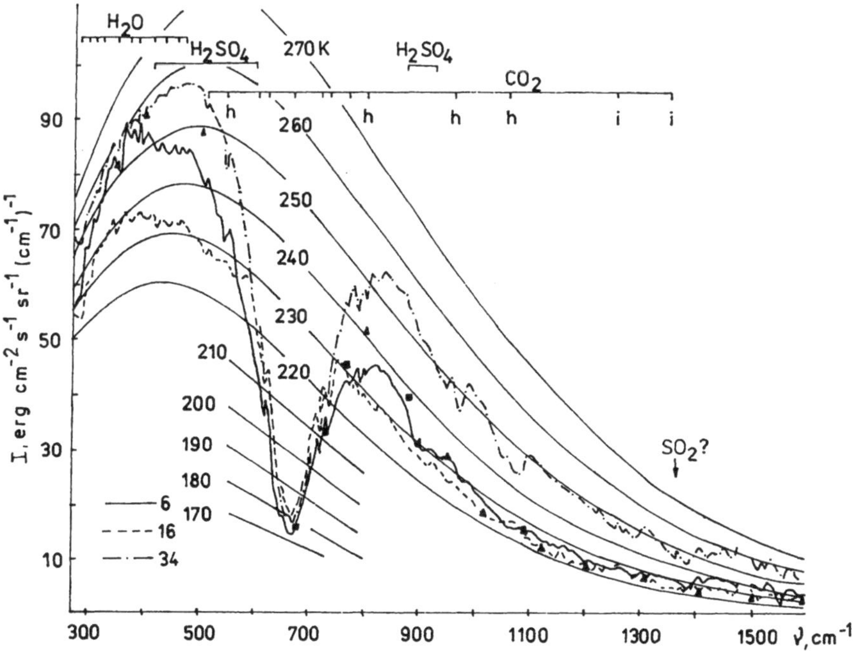

Fig. 7. Characteristic patterns of spectra obtained on October 14
No. 6 ($\varphi = 52°$) — example of a spectrum of first group (latitudes less than 60°), No. 16 ($\varphi = 70°$) — of second group (latitudes 70—75°, night side), No. 34 ($\varphi = 79°$) of third group ("hot") region (latitudes 77—82°, day). The position of absorption bands of gases (CO_2, H_2O, SO_2) and of aerosol (water solution H_2SO_4) are shown. Squares — radiance densities, measured in the infrared experiment aboard spacercraft Pioneer-Venus (orbit 33, $\varphi = 50°$), see Fig. 6 in [5]). Triangles — synthetic radiance spectrum of aerosol, obtained in modelling spectra of first group (see text). Letters *h* denote the hot bands CO_2, *i* — bands of isotope molecules $^{12}C^{16}O^{18}O$

Group 1 — these are spectra obtained at latitudes $\varphi \lesssim 60°$ both on the night and on the day side — is characterized by a flat shape of the ν_2CO_2 band wings, by a high brightness temperature T_B (365) ≈ 260 K, a high temperature difference T_B (365) — T_B (870) ≈ 30 K and by distinct H_2SO_4 bands. Radiation of clouds in these spectra is obviously formed in the altitude region with a maximum temperature gradient.

Group 2 — these spectra are obtained in the "cold" zone at latitudes 70—75° at night (cf. Fig. 8) — is characterized by a steep turn in the wings of the ν_2CO_2 band, by a low brightness temperature T_B (365) ≈ 240 K, small difference T_B (365) — T_B (870) < 10 K, failing or even inversion of the H_2SO_4 absorption band at 900 cm^{-1} and a small depth of the band at 450 cm^{-1}. It can be supposed that in this case the radiation is formed in the isotherm or inversion zone of the temperature profile.

Group 3 — these are spectra obtained in the 'hot' zone at latitudes 77—82° on the day side (see Fig. 8) — shows a smooth shape of the wings, a maximum depth of the hot bands at 961 and 1064 cm^{-1}, a low depth of the H_2SO_4 band. Obviously, the radiation of the cloud layer is in this case formed in the zone with a distinct temperature gradient, and the upper boundary of the cloud layer is very deep.

A more detailed classification can be done, of course. From the brightness temperatures $T_B(\nu)$ of Fig. 8, the two wave numbers 365 and 872 cm^{-1} represent spectral regions free or nearly free of gaseous absorption, whereas the two others (667 and 626 cm^{-1}) are

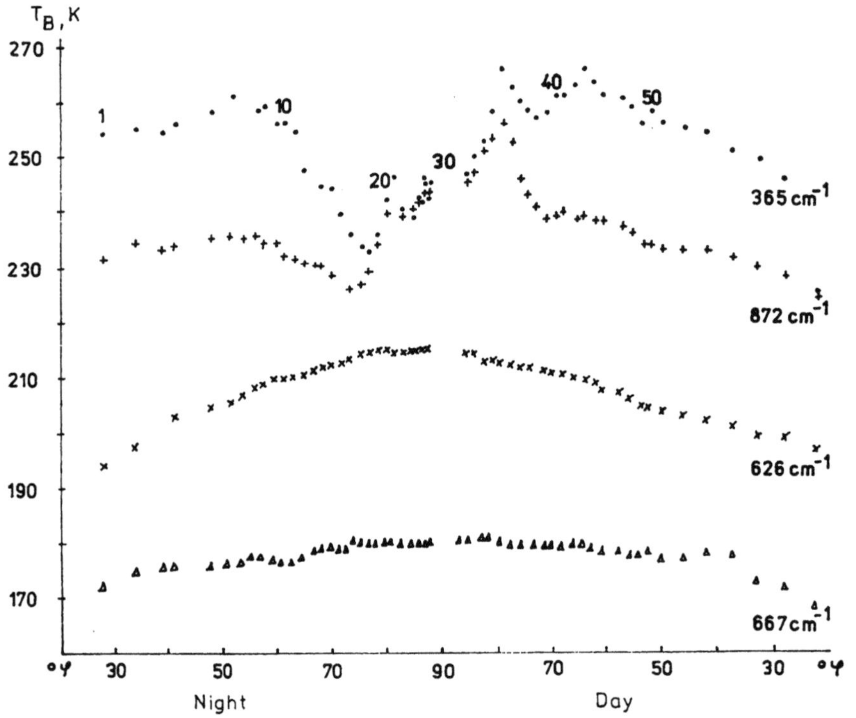

Fig. 8. Meridional profiles of brightness temperatures in selected regions of 57 spectra, obtained on October 14

On the abscissa axis — latitudes, on the ordinate axis — brightness temperatures T_B. Digits above the profile of $\nu = 365$ cm^{-1} — number of spectra within the ν_2CO_2 band. The behaviour of these two curve pairs is quite different. The curves T_B (667) and T_B(626) are smooth, the brightness temperatures increase continuously with increasing latitude. At these wave numbers the radiation originates above the clouds at levels of about 0.5 and 10 hPa. The curves T_B (365) and T_B (872) are far from each other at latitudes $\varphi \lesssim 60°$, approach quickly towards the pole, form a minimum at 75° latitude, and are nearly identical at latitudes 77—87° (on the night side) with continuous increase in T_B. This increase continues on the dayside reaching a maximum at the latitude of about 80°. Then the curves quickly diverge. A comparison with maps

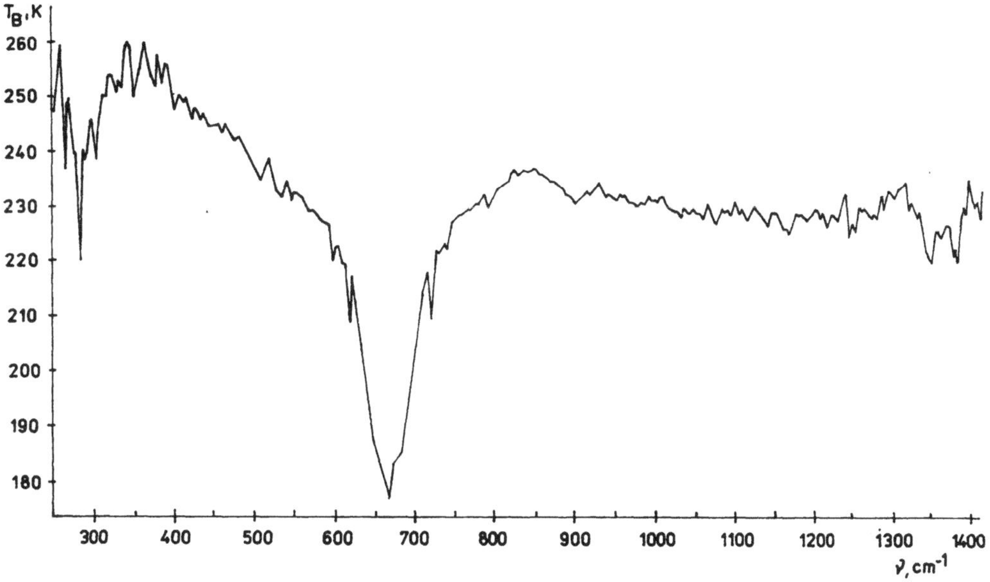

Fig. 9. One of the spectra obtained after Earth based processing of the interferogram. Venera 15, direct transfer mode, November 17, 1983

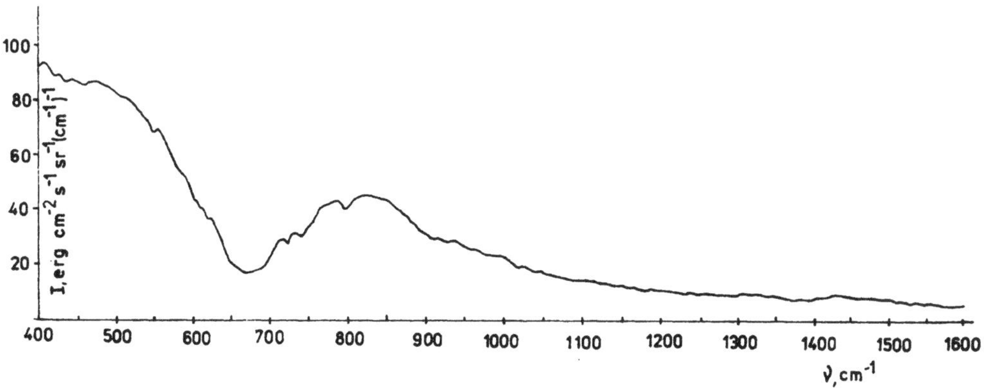

Fig. 10. One of the spectra obtained in memory mode aboard Venera 16 with onboard Fourier transformation (October 16, 1983)

obtained by the Pioneer-Venus infrared experiment in the 872 cm^{-1} channel show that the minimum at 75° corresponds obviously to the intersection of the so called 'cold collar', and the maximum the passage through the 'warm dipole'. We note that in some of the Venera-15-measurements T_B-curves with 2 maxima were obtained, i.e. the hot region near the pole has not only kept its position approximately but also its shape in the course of 5 years having passed since the Pioneer-Venus measurements.

The integrated radiation of the spectral range covered from a fast Fourier transformation is about 20% higher in the 'dipole region' than in middle latitudes.

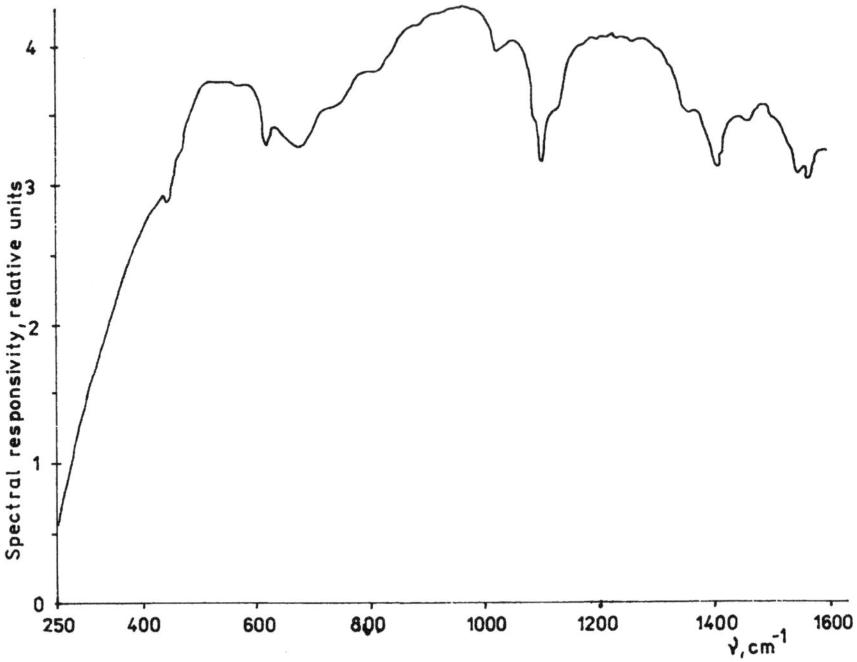

Fig. 11. Curve of spectral responsivity of the instrument installed aboard Venera 15 obtained after processing of measurement data from the orbit in October 1983

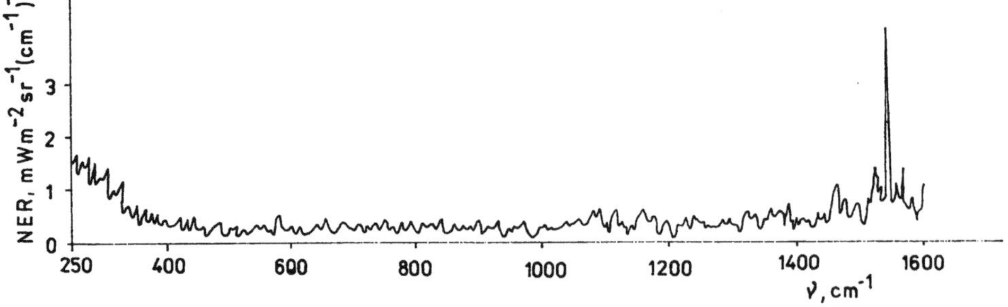

Fig. 12. Dependence of noise equivalent radiation on the wave number for the instrument, installed aboard Venera 15, obtained after processing of measurement data from the orbit in October 1983

Fig. 9 presents a spectrum processed on Earth. Such spectra have a higher spectral resolution and comprise a somewhat wider spectral region — their real long-wave border is approximately at 40 μm (250 cm^{-1}). In Fig. 10 one of the spectra obtained with Venera 16 is given. The long-wave border for these spectra is displaced in direction to smaller wave lengths, it is approximately at 25 μm (400 cm^{-1}). The quality of the spectra was analyzed directly on the basis of the measurements by means of curves of spectral responsivity and spectral noise power density. Fig. 11 shows the responsivity curve of the FS 1/4 instrument aboard the automatic interplanetary station Venera 15, obtained from onboard difference spectra S_{bb} ('black body'), and S_c ('cosmos') during measurements in October 1983. The best quality measure for Fourier spectrometer measurements is the curve of spectral density of the noise power, or the noise equivalent radiation, determined by the equation

$$\text{NER} = \frac{\Delta a \times B_{bb}}{\bar{a} \times \sqrt{2}} \tag{12}$$

and obtained from onboard measurements, where $\Delta a = |a - \bar{a}|$ is the dispersion of the spectral responsivity and $a(\nu)$ is given by formula (9). To obtain $\bar{a}$, 15—50 responsivity curves of different pairs S_{bb} and S_c are averaged, with identical temperature conditions in the instrument. Fig. 12 shows the noise equivalent radiation of FS 1/4, i.e. the estimation of the real mean quadratic random error of the measured absolute planetary spectra [6] aboard the automatic interplanetary station Venera 15 obtained during the experiment from 12 pairs of S_{bb} and S_c. This error is in the limits $(2-3) \times 10^{-8}$ W/cm^2 $\times$ cm^{-1} $\times$ sr in the wave number interval 500—1000 cm^{-1} and does not exceed 1×10^{-7} W/cm^2 $\times$ cm^{-1} $\times$ sr for 300—500 cm^{-1} and 1000—1500 cm^{-1}. The burst at 1540 cm^{-1} is caused by systematic additive interference from an unknown source which influences the electronic part of the IR channel with a frequency of ~110 Hz. The real random error in the centre of the measuring range is 2—3 times greater than the theoretical value $\varrho_{min} = 1 \times 10^{-8}$ W/cm^2 $\times$ cm^{-1} $\times$ sr by formula (6). This discrepancy can be explained by the fact, that the given theoretical estimation for ϱ_{min} does not consider temperature fluctuations in the instrument during the time between subsequent measurements, and that the quality of S_{bb} and S_c pairs, used for the calculation of NER, is at the lower boundary of the recommended value. The spectral responsivity and the spectral noise density did not vary during the experiment.

The spectra obtained in the present experiment are used for new data on the chemical composition of the atmosphere at altitudes 60—70 km [17], on its temperature above the clouds [15], on the composition of the upper cloud layer [16], and on thermal fluxes of the outgoing radiation [18]. It is suggested to carry out investigations on dynamic atmospheric phenomena based on the data analysis of atmospheric temperature and cloud morphology.

References

[1] Kunde, V. G., Hanel, R. A., Herath, L. W.: High spectral resolution ground based observation of Venus in the 450—1250 cm^{-1}. Icarus **32** (1977) 2, 210.

[2] Chase, S. C., Kaplan, L. D., Neugebauer, G.: The Mariner 2 infrared radiometer experiment. J. Geophys. Res. **68** (1963), 6157.

[3] Chase, S. C., Miner, E. D., Morrison, D. et al.: Preliminary infrared radiometry of Venus from Mariner 10. Science **183** (1974), 1291.

[4] Ksanfomaliti, L. V., Dedova, E. V., Obukhova, L. F. et al.: Infrakrasnoe izluchenie oblakov Venery. Kosmich. Issled. **14** (1976) 5, 768.

[5] Taylor, F. W., Beer, R., Chahine, M. T. et al.: Structure and meteorology of the middle atmosphere of Venus: infrared remote sensing from the Pioneer Orbiter. J. Geophys. Res. **85** (1980) A 13, 7963.

[6] Bell, R.: Vvedenie v fur'e spektroskopiju/perevod s angl. G. N. Zhizhina. Mir, Moskau, 1975.

[7] Oertel, D.: Sputnikovye infrakrasnye fur'e-spektrometry dlja issledovanija planet. Kosmicheskoe priborostroenie. Nauka, Moskau 1982, 153.

[8] Hanel, R. A. and Conrath, B.: Preliminary results from the interferometer experiments on Nimbus III. Science **165** (1969), 1258.

[9] Hanel, R., Conrath, B., Hovis, W., Kunde, V.: Investigation of the Martian environment by infrared spectroscopy on Mariner 9. Icarus **17** (1972), 2, 423.

[10] Hanel, R., Conrath, B., Flasar, F. et al.: Infrared observation of the Jovian system from Voyager 1. Science **204** (1979), 972.

[11] Hanel, R., Conrath, B., Flasar, F. et al.: Infrared observations of the Saturnian system with Voyager 1: preliminary results. Science **212** (1981), 192.

[12] Hanel, R., Conrath, B., Flasar, F. M. et al.: Infrared observations of the Uranian System. Science **233** (1986), 70–74.

[13] Distancionnoe zondirovanie atmosfery so sputnika "Meteor"/pod. red. L. A. Nakhomova. Gidrometeoizdat, Leningrad 1979.

[14] Oertel, D., Moroz, V. K., Linkin, V. M. et al.: "Venera-15" i "Venera-16": Pervye rezul'taty ehksperimenta po infrakrasnojj spektrometrii. Pis'ma v Astron. Zh. **10** (1984) 4, 243.

[15] Spänkuch, D., Zasova, L. V., Schäfer, K. et al.: Infrared experiment aboard the automatic interplanetary stations Venera 15 and Venera 16. 2. Preliminary results of temperature profile retrieval. Veröffentlichungen des Forschungsbereichs Geo- und Kosmoswissenschaften, Heft 18, 1990.

[16] Zasova, L. V., Spänkuch, D., Moroz, V. I. et al.: Infrared experiment aboard the automatic interplanetary stations Venera 15 and Venera 16. 3. Some conclusions on the structure of clouds, based on the analysis of spectra. Veröffentlichungen des Forschungsbereichs Geo- und Kosmoswissenschaften, Heft 18, 1990.

[17] Moroz, V. I., Döhler, W., Ustinov, E. A. et al.: Infrared experiment aboard the automatic interplanetary stations Venera 15 and Venera 16. 4. Preliminary results of spectral analysis in the region of H_2O and SO_2 absorption bands. Veröffentlichungen des Forschungsbereichs Geo- und Kosmoswissenschaften, Heft 18, 1990.

[18] Linkin, V. M., Schäfer, K., Macygorin, I. A. et al.: Infrared experiment aboard the automatic interplanetary stations Venera 15 and Venera 16. 5. Preliminary results of the analysis of fields of brightness temperatures and thermal fluxes. Veröffentlichungen des Forschungsbereichs Geo- und Kosmoswissenschaften, Heft 18, 1990.

[19] Jahn, H.: Zur Rekonstruktion von Abtastwerten. Zeitschrift für elektrische Informations- und Energietechnik, Leipzig **7** (1977) 3, 280.

[20] Hanel, R. et al.: Nimbus-4-Michelson-interferometer. Appl. Optics **10** (1971), 1376.

[21] Röseler, A., Wieczorek, L. W.: Optische Dimensionierung eines IR-Satelliten-Fourier-Spektrometers. Experimentelle Technik der Physik **26** (1978) 4, 368.

[22] Oertel, D.: Metodicheskie predlozhenija po spektrometrirovaniju i specificheskie trebovanija po meteorologicheskomu obespecheniju s'emok poverkhnosti Zemli i okeana v teplovojj IK-oblasti. Preprint IKI AN SSSR, Moskau (1980) 566.

[23] Becker-Ross, H., Stadthaus, W.: Aufbau des Optikblockes des Infrarot-Fourier-Spektrometers FS-1/4. Feingerätetechnik **33** (1984) 8, 341.

2. Infrared experiment aboard the automatic interplanetary stations Venera 15 and Venera 16

2. Preliminary results of temperature profile retrieval

D. Spänkuch, L. V. Zasova, K. Schäfer, E. A. Ustinov, J. Güldner, V. I. Moroz, W. Döhler, V. M. Linkin, R. Dubois, V. D'jachkov, H. Becker-Ross, A. N. Lipatov, W. Stadthaus, I. A. Macygorin, D. Oertel, V. V. Kerzhanovich, J. Nopirakowski, H. Jahn, G. Fellberg, R. Schuster, A. A. Shurupov

Abstract

The Venusian infrared radiance spectra obtained aboard Venera 15 and Venera 16 were used for temperature retrieval in the Venusian atmosphere in the altitude range 60—95 km. The profiles show that the temperature diminishes monotonously with rising altitude within the latitude range $\pm 60°$. At higher latitudes, the atmosphere is warmer, and generally an inversion or isothermy is found at the pressure range $10 \lesssim p \lesssim 100$ hPa. Significant diurnal temperature variations have not been observed.

In the Fourier spectrometer infrared experiment aboard the spacecrafts Venera 15 and Venera 16 about 1500 Venusian spectra were received in the range 6—40 μm [1]. One of the important information of analyzing the obtained spectra are temperature profiles of the Venusian atmosphere at altitudes of approximately 60 to 95 km. Preceding investigations on the structure of the atmosphere in these altitudes can be subdivided into three groups:

a) measurements of descend probes of spacecrafts in the atmosphere (Venera 8 [2], Venera 9, 10 [3], Venera 11, 12 [4], Venera 13, 14 [5], Pioneer-Venus Probes [6]).
b) measurements of radiosignal parameters on Earth during the occultation phase of spacecrafts (Mariner 5 [7], Mariner 10 [8], Venera 9, 10 [9], Pioneer-Venus Orbiter [10]).
c) intensity measurements in the 15 μm CO_2 band by means of filters (Pioneer-Venus Orbiter [11]).

The results of these investigations together with data for other altitudes are generalized in some models of the atmosphere [12—16]. Nevertheless, more detailed measurements are necessary, because the results of investigations a), b) were obtained by soundings in a limited number of points over the planet and at very different times in most cases. Only the spectral analysis in the 15 μm CO_2 band is able to guarantee simultaneous measurements of the atmospheric structure in a great latitude range.

The first series of those measurements were carried out aboard the Pioneer-Venus Orbiter [11], but the number of spectral channels within the band was small — altogether 4. In our case, a greater number of channels is being used for analyzing, so that more detailed information on temperature profiles can be obtained. In principle, the fact that the whole spectrum is recorded and not some single points ('spectral channels') is in the case of Venus very important because only this makes it possible to consider the influence of clouds in planetary radiation exactly.

1. Physical bases of retrieval approaches

The radiance density for a semi-infinite pure absorbing medium is in dependence on wave number ν determined by the well known solution of the transport equation

$$I_\nu = \sec\theta \times \int_0^\infty B_\nu[T(z)] \times \varkappa_\nu[p(z), T(z)] \times \exp\left\{-\sec\theta \times \int_0^\infty \varkappa_\nu[p(z'), T(z')] \times \mathrm{d}z'\right\} \times \mathrm{d}z, \quad (1)$$

where z — altitude, $T(z)$ and $p(z)$ — temperature and pressure at altitude z, $B_\nu(T)$ — Planck function, $\varkappa_\nu(p, T)$ — volume absorption coefficient, θ — angle between normal and outgoing radiance direction.

It is convenient to use instead of altitude z pressure p — both parameters are connected by the hydrostatic equilibrium equation,

$$\frac{\mathrm{d}p}{p} = -\frac{\mathrm{d}z}{H}, \quad (2)$$

where H — altitude scale and to introduce the atmospheric transmission function from the top of the atmosphere to the pressure level p

$$t_\nu = \exp\left[\sec\theta \times \int_{\ln p}^{-\infty} H \times \varkappa_\nu(p', T) \times \mathrm{d}\ln p'\right]. \quad (3)$$

Then, (1) can be rewritten in shorter form

$$I_\nu = \int_{-\infty}^{\infty} B_\nu[T(p)] \times K_\nu(p, T) \times \mathrm{d}\lg p, \quad (4)$$

where

$$K_\nu = -\frac{\partial t_\nu}{\partial \lg p}. \quad (5)$$

This quantity is called weighting function. It determines the contribution of the given atmospheric level p to the density of outgoing radiation at the given wave number ν.

The task is the determination (retrieval) of function $T(p)$ or, what is the same, $T(z)$ for the measured spectrum $I(\nu)$. One of the necessary conditions for its solution is the knowledge of volume absorption coefficients in dependence on pressure and temperature $\varkappa_\nu(p, T)$ or on transmission function t_ν. The absorption coefficient is the sum of two quantities

$$\varkappa_\nu = \varkappa_{\nu,\mathrm{g}} + \varkappa_{\nu,\mathrm{a}}, \quad (6)$$

where $\varkappa_{\nu,g}$ — gas absorption, $\varkappa_{\nu,a}$ — aerosol absorption. In our case, $\varkappa_{\nu,g}$ is determined for the spectrometer working range mainly by CO_2, the main component of the atmosphere. Its absorption spectrum is sufficiently known, and the retrieval task does not contain principal difficulties if the CO_2 band wings, where the quantity $\varkappa_{\nu,a}$ becomes comparable with $\varkappa_{\nu,g}$, are not used. In the course of moving to the edges of the band, aerosol contributes more and more to radiation. Consequently, it is necessary to consider cloud extinction in t_ν for retrieval of the temperature profile in its lower part. In our case this information can on principle be obtained by the measurements themselves since we get the cloud spectrum outside the CO_2 absorption bands in a sufficient great region.

2. Selection of spectral ranges

We will now consider the selection of wave numbers ('spectral channels') used for retrieving tasks. The spectra of the present experiment give the radiance density with a wave number step of 3.472 or 1.736 cm^{-1} for a spectral response function width (level 0.5) of 7.5 and 5 cm^{-1} respectively [1]. The 15 μm band covers about 150 cm^{-1}, and at first sight it seems to be possible to use up to 100 wave numbers for temperature retrieval. But this is not purposeful since the wave numbers to be selected must guarantee independent information about atmospheric temperature. However, the weighting functions t_ν are 'blurred' in the altitude, and they are, as a rule, similar for a wave number step of the order of 5 cm^{-1} (see below). There are quantitative criteria for the selection of the optimum number of spectral wave numbers, which guarantee maximum retrieval accuracy (see, e.g. [17]). These criteria take into consideration measuring errors but also errors of data to be used (t_ν and retrieval algorithm). Selecting the position of the 'channels' within the 15 μm CO_2 band it is necessary to avoid overlapping with bands belonging to other gases. For instance, because of overlapping with the rotational lines of the H_2O band in sounding the Earth's atmosphere it is not recommended to use the long-wave wing of 15 μm CO_2 band.

Tabelle

Selected wave numbers in the 15 μm CO_2 band for retrieval of temperature profiles

Wave numbers cm^{-1}	pressure level for maximum weighting function hPa	wave numbers cm^{-1}	pressure level for maximum weighting function hPa
555.63	200	677.17	1
572.99	100	691.06	2
586.88	45	701.48	5
600.77	20	711.90	10
614.66	12	732.73	16
632.02	5	753.57	45
642.44	2	770.93	90
656.33	1	802.18	200
666.75	0.5		

In our case this limitation does not occur since the H_2O mixing ratio in the Venusian atmosphere is small, and we can, consequently, use both wings for retrieving. The short-wave wing, however, should be preferred because of smaller influence of aerosol extinction. The table gives a specification of selected spectral wave numbers, and their position are also given in Fig. 1.

3. Transmission function t_ν and weighting functions K_ν for standard models

For the solution of the integral equation (4) it is necessary to know the wighting functions K_ν for a certain set of a-priori-profiles $T(p)$. We employed three models (mean, maximum, and minimum) given in Fig. 2. These are empirical profiles, based on the analysis of results of the Venusian atmosphere published till 1979 (begin of work on transmission functions calculation). In Fig. 2, model profiles taken from the paper of Kerzhanovich et al. [15] considering the information obtained till 1983 are given for comparison. Whereas the new profile is within the range of our assumed temperature profiles, that range is exceeded below 67 km at high latitudes. Nevertheless this deviation can be accepted for the first approximation t_ν and K_ν.

The CO_2 mass mixing ratio has been considered equal to 0.98 (remain — N_2). For different zenith angles and pressure levels the quasimonochromatic atmospheric trans-

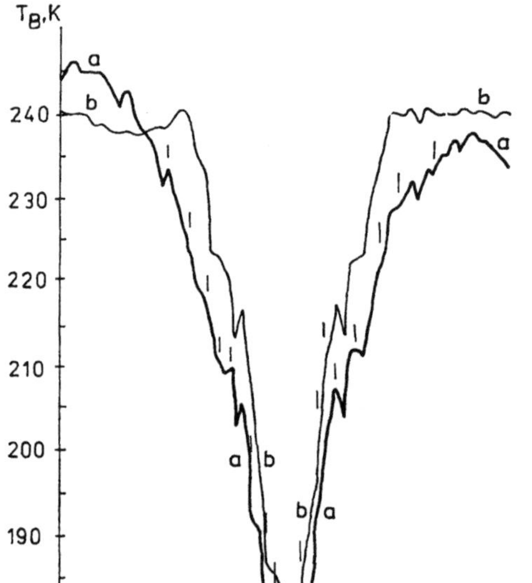

Fig. 1. The 15 µm CO_2 band shape in the Venusian spectrum for 2 individual spectra obtained in the orbit on October 14 (Venera 15). *a* — No. 5, $\varphi = +44°$; *b* — No. 25, $\varphi = +82°$. Both spectra belong to the night side of the planet (zenith distances of the sun 114° and 96°). The locations of the spectral ranges which were used for the temperature profile retrieval are marked by strokes

mittances were calculated (from level 600 hPa to 0.006 hPa) and then averaged in the interval $\Delta_\nu = 0.1$ cm^{-1} with a step of 0.1 cm^{-1}. Calculations were carried out line-by-line, using spectroscopic informations (AFGRL line compilation version 1976 [18]). The Voigt line profile was calculated by the analytical expression proposed by Matveev [19]. For the Lorentz wings, the effect of exponential "cutting" at great distances from the line centre was taken into consideration [20]. Hereby the Lorentz line profile was supposed in a modified form

$$f_\nu = f_{\nu,L} \times \exp\{-a \times [|\nu - \nu_0| - \nu_{min}]^b\}, \tag{7}$$

where $f_{\nu,L}$ — the Lorentz profile, $a = 1.4$, $b = 0.25$ and $\nu_{min} = 3.5$ cm^{-1}. The coefficient of self-broadening of the CO_2 line was taken as 1.296. The transmittances calculated in the same manner of the Earth atmosphere [21], coincide in the 640—800 cm^{-1} range with the measured ones in the boundaries of $\pm 5\%$; this error is smaller than in analogous papers of other authors [22].

The next stage was the convolution of transmission functions (calculated for $\Delta_\nu = 0.1$ cm^{-1}) with the spectral response function of the Fourier spectrometer with a resolution of 5 cm^{-1} [23]. Examples of convoluted transmission functions are represented in Fig. 3. Then, the weighting functions K_ν used in the present paper were calculated. These weighting functions do not consider aerosol extinction.

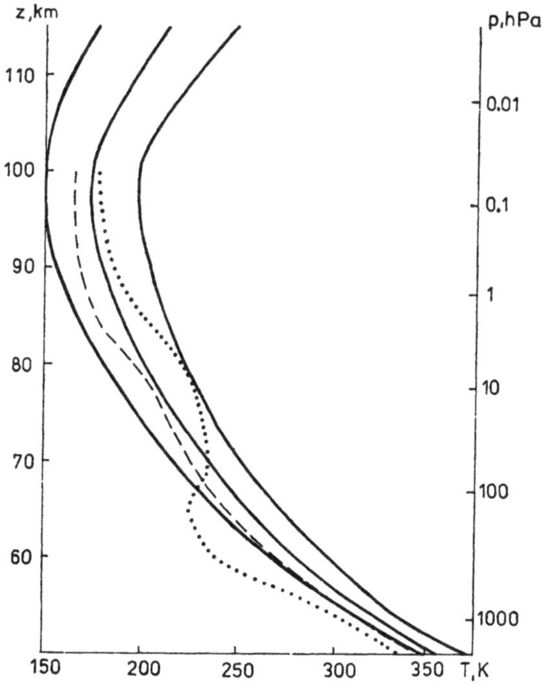

Fig. 2. A-priori-models of the vertical temperature profile of Venusian atmosphere, which were used for calculations of the transmission function (three full lines — maximum, mean, and minimum model). For comparison two model profiles from [15] are given for latitudes $|\varphi| < 50°$ (dashed curve) and $|\varphi| = 75°$ (dotted curve)

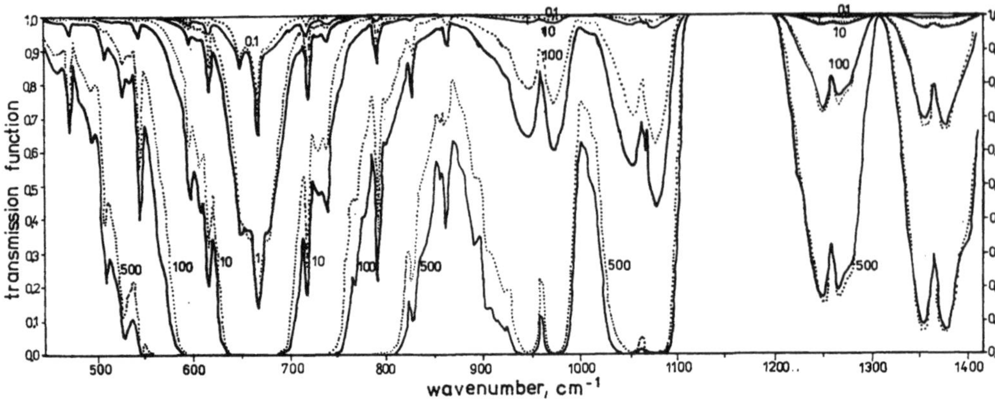

Fig. 3. CO_2 spectral transmission function t_ν with spectral resolution $\Delta_\nu = 5\ cm^{-1}$ for zero zenith angle from the top of the atmosphere to the pressure level indicated: 0.1; 1; 10; 100 and 500 hPa. Full line — maximum atmospheric model (see Fig. 2), dotted line — minimum model

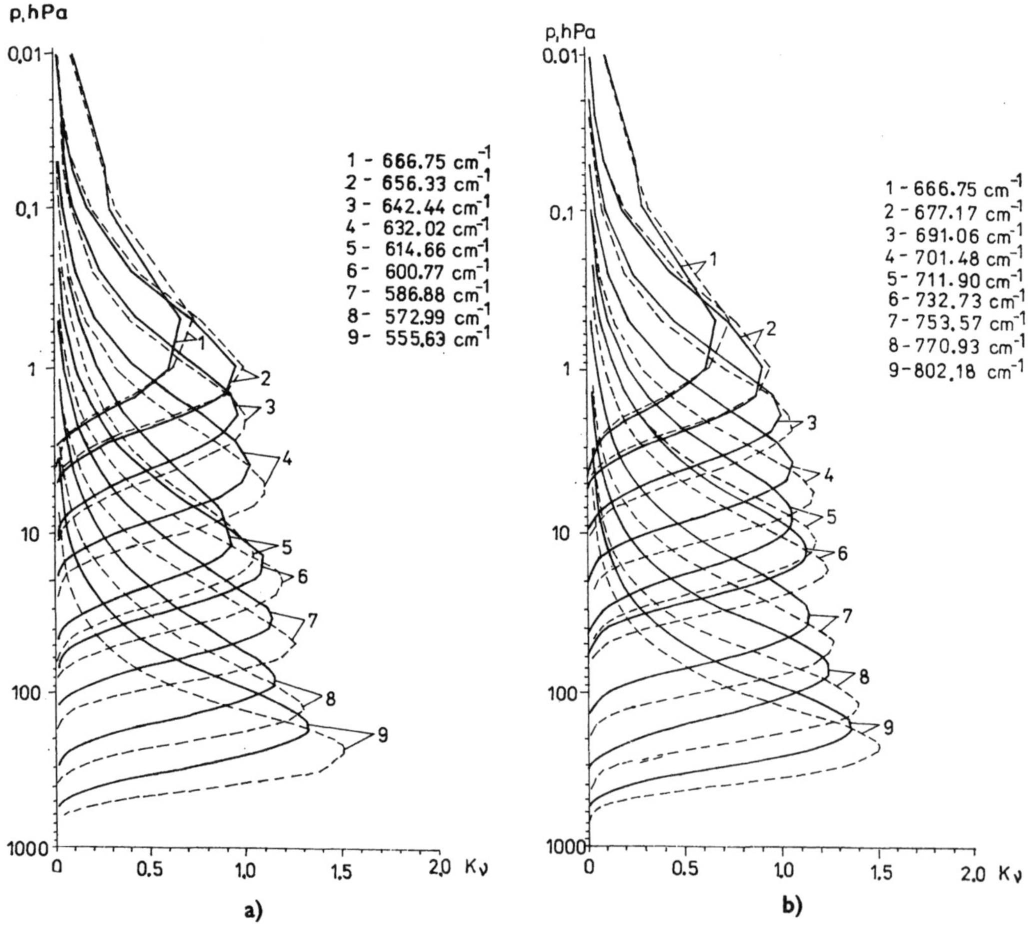

Fig. 4. Weighting functions for selected wavenumbers (4a) longwave wing and 4b) shortwave wing), used for thermal sounding. Full line — maximum model, dotted line — minimum model

The table gives the wave numbers used for temperature retrieval and also the levels where the weighting functions $K_\nu(p)$ of the mean model $T(p)$ have their maximum. In Fig. 4 the weighting functions $K_\nu(p)$ are shown for the minimum and the maximum temperature model. Later on, additional wave numbers were introduced with their maxima of $K_\nu(p)$ between 10 and 200 hPa, where inversions of the $T(p)$ profile are observed. The calculation of aerosol extinction coefficients and aerosol transmission functions, was made in two steps starting with an a-priori-model, and using afterwards data directly retrieved on the base of measurements outside the CO_2 band [24]. The transmission function of the gas and aerosol mixture is equal to the product of functions calculated separately for both components.

Calculations of gas transmission functions were performed in the GDR, that of aerosol functions in the GDR and USSR.

4. Retrieval algorithms

Two different solution methods of equation (4) were used: the method of statistical regularization [25, 26] and the relaxation method [27]. We describe their main characteristics:

a. Method of statistical regularization

The wanted profile $T(p)$ is considered as one element of a certain statistic ensemble of functions $T(p)$ described by the mean profile $\langle T(p)\rangle$ and the correlation function

$$C_T(p, p') = \langle [T(p) - \langle T(p)\rangle] \times [T(p') - \langle T(p')\rangle] \rangle .$$

As is known [25, 28] such a-priori-information is sufficient for the determination of $T(p)$ from radiance density measurements in selected frequencies ν_i ($i = 1, \ldots, m$, m — frequency number).

By linearization of relation (4) we have

$$\Delta I_\nu = \int_{-\infty}^{\infty} K_\nu(p) \times \frac{\partial B_\nu(T)}{\partial T} \times \Delta T(p) \times \mathrm{d} \lg p .$$

The temperature at given pressure levels p_j ($j = 1, \ldots, n$, where n — number of levels) is determined from the solution of the linear algebraic equation system of the kind

$$(\boldsymbol{K}^+ \times \boldsymbol{W} \times \boldsymbol{K} + \boldsymbol{C}^{-1}) \times \varphi = \boldsymbol{K}^+ \times \boldsymbol{W} \times \boldsymbol{f}, \tag{8}$$

and the error is found by the diagonal elements of the a-posteriori correlation matrix of the solution

$$\boldsymbol{S} = (\boldsymbol{K}^+ \times \boldsymbol{W} \times \boldsymbol{K} + \boldsymbol{C}^{-1})^{-1} . \tag{9}$$

Here, the following expressions are introduced: φ — n-vector of differences $\Delta T_j = T(p_j) - \langle T(p_j)\rangle$ between unknown and mean profile, $\boldsymbol{f}$ — m-vector of differences $\Delta I_i = I_{\nu_i}^{(m)} - I_{\nu_i}^{(c)}[\langle T(p)\rangle]$ between measured and calculated spectral radiance densities for the mean profile, $\boldsymbol{K}$ — $n \times m$ matrix, determined from the weighting functions $K_\nu(p)$ for selected frequencies at given levels, $\boldsymbol{C}$ — $n \times n$ matrix of values of the a-priori correlation function $C_T(p, p')$, $\boldsymbol{W} = \mathrm{diag}\,(1/s_i^2)$ — diagonal matrix with non-zero elements, to be determined by s_i-values of the error of I_ν-measurements at selected wave numbers.

In case of thermal sounding of the Earth's atmosphere from artificial satellites, extensive a-priori-information can be used. For different latitudes and seasons both the mean temperature profiles and their correlation functions are well determined. The differences between unknown and mean profiles are not large and do normally not exceed some degrees allowing, therefore, to confine to linear approximation (8). For Venus, significant data have been stored up to now so that atmospheric models of the latitudinal dependence of the thermal structure could be defined [15]. The first temperature retrievals from real spectra, obtained by the FS 1/4 instrument showed, however, that the differences between retrieved and model temperatures can reach dozens of degrees. Therefore, the temperature retrieval relation (8) was used iteratively. Hereby, the profile $T(p)$ obtained in the preceding approximation was used instead of the mean profile $\langle T(p)\rangle$. As a rule, three iterations were sufficient for coincidence of the calculated spectrum using the retrieved profile with the measured spectrum within the limits of the measuring error. The application of C as a matrix obtained by discretization of any real correlation function $C_T(p, p')$ would be sensible for the first approximation only. Also other kinds of correlation matrixes had been used in the first calculations for example, in the form

$$C_T(p, p') = \sigma^2(p) \times \exp\left\{-\frac{(\lg p - \lg p')^2}{2 \times R_{cor}^2}\right\} \tag{10}$$

for different values of the a-priori-error of temperature $\sigma(p)$ and correlation radius R_{cor}. The numerical experiments carried out showed that the sensitivity of the retrieved temperatures to the numerical values of (10) is extremely weak in a wide range of variations in σ and R_{cor}, namely 10—50 K in σ and 0.5—2 km in R_{cor}. For data processing R_{cor} was chosen corresponding to the half-width of weighting functions in the lg p-coordinate. Fig. 5 gives an example of the retrieved profile $T(p)$ for a spectrum with pure gaseous transmission functions. The dotted region corresponds to the formal retrieval error, determined by relation (9). This error is small and does not exceed $\sim$2 K in the altitude region between the maxima of the 'upper' and the 'lower' weighting function (0.5 to 200 hPa). However, the real error is larger than the one given by (1) at lower levels due to the indeterminancy in the aerosol absorption.

b. Relaxation method

The radiance density, measured at wave number ν_i, is determined by formula (4). It corresponds to the brightness temperature

$$T_{\nu_i} = \frac{1.4388 \times \nu_i}{\ln(1.1911 \times 10^{-5} \times \nu_i^3/I_{\nu_i} + 1)}, \tag{11}$$

where $i = 1, \ldots, m$, m — number of wave numbers, used for temperature retrieval. The iteration method supposed by Chahine [27] and applied by Taylor et al. [11] consists in the following:

1. Starting with a certain initial temperature profile $T_j^{(1)}$, where $j = 1, \ldots, n$, n — number of levels in the atmosphere, the outgoing radiance density $\tilde{I}_i^{(1)}$ and the corresponding brightness temperature $\tilde{T}_i^{(1)}$ are calculated according to (4) and (11), respectively.

2. The root mean square error is calculated

$$\sigma = \sqrt{\sum_{i=1}^{m} \left(\frac{I_{\nu_i} - \tilde{I}_i^{(1)}}{I_{\nu_i}} \right)^2 \Big/ m} . \tag{12}$$

The value σ is compared with the error in measurement

$$\sigma_0 = \sqrt{\sum_{i=1}^{m} \tilde{\sigma}_i^2 / m} \tag{13}$$

where $\tilde{\sigma}_i = s_i / I_{\nu_i}$ — relative measuring error in channel i.

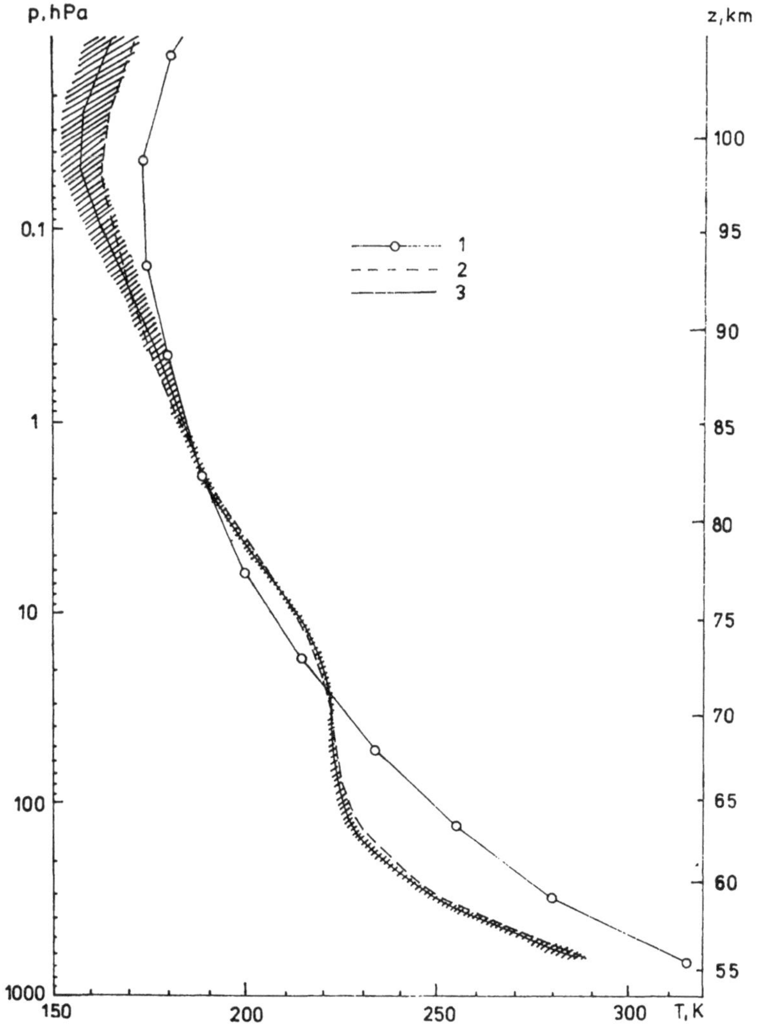

Fig. 5. Results of $T(p)$ profile retrieval by the statistical regularization method with pure gaseous transmission functions for spectrum No. 19 of the orbit on October 14, 1983 ($\varphi = 75°$, $z_{\odot} = 100°$). 1 — initial temperature profile, 2 — first iteration, 3 — second and third iteration. Pressure scale p — exact, altitude scale z — approximated

3. If the condition $\sigma \leqq \sigma_0$ is not fulfilled, a new temperature profile

$$T_j^{(2)} = T_j^{(1)} \times \left[\sum_{i=1}^{m} T_{\nu_i}/\bar{T}_i^{(1)} \times \left(\frac{\partial t_\nu}{\partial \lg p}\right)_{ij} \times \bar{\sigma}_i \Big/ \sum_{i=1}^{m} \left(\frac{\partial t_\nu}{\partial \lg p}\right)_{ij} \times \bar{\sigma}_i \right] \tag{14}$$

is calculated and the rerrieval process returns to step 1. This procedure is repeated until the condition $\sigma \leqq \sigma_0$ is fulfilled. At the n-th step

(15)

$$T_j^{(n+1)} = T^{(n)} \times \left[\sum_{i=1}^{m} T_{\nu_i}/\bar{T}_i^{(n)} \times \left(\frac{\partial t_\nu}{\partial \lg p}\right)_{ij} \times \bar{\sigma}_i \Big/ \sum_{i=1}^{m} \left(\frac{\partial t_\nu}{\partial \lg p}\right)_{ij} \times \bar{\sigma}_i \right]. \tag{16}$$

Both retrieval programmes were checked at first by means of numerical experiments. In both cases the retrieved temperatures differed from the given temperature profile

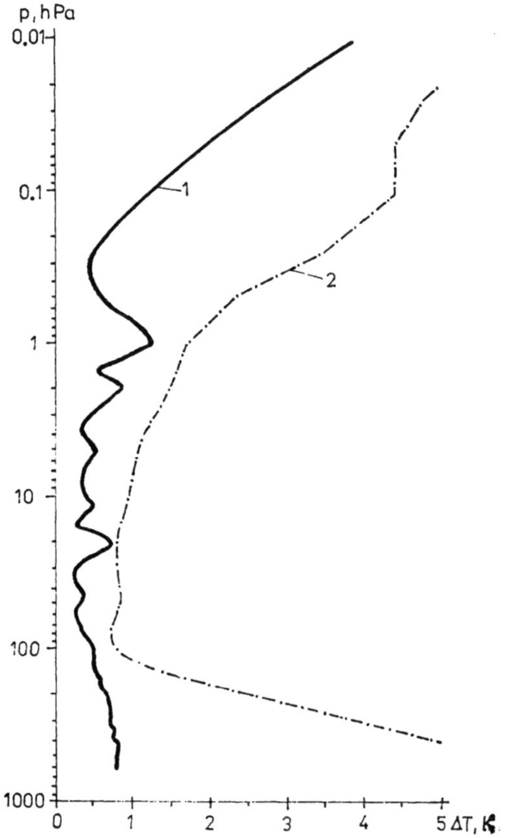

Fig. 6. Estimation of retrieval errors by the relaxation method on the base of numerical experiments. Mean absolute error of temperature profiles retrieved from a synthetic spectrum (for 30 independent retrievals) with errors, given by means of a random number generator. — 1. For comparison a formal error of the retrieval is given for the statistical regularization method — 2 (equation (9))

only by 1—2 K in the altitude range defined by the weighting functions. Examples of retrieval errors for both methods are given in Fig. 6 showing nearly identical retrieved temperature profiles with differences of less than 3 K below the level of 0.5 hPa in general. An example of such a comparison is give in Fig. 7. The gaseous transmission functions depend on the atmospheric temperature profile, therefore, they have to be corrected for according to the calculated profile $T(p)$ within the iteration cycle.

The calculations show that the sensitivity of the transmission function to the choice of the a-priori-model $T(p)$ is not very great. Fig. 8 presents the results of retrieved profiles $T(p)$ obtained by the relaxation method using transmission functions for the minimum and mean model (without consideration of aerosol). The retrieved profiles $T(p)$ do not differ by more than 3 K.

One possibility to check the reliability of the retrieval method is the comparison of temperature profiles obtained with spectral channels of the short-wave and the long-wave wing. Such a comparison is given in Fig. 9. The profiles for two wings differ by not more than 2 K.

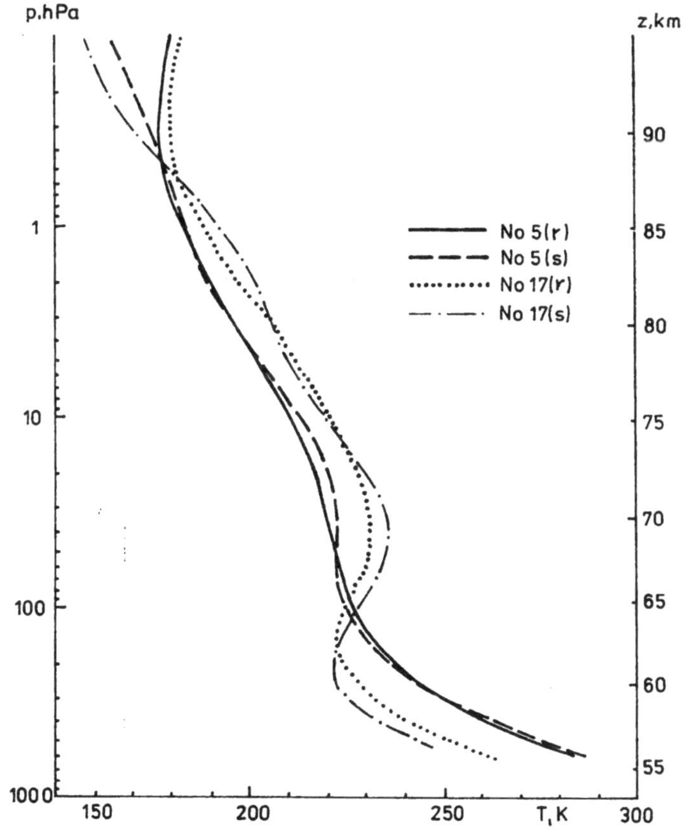

Fig. 7. Comparison of $T(p)$ profile retrieval results by the relaxation method (r) and the statistical regularization method (s) for spectra No. 5 and 17 of the orbit on October 14, 1983 (latitudes 48 and 73°, night side, zenith distances of the sun 112 and 100° respectively). Pressure scale p — exact, altitude scale z — approximated

5. $T(p)$ profiles in Venusian atmosphere

The estimation of temperature profiles from the measured spectra is still in the initial stage, but a relatively great number of examples has already been gathered. All temperature retrievals based on onboard-processed spectra from Venera 15. Calculations are carried out in both countries by both methods.

The measurements of October 14, 1983 are completely processed with pure gaseous transmission functions by both methods and considering aerosol by the relaxation method. All other data are processed with pure gaseous transmission functions by the relaxation method.

In paper [1] it was mentioned that some groups of spectra can be separated which differ in the form of the wings of the 15 μm CO_2 band and several other parameters. A more comprehensive discussion of this problem is given in [24]. These groups are characteristic for certain latitude zones, and their localization does not vary with time. In Fig. 10 temperature profiles are given averaged over latitudinal zones (orbit on October 14). They all belong to latitudes higher than 57° and have a characteristic S-shape with inversion or isothermy in the pressure range from 200 to 20 hPa.

The temperature profiles are flatter in lower latitudes (Fig. 10). Fig. 11, compares profiles obtained with pure gaseous transmission functions and under consideration of

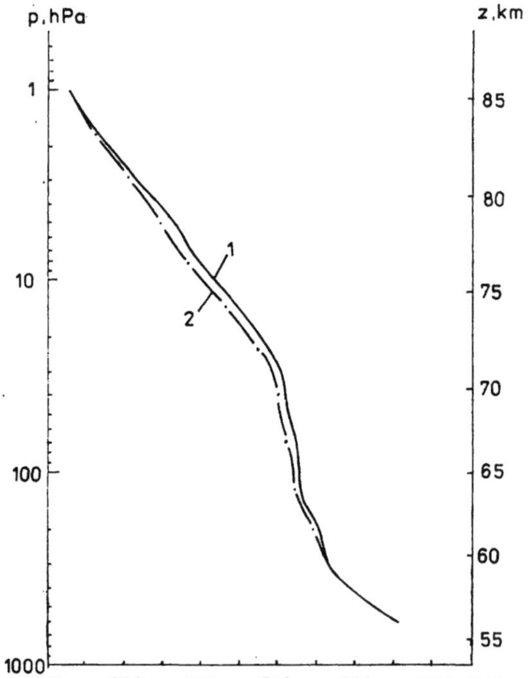

Fig. 8. Examples of $T(p)$ temperature profile retrieval results (October 14, 1983, $\varphi = 87°$, $z_{\odot} = 91°$) by the relaxation method with use of transmission functions for the mean model (1) and the minimum model (2) in the longwave wing of the 15 μm CO_2 band. Pressure scale p — exact, altitude scale z — approximated

aerosol (a-priori-model, described in paper [24]). There is a divergence between the retrievals of the two wings below the level of 30 hPa if aerosol is not considered. The divergence disappears and the temperature below 30 hPa rise when aerosol is taken into account. It is important, however, that this rising is not significant, about 10 K. Above the 30 hPa level the aerosol effect is negligible.

A great part of spectra was optained on the northern hemisphere, but there were also some cycles with a turn of the viewing direction (see [1]) to investigate the southern hemisphere (up to a latitude of about 70°), too. In Fig. 12 a comparison of temperature profiles in 65° latitude on the southern and northern hemispheres is given. The differences between the profiles of both hemispheres are not great, and do not exceed normal local temporal variations (see below).

At present, the temperature retrieval method with consideration of aerosol has been worked out for low latitudes only ($\varphi < 60°$) as in higher latitudes the clouds have another structure [24] so that it is necessary to introduce more detailed aerosol models for latitudes $\varphi > 60°$. Therefore, the greater part of the temperature profiles are retrieved with pure gaseous transmission functions.

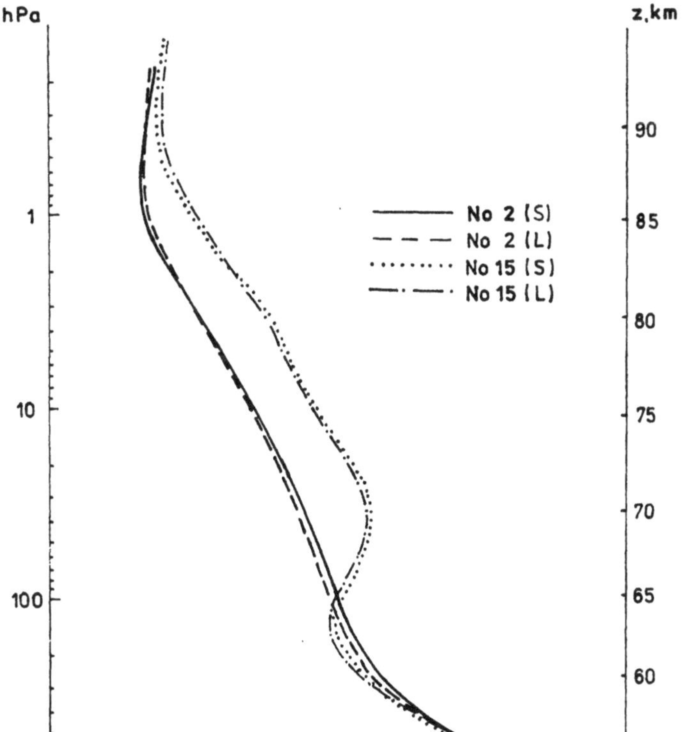

Fig. 9. Comparison of profiles obtained for both wings of the 15 μm CO_2 band. Orbit on October 14, spectra No. 2 ($\varphi = 34°$, $z_\odot = 118°$) and No. 15 ($\varphi = 68°$, $z_\odot = 103°$), s — shortwave wing, l — longwave wing. Relaxation method, pure gaseous transmission functions. Pressure scale p — exact, altitude scale z — approximated

In Fig. 13 isotherms in pressure-latitude coordinates are given for October 14. One part of this information is presented as profiles in Fig. 14. Fig. 14 contains, too, temperature profiles obtained by analysis of Orbiter Infrared Radiometer measurements of Pioneer-Venus [11] for comparison demonstrating the information gain in vertical temperature structure when more spectral channels can be used for the retrieval procedure.

The analysis of the retrieved temperature profiles allows the following conclusions:

1. Within the latitude range of $\pm 60°$ the temperature decreases monotonously with altitude. Typical temperature values are about 230 K at 70 km altitude ($p \simeq 30$ hPa), about 200 K at 80 km altitude ($p \simeq 3$ hPa) and about 180 K at 90 km altitude ($p \simeq 0.3$ hPa).

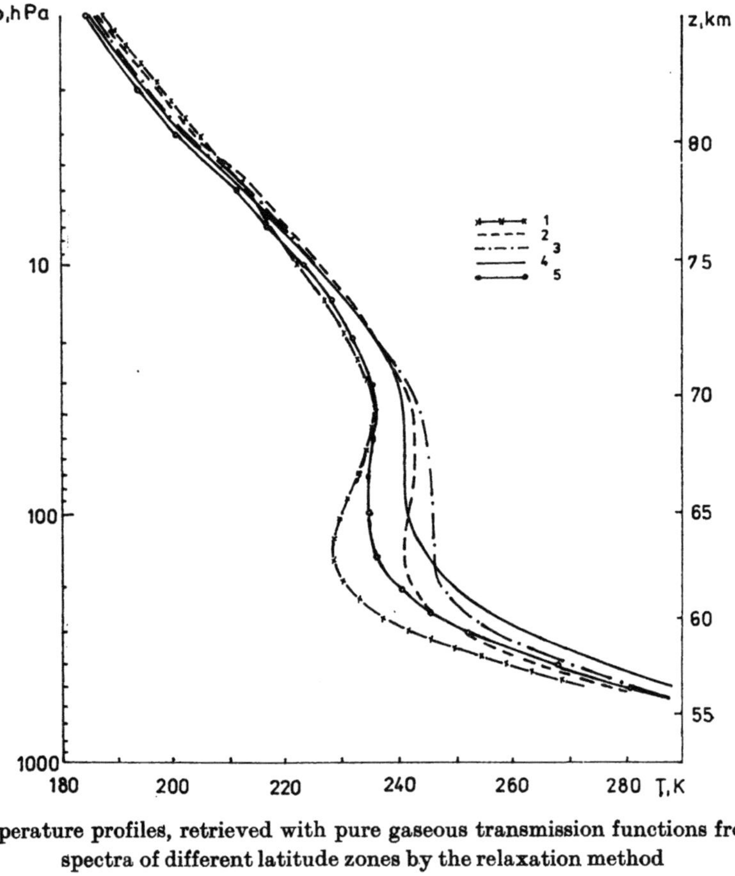

Fig. 10. Temperature profiles, retrieved with pure gaseous transmission functions from averaged spectra of different latitude zones by the relaxation method

1 — from 63 to 70° ($z_\odot = 106-102°$),
2 — from 79 to 85° ($z_\odot = 98-94°$),
3 — from 86 to 87° ($z_\odot = 93-91°$),
4 — from 85 to 81° ($z_\odot = 90-87°$),
region of the hotspot — of the "dipole", see [1]),
5 — from 78 to 71° ($z_\odot = 85-82°$).

Pressure scale p — exact, altitude scale z — approximated

2. At latitudes $\varphi > +60°$ there is normally an inversion or an isothermy in the pressure range $10 < p < 100$ hPa.
3. The atmosphere is warmer at latitudes $\varphi > +60°$ than at middle or lower latitudes. The temperature difference between high and low latitudes reaches 20 K at 70 km altitude and decreases to 5 K at 90 km altitude.
4. Temperatures at day nearly coincide with temperatures at night.

We note that several publications mention diurnal atmospheric temperature variations with amplitudes up to 20 K [9, 29, 30], but according to other data their amplitude does not exceed some degrees [11, 31]. In models [14, 15] the latter view has been taken, and obviously this was right. The standard geometry of our measurements is very favourable for investigations of diurnal variations (the measuring line passes the pole from latitudes ~20° on the night side till 20° on the day side) and they show convincingly that diurnal variations are small or even absent at all altitudes. These are local-temporal variations of temperature profiles at one latitude, but they are not connected with local

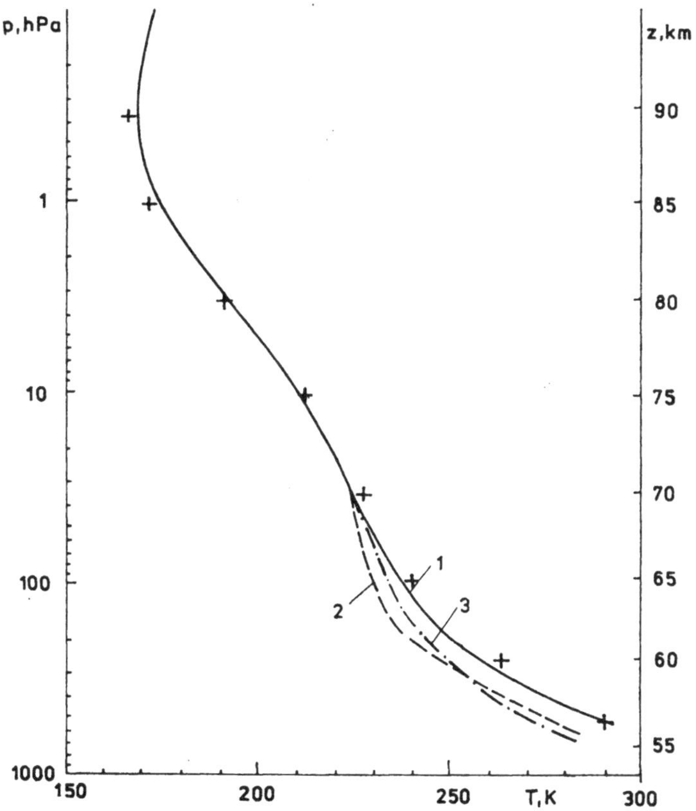

Fig. 11. Comparison of profiles calculated with (curve 1) and without (2 — longwave wing, 3 — shortwave wing) consideration of aerosol, equatorial region, orbit with turn of the spacecreaft-orientation on November 11, spectrum No. 13 ($\varphi = 0.5°$, $z_{\odot} = 76°$). Crosses — profile in low latitudes corresponding to model [15]. Pressure scale p — exact, altitude scale z — approximated

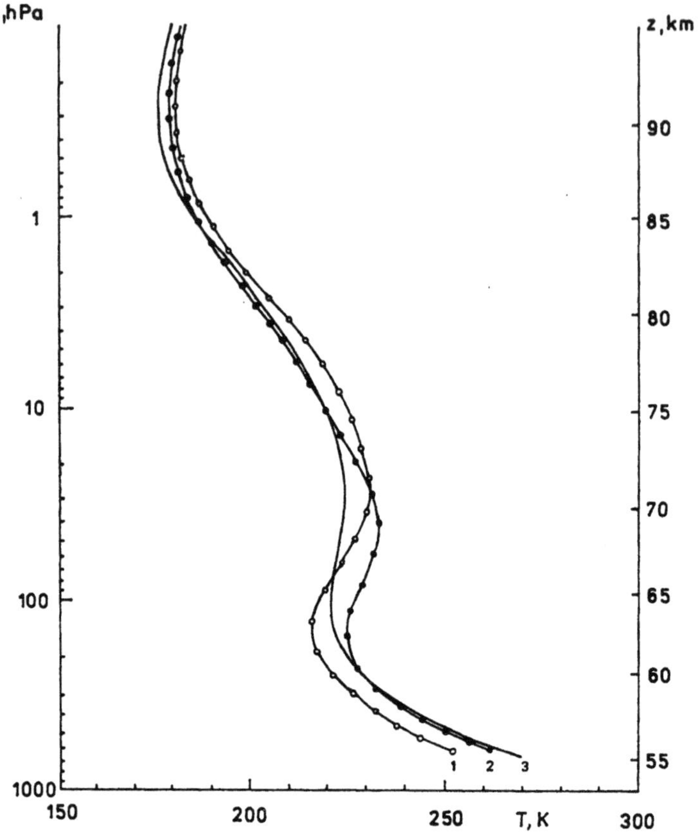

Fig. 12. Comparison of southern and northern hemispheric tempeature profiles

$1 - \varphi = +66°$, $z_\odot = 90°$ (orbit on November 10),
$2 - \varphi = +66°$, $z_\odot = 100°$ (orbit on November 10),
$3 - \varphi = -65°$, $z_\odot = 80°$ (orbit on November 11).

Pressure scale p — exact, altitude scale z — approximated

Fig. 13. Temperature retrieval results for October 14 with pure gaseous transmission functions by the statistical regularization method. The isotherms in the coordinates pressure-latitude are represented

Fig. 14. Temperature profiles of the atmosphere above the clouds, obtained from spectra No. 6 ($\varphi = 52°$, $z_\odot = 111°$), No. 16 ($\varphi = 70°$, $z_\odot = 102°$), No. 34 ($\varphi = 79°$, $z_\odot = 87°$), and No. 51 ($\varphi = 46°$, $z_\odot = 70°$) of the orbit on October 14, 1983 with pure gaseous transmission functions by the statistical regularization method. Squares and circles — temperatures of the atmosphere according to infrared experiment data of Pioneer-Venus orbiter [11] at latitudes 50 and 70° correspondingly. Pressure scale p — exact, altitude scale z — approximated

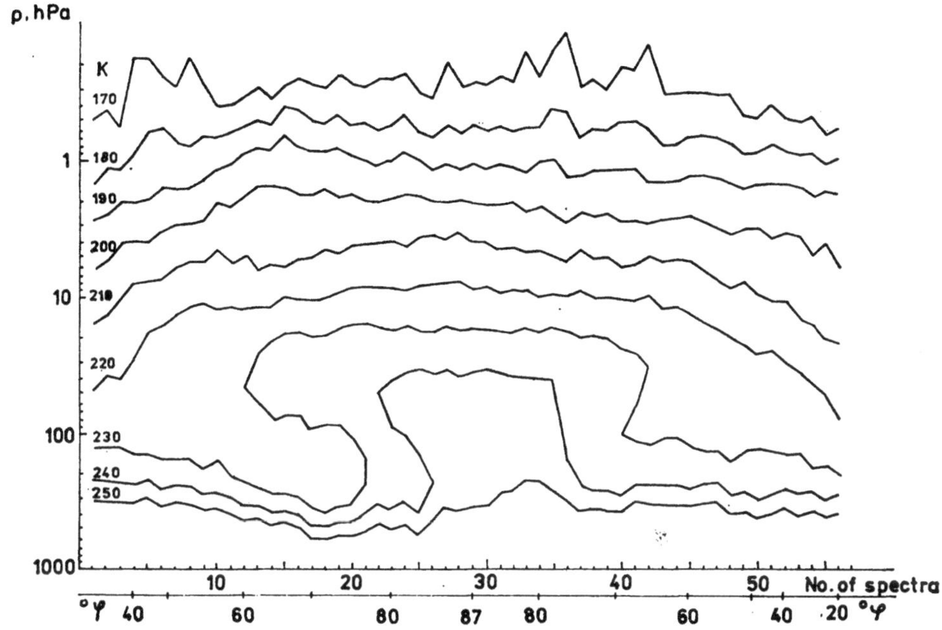
p, hPa
K
170
180
190
200
210
220
230
240
250
1
10
100
1000
10
20
30
40
50
No. of spectra
°φ 40
60
80
87
80
60
40
20 °φ

Fig. 13

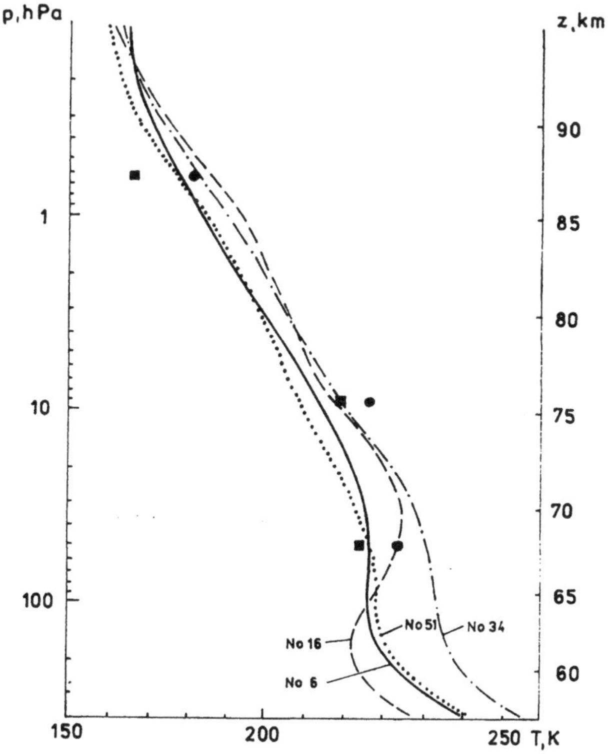
p, hPa
z, km
1
10
100
90
85
80
75
70
65
60
No 16
No 6
No 51
No 34
150
200
250 T, K

Fig. 14

solar time in a dinstinguishable way. These variations are most intense in high latitudes. In Fig. 15 some temperature profiles are given for latitudes $66° < \varphi < 70°$, showing the wide range of the inversion depth at the 150 hPa level. In some cases it reaches 15 K, in other ones it disappears completely.

The analysis of the material obtained by the infrared experiment aboard Venera 15 and Venera 16 is only in the very first stage. It is important that from the methodical view the temperature retrieval procedure has been worked out and is ready for general processing. Some improvements have still to be made (consideration of aerosol extinction at any latitudes — not only at low ones, iteration of transmission functions), but it can be expected that they will not lead to significant changes of the temperature profiles.

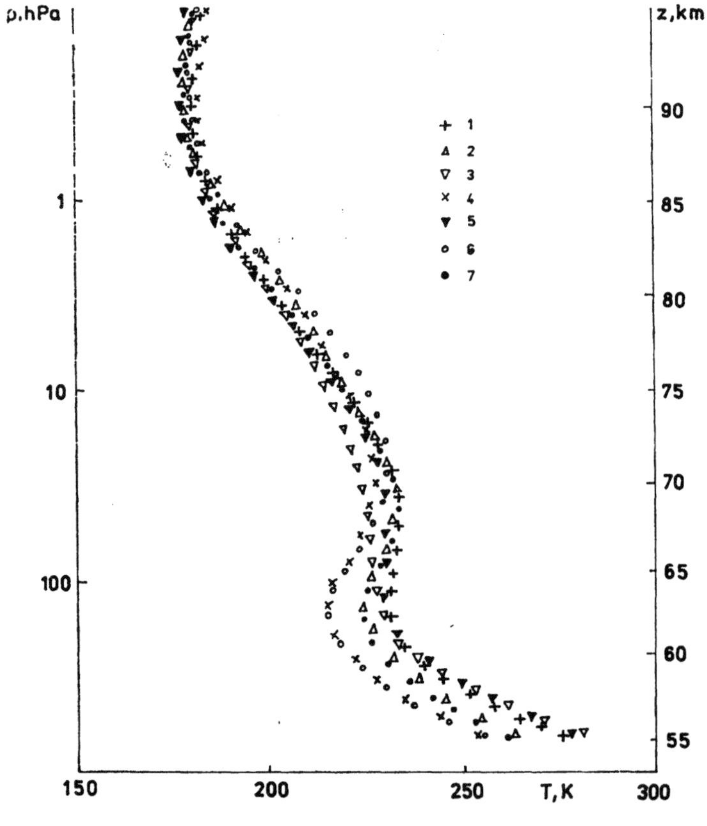

Fig. 15. Local-temporal variation of $T(p)$ profiles at latitudes $\varphi \simeq 65-70°$:

1 — October 12, 1983 $\varphi = 70°$, $z_\odot = 103°$,
2 — October 14 $\varphi = 66°$, $z_\odot = 104°$,
3 — October 14 $\varphi = 66°$, $z_\odot = 80°$,
4 — October 25 $\varphi = 66°$, $z_\odot = 99°$,
5 — October 25 $\varphi = 66°$, $z_\odot = 88°$,
6 — November 10 $\varphi = 65°$, $z_\odot = 90°$,
7 — November 10 $\varphi = 66°$, $z_\odot = 100°$.

Pressure scale p — exact, altitude scale z — approximated

References

[1] Oertel, D., Moroz, V. I., Nopirakowski, J. et al.: Infrared experiment aboard the automatic interplanetary stations Venera 15 and Venera 16. 1. Methodics and first results. Veröffentlichungen des Forschungsbereiches Geo- und Kosmoswissenschaften, Heft 18, 1990.

[2] Cheremukhina, Z. P., Morozov, S. F., Borodin, N. F.: Ocenka temperatury stratosfery Venery po dannym o peregruzkakh avtomaticheskoj stancii "Venera-8". Kosmich. Issled. **12** (1974) 2, 264.

[3] Avduevskijj, V. S., Borodin, N. F., Burcev, V. N. et al.: Avtomaticheskie stancii "Venera-9", i "Venera-10" — funkcionirovanie spuskaemykh apparatov i izmerenie parametrov atmosfery. Kosmich. Issled. **14** (1976) 5, 655.

[4] Avduevskijj, V. S., Borodin, N. V., Vasil'ev, V. N. et al.: Analiz rezul'tatov izmerenija parametrov atmosfery Venery v mestakh posadki AMS "Venera-11" i "Venera-12". Kosmich. Issled. **17** (1979) 5, 655.

[5] Avduevskijj, V. S., Godnev, A. G., Semenchenko, V. V. et al.: Issledovanie kharakteristik stratosfery Venery po izmerenijam peregruzok pri tormozhenii stancijj "Venera-13" i "Venera-14". Kosmich. Issled. **21** (1983) 2, 205.

[6] Seiff, A., Kirk, D. B., Young, R. E. et al.: Measurements of thermal structure and thermal contrasts in the atmosphere of Venus and related dynamical observations: results from the four Pioneer Venus Probes. J. Geophys. Res. **85** (1980) A 13, 7903.

[7] Kliore, A. J., Levy, G. S., Cain, D. L. et al.: Atmosphere and ionisphere of Venus from the Mariner 5 S-band radio occultation measurement. Science **158** (1967) 105.

[8] Howard, H. T., Tyler, G. L., Fjeldbo, G. et al.: Venus: mass, gravity field, atmosphere and ionosphere as measured by Mariner 10 dualfrequency radio system. Science **183** (1974) 1297.

[9] Jakovlev, O. I., Efimov, A. I., Timofeeva, T. S. et al.: Atmosfera Venery po predvaritel'nym dannym ehksperimentov radioprosvechivanija s pomoshh'ju apparatov "Venera-9" i "Venera-10". Kosmich. Issled. **14** (1976) 5, 722.

[10] Kliore, A. J., Patel, I. R.: Vertical structure of the atmosphere of Venus from Pioneer Venus orbiter radio occulations. J. Geophys. Res. **85** (1980) A 13, 7957.

[11] Taylor, F. W., Beer, R., Chahine, M. T. et al.: Structure and Meteorology of the Middle Atmosphere of Venus: infrared remote sensing from the Pioneer Orbiter. J. Geophys. Res. **85** (1980) A 13, 7963.

[12] Moroz, V. I.: Rabochaja model atmosfery Venery. Preprint IKI AN SSSR **162** (1973).

[13] Marov, M. Ja., Rjabov, O. L.: Model' atmosfery Venery. Preprint IPM AN SSSR **112** (1974).

[14] Moroz, V. I.: The atmosphere of Venus. Space Sci. Rev. **29** (1981) 1, 3.

[15] Kerzhanovich, V. V., Marov, M. Ya., Moroz, V. I.: Proposals for VIRA (Venus International Reference Atmosphere): dynamics and structure below 100 km. Preprint IKI AN SSSR Pr 831 (1983).

[16] Seiff, A.: Models of the lower and middle atmosphere of Venus from the surface to 100 kilometers. Report for the Venus International Reference Atmosphere Workshop, Hamburg, 1983.

[17] Kondrat'ev, K. Ja., Timofeev, Ju. M.: Meteorologicheskoe zondirovanie atmosfery iz Kosmosa. Gidrometeoizdat. Leningrad, 1978.

[18] Rothman, L. S. and McClatchey, R. A.: Updating of the AFCRL atmospheric absorption line parameters compilation. Appl. Opt. **15** (1976) 11, 2616.

[19] Matveev, V. S.: Priblizhennye predstavlenija koehfficienta pogloshhenija i ehkvivalentnykh shirin linijj s fojjgtovskim konturom. Zh. Prikl. Spektr. **16** (1972) 2, 228.

[20] Winters, B. H., Silverman, S., Benedict, W. S.: Line shape in the wing beyond the band head of the 4.3 μm band CO_2. J. Quant. Spectrosc. Radiat. Transfer **4** (1964) 3, 527.

[21] SPÄNKUCH, D.: Arbeiten des Meteorologischen Dienstes der DDR auf dem Gebiet der indirekten Sondierungen. Zeitschrift für Meteorologie **30** (1980), 205.

[22] KUNDE, V. G., MAGUIRE, W. C.: Direct integration transmittance model. J. Quant. Spectrosc. Radiat. Transfer **14** (1974) 6, 803.

[23] DÖHLER, W., SCHÄFER, K., SPÄNKUCH, D.: Line-by-line gaseous transmittance for Venus standard atmospheres in the 250 to 1400 cm^{-1} region. Report for the Venus International Reference Atmosphere Workshop, Hamburg, 1983.

[24] ZASOVA, L. V., SPÄNKUCH, D., MOROZ, V. I. et al.: Infrared experiment aboard the automatic interplanetary stations Venera 15 and Venera 16. 3. Some conclusions on the structure of clouds, based on the analysis of spectra. Veröffentlichungen des Forschungsbereichs Geo- und Kosmoswissenschaften, Heft 18, 1990.

[25] TURCHIN, V. F., MALKEVICH, M. S., GORCHAKOVA, I. A.: Primenenie statisticheskojj reguljarizacii k opredeleniju vertikal'novo profilja temperatury atmosfery. Izv. AN SSSR. ser. fiziki atmosfery i okeana **5** (1969) 5.

[26] MALKEVICH, M. S.: Opticheskie izmerenija atmosfery so sputnikov. Nauka, Moskau 1973.

[27] CHAHINE, M. T.: Inverse Problems in radiative transfer: determination of atmospheric parameters. J. Atmos. Sci. **27** (1970) 6, 960.

[28] WESTWATER, E. R. and STRAND, O. N.: Statistical information content of radiation measurements used in indirect sensing. J. Atm. Sci. **25** (1968) 5, 750.

[29] JAKOVLEV, O. I., EFIMOV, S. S., MATJUGOV, T. S. et al.: Nochnaja atmosfera Venery po rezul'tatam radioprosvechivanija s pomoshh'ju sputnikov "Venera-9" i "Venera-10". Kosmich. Issled. **16** (1978) 1, 117.

[30] KOLOSOV, M. A. et al.: Dnevnaja atmosfera Venery po rezul'tatam radioprosvechivanija s pomosh'ju sputnikov "Venera-9" i "Venera-10". Kosm. Issled. **16** (1978) 2, 278.

[31] KLIORE, A. J., PATEL, I. R.: Thermal structure of the atmosphere of Venus from Pioneer Venus radio occultations. Icarus **52** (1982) 2, 320.

3. Infrared experiment aboard the automatic interplanetary stations Venera 15 and Venera 16

3. Some conclusions on the structure of clouds, based on the analysis of spectra

L. V. Zasova, D. Spänkuch, V. I. Moroz, K. Schäfer, E. A. Ustinov, W. Döhler, V. M. Linkin, D. Oertel, A. V. D'jachkov, H. Becker-Ross, I. A. Macygorin, J. Nopirakowski, V. V. Kerzhanovich, R. Dubois, A. N. Lipatov, W. Stadthaus, A. A. Shurupov

Abstract

Spectra of different Venusian latitudinal zones, obtained by Venera 15 and Venera 16 in the spectral range 250—1650 cm^{-1}, are analyzed. It is shown that the differences in the spectra can be explained at least to some extent by variations of the cloud layer structure. At low latitudes ($\varphi < 60°$) the cloud layer is not so dense as at polar latitudes, and diurnal variations are not observed. A region of enhanced particle concentration, extended in longitude, but not closed, exists at latitudes 65—78°. The cloud top within the "hot spot" ($\varphi \approx 78—82°$) is lower than in other regions of the planet.

The infrared experiment aboard Venera 15 and Venera 16 showed the variability of the spectra of the Venusian thermal radiation in the range 250—1600 cm^{-1} in different regions of the planet, especially in high latitudes. In [1—3] some of these variations are explained by different structure of the Venusian clouds in different regions of the planet. In the following this conclusion will be discussed in more detail, and some quantitative characteristics of the aerosol medium are given, based on spectral analysis of onboard processed spectra from Venera 15 in the range 280—1600 cm^{-1}.

1. Classification of measured spectra

In [1] three main groups of spectra were introduced, differing in their characteristics: spectra at latitudes $\varphi < 60°$, spectra in the 'cold' zone at latitudes 65—75° and some spectra in the 'hot' zone near the pole. A more detailed classification for each individual measuring cycle is useful to define characteristic averages and limits of their variability (Fig. 1, 2). One example of classification for a sequence of spectra, measured on the October 14, 1983 orbit is given in Tab. 1.

The averaged spectra are useful for a great number of tasks, among them for the analysis of pecularities of cloud structure. Spectra with 'transition' character were not included in the average procedure. Since no quantitative criteria of homogeneity had

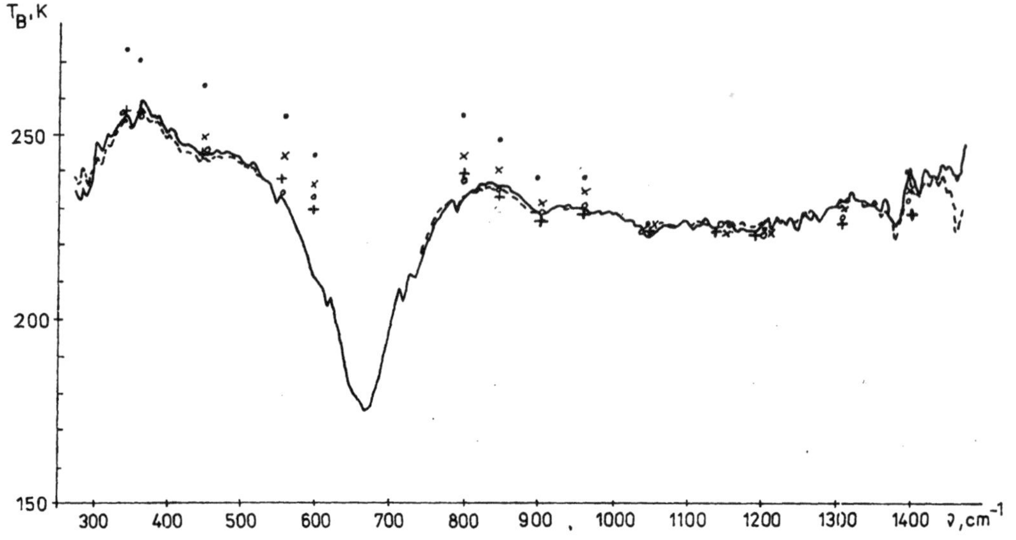

Fig. 1a

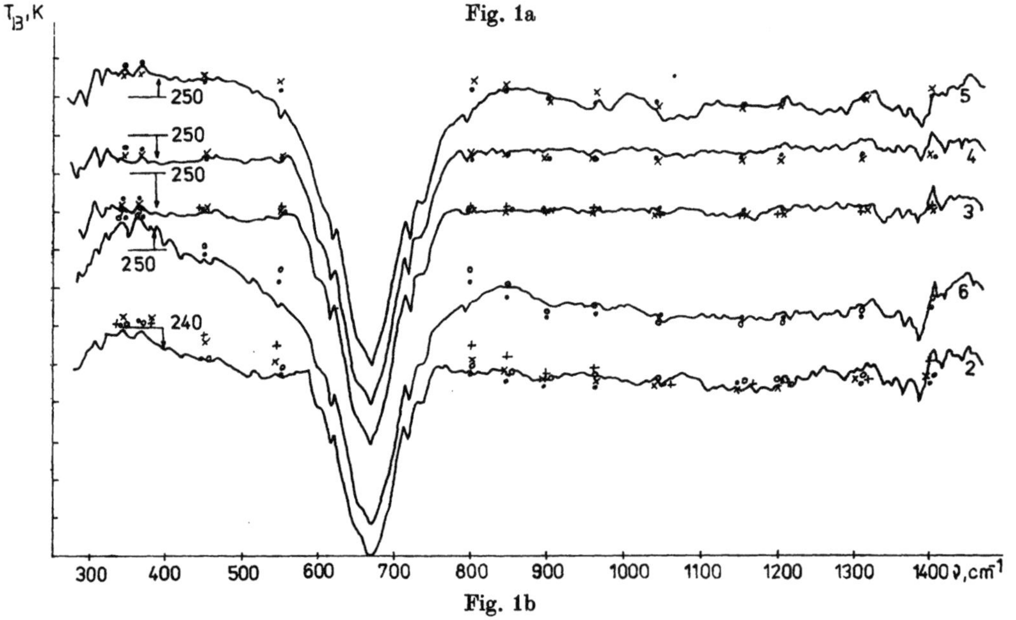

Fig. 1b

Fig. 1. Averaged spectra classified in groups for the orbit on October 14 (see Tab. 1).
a) Group 1 — full line, group 7 — dotted line. For comparison, synthetic spectra are given calculated at 14 wave numbers — points for the a-priori model; — circles for the modified a-priori model (see Tab. 5); model with constant concentration (variant A) with $H_0 = 72$ km and $N = 109$ cm^{-3} — upright crosses; model with exponential dependence of concentration on altitude (variant B) with $H_a = 3.85$ km and $H_1 = 68.5$ km — crosses

b) Curves for groups 2—6 are given with a shifted vertical axis. Right the group number is given, left the brightness temperature 250 K or 240 K. Scale division on the ordinate 10 K.

Synthetic spectra:
Group 2: point — variant A with $H_0 = 64.5$ km and $N = 315$ cm^{-3}, circles — variant A with

been applied such subdivision into groups contains some equivocalness. For instance, groups 3 and 4, but also 1 and 7 could obviously be combined. A qualitative spectral analysis for 10 orbits showed, however, the representativeness of the characteristic groups found for the orbit on October 14, 1983 (Fig. 1). One has to add one class of spectra typical for the equatorial region of the planet which was observed during the orbit on November 11, 1983. The number of characteristic main groups can be confined to 5 (Tab. 2, Fig. 2).

In [4], temperature profiles are given for some of the groups for the orbit on October14. They all were obtained without consideration of aerosol extinction, and their direct application for retrieval of cloud characteristics can cause incorrect results, especially in the region of wave numbers $\nu > 500\ \mathrm{cm}^{-1}$ where (as we will see below) the cloud transmission is much greater than in the region of the 15 μm band. In this paper temperature

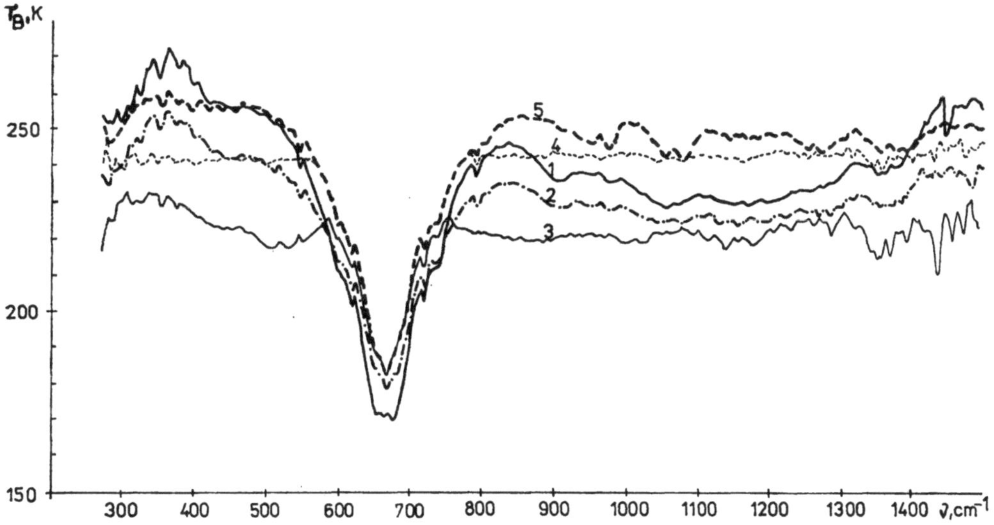

Fig. 2. Mean spectra of groups corresponding to Tab. 2. The number at the curves denote the group number

---

$H_0 = 73$ km and $N = 180\ \mathrm{cm}^{-3}$, crosses — variant B with $H_a = 4.2$ km and $H_1 = 69$ km, upright crosses — variant B with $H_a = 1.45$ km and $H_1 = 62$ km. Similar to the last model is the type A model with $H_0 = 61$ km and $N = 615\ \mathrm{cm}^{-3}$, which is not given here. H_1 = corresponds to $\nu = 1152\ \mathrm{cm}^{-1}$

Group 3: points — variant A with $H_0 = 70$ km and $N = 370\ \mathrm{cm}^{-3}$, crosses — variant B with $H_a = 1.1$ km and $H_1 = 69$ km, upright crosses — variant B with $H_a = 3$ km and $H_1 = 67$ km

Group 4: points — variant A with $H_0 = 65$ km and $N = 595\ \mathrm{cm}^{-3}$, crosses — variant B with $H_a = 2$ km and $H_1 = 65$ km.

Group 5: points — variant A with $H_0 = 60.5$ km and $N = 745\ \mathrm{cm}^{-3}$, crosses — variant B with $H_a = 1.1$ km and $H_1 = 60.5$ km.

Group 6: points — variant A with $H_0 = 72.5$ km and $N = 124\ \mathrm{cm}^{-3}$, circles — variant B with $H_a = 4.5$ km and $H_1 = 67$ km.

Table 1

Groups of similar spectra measured in the orbit on October 14, 1983

No. groups	Number of spectra	latitudes °φ	zenith distances of sun $z_\odot$	zenith distances of satellite $z_\bigcirc$	characteristics of groups
1	5	34–50°	118–112°	57–41°	Brightness temperature T_B strongly varies with ν outside gaseous absorption bands. Absorption near 900 cm^{-1} (H_2SO_4) very strong ($T_B = 224$ K at 1152 cm^{-1} and 258 K at 365 cm^{-1})
2	5	63–70	106–102	28–21	T_B varies with ν outside gaseous absorption bands weaker as in group 1. Absorption near 900 cm^{-1} weaker. Inversion like emission in the edges of the 15 µm CO_2 band ($T_B = 224$ K at 1152 cm^{-1} and 239 K at 365 cm^{-1}).
3	5	79–85	98–94	12–6	T_B constant with 2 K outside gaseous absorption bands. Absorption near 900 cm^{-1} is absent. Mean T_B about 240 K.
4	4	86–87	93–91	4–2	Like group 3 but T_B about 4 K higher.
5	4	85–81	90–87	3–8	$T_B = 248$ K at 1152 cm^{-1} and 258 K at 365 cm^{-1}. Very strong absorption of CO_2 near 960 and 1060 cm^{-1}. Absorption at 900 cm^{-1} visible.
6	5	78–71	86–82	11–18	$T_B = 232$ K at 1152 cm^{-1} and 259 K at 365 cm^{-1}. Similar as group 1. Absorption near 900 cm^{-1} weak. Hot CO_2 bands at 960 and 1060 cm^{-1} visible.
7	5	59–50	76–72	30–39	Like group 1.

profiles are corrected (at altitudes where the clouds can influence temperature retrieval results) by means of one of the published models [5]. These corrected profiles are given in Fig. 3. In the following the method of subsequent approximations is used which allows the calculation of the temperature profile and the parameters of a simplified aerosol model in a self-consistent manner. One of the tasks solved in this paper is the selection of the type of this aerosol model, though the results obtained are more general and represent themselves some new information on the cloud structure of Venus. For each spectral group such characteristics of vertical cloud structure were found which allowed outside the gaseous absorption to reproduce the measured spectrum bands nearly exactly (up to 1–2 K in brightness temperature).

Table 2
Groups of similar spectra based on the analysis of 10 orbits

No. of groups	Date of 1983	Number of spectra	latitudes °φ	zenith distance		characteristics of groups
				of sun $z_{\odot}$	of satellite $z_{\bigcirc}$	
1	Nov. 11	9	0.5–9°	76–77°	4–13°	Brightness temperature T_B strong varies with ν ($T_B = 230$ K at 1152 cm^{-1} and 272 K at 365 cm^{-1}). Strong H_2SO_4 bands near 960 and 1060 cm^{-1}.
2	Oct. 22	10	36–71	108–99	56–20	Analogous to group 1 of Table 1. $T_B = 255$ K at 365 cm^{-1} and 223 K at 1152 cm^{-1}. H_2SO_4 absorption near 450 and 900 cm^{-1} visible. Distinct hot CO_2 bands near 960 and 1060 cm^{-1}.
3	Oct. 25	3	66–73	99–98	24–17	Analogous to group 2 of Table 1. $T_B = 232$ K at 365 cm^{-1} and 220 K at 1152 cm^{-1}. Inversion like emission in the edges of the 15 µm CO_2 band.
4	Oct. 25	3	85–87	94–93	5–3	Polar group of spectra, analogous to group 4 of Table 1. Mean brightness temperature 242 K.
5	Oct. 23	2	77–80	97–96	15–12	Analogous to group 5 of Table 1. $T_B = 248$ K at 1152 cm^{-1} and 260 K at 365 cm^{-1}. Strong absorption of hot CO_2 bands near 960 and 1060 cm^{-1}. Distinct H_2SO_4 absorption near 900 cm^{-1}.

2. The a-priori aerosol model and its exact determination

The Earth based observations and the direct measurements already gave relatively extensive information on the structure and the composition of Venusian clouds [6–13]. These data refer chiefly to latitudes $\varphi < 65°$, but their basic characteristics (e.g. composition) can be assumed also at higher latitudes. These characteristics can be summarized shortly as follows:

a) The lower boundary of the clouds is relatively distinct, the particle concentration decreases approximately for two orders of magnitude within one km. The altitude of the lower boundary is in the range 47–50 km but possibly with some local variation.

b) The upper boundary is not distinct, the concentration decreases above ~65 km altitude exponentially with the scale height 2–3 km.

c) The particle size distribution is complicated and can formally be described by dif-

Table 3

Parameters of the a-priori size distribution of cloud particles. Each mode is described by a log normal distribution

$$n(r) \sim \frac{1}{r} \times \exp\left\{-\left[\ln\left(\frac{r}{\bar{r}}\right)\right]^2 \Big/ 2 \times (\ln\sigma)^2\right\}$$

in the dominating altitude regions

	mode 1 $H > 48$ km	mode 2 $H > 57$ km	mode 2′ $H < 57$ km	mode 3 $H < 57$ km
$\bar{r}$, μm	0.15	1.05	1.40	3.85
σ	1.91	1.21	1.23	1.30

ferent methods. One of them is the so-called three mode distribution. We will principally confine to this description, considering that each mode is represented by a log normal distribution. Hereby, for mose 2 we will suppose the existence of two variants (2 and 2′), i.e. actually a four-mode size distribution is used. The parameters of these 4 distributions are given in Tab. 3. These parameters are recommended by Pollack et al. [7]. The particles of modes 1 and 2 (or 2′) are present in all altitudes, of mode 3 only lower than 57 km altitude.

Regarding the chemical nature of mode 1 and mode 2 there are serious arguments for water soluble sulphuric acid droplets [11—14]. We suppose the mode 3 particles as of the same composition noting that this supposition is for our purposes not very critical due to relatively small contribution of these particles to the outgoing infrared radiation. Tab. 4. compiles data on extinction cross sections and single scattering albedo used in our calculations. The share of mode 1 particles is not great due to their small extinction cross section. In some cases only mode 2 particles were considered in the claculation of synthetic spectra. Nevertheless, for more exact calculations all three modes should be taken into account.

The particle concentrations for all modes applied in the basic a-priori model are given in the left part of Tab. 5. The original model of Pollack et al. [7] was applied at altitudes $H < 65$ km. In higher altitudes the distribution was chosen according to the following considerations: the mode 2 particle concentration is 100 cm^{-3} at an altitude of 68 km, and the ratio of the number of particles of modes 1 and 2 is assumed as 5. Below 68 km altitude the number of particles increase reaching the value obtained in direct measurements [6] aboard Pioneer Venus at an altitude of 65 km. Above 68 km altitude the number of particles decreases exponentially with 2.5 km scale height. The optical depth $\tau \simeq 1$ is at 70 km altitude for the visible part of the sepctrum and at 66 km altitude for the infrared region ($\lambda = 11.5$ μm). Such a model gives a brightness temperature $T_B = 225$ K at 11.5 μm wave length, obtained in Earth based observations and in the infrared experiment aboard Pioneer Venus [15] at low latitudes.

The outgoing thermal radiance density of a plane-stratified medium,

$$I_\nu = \int_0^\infty B_\nu(T) \times e^{-\tau_\nu} \times d\tau_\nu, \tag{1}$$

Table 4
Parameter $\langle Q \times r^2 \rangle_{i,\nu}$ for the size distribution of Table 3 and ω_0 for mode 2

ν, cm^{-1}	$\langle Q \times r^2 \rangle_i$				ω^0
	1	2	2′	3	for mode 2
312	0.0025	0.18	0.45	18.5	0.010
344	0.0022	0.15	0.37	20.0	0.018
365	0.0022	0.15	0.37	23.0	0.020
451	0.0033	0.22	0.62	35.6	0.043
500	0.0027	0.18	0.56	38.0	0.072
549	0.0042	0.30	0.93	46.5	0.089
601	0.0080	0.54	1.45	37.0	0.045
798	0.0036	0.31	1.10	54.0	0.260
844	0.0048	0.42	1.58	55.7	0.290
899	0.0112	0.88	2.68	48.0	0.170
958	0.0076	0.67	2.30	52.0	0.260
1042	0.020	1.59	4.24	47.0	0.220
1152	0.0258	1.88	4.37	45.0	0.198
1202	0.0282	1.76	3.90	41.0	0.160
1312	0.0160	1.00	2.30	33.0	0.102
1399	0.0084	0.59	1.52	33.0	0.150
1500	0.0073	0.60	1.70	42.0	0.272
1601	0.0090	0.81	2.37	48.0	0.360
18000*)	0.136	2.90	4.92	36.6	1.000

*) corresponds to 0.55 μm, useful for control of calculated and measured optical depths available from measurements of scattered solar radiation

where $B_\nu(T)$ is the Planck function, T — temperature,

$$\tau_\nu = \sec \Theta \times \int_z^\infty k_\nu \times \mathrm{d}z \tag{2}$$

τ — optical depth corresponding to altitude z, Θ — angle between normal and direction of sounding, k_ν — extinction coefficient.

In case of aerosol extinction outside the gaseous absorption bands

$$k_\nu = 10^{-3} \times \pi \times \sum_{n=1}^{N} n_i(z) \times \langle Q \times r^2 \rangle_i,\ \mathrm{km}^{-1} \tag{3}$$

where

$$n_i(z) = \int_0^\infty n_i(z, r) \times \mathrm{d}r,\ \mathrm{cm}^{-3}; \tag{4}$$

$n_i(z, r) \times \mathrm{d}r$ — number density of particles of mode i with the radius from r to $r + \mathrm{d}r$ at the altitude z.

Table 5
Particle number densities for the a-priori model at lower latitudes ($\varphi < 60°$)

altitude km	basic model; N, cm^{-3}				modified model; N, cm^{-3}			
	mode 1	2	2′	3	1	2	2′	3
85	0	0	0	0	1.0	0	0	0
84	0	0	0	0	1.8	0	0	0
83	0	0	0	0	2.5	0	0	0
82	1	0	0	0	4	0	0	0
81	2	0	0	0	6	0	0	0
80	3	0	0	0	10	1	0	0
79	4	0	0	0	15	1.5	0	0
78	9	1	0	0	20	2	0	0
77	14	3	0	0	30	3	0	0
76	20	4	0	0	50	5	0	0
75	30	6	0	0	70	7	0	0
74	45	9	0	0	110	11	0	0
73	68	14	0	0	160	16	0	0
72	100	20	0	0	240	24	0	0
71	150	30	0	0	360	36	0	0
70	225	45	0	0	530	53	0	0
69	335	68	0	0	800	80	0	0
68	500	100	0	0	1200	120	0	0
67	470	80	0	0	1800	180	0	0
66	280	27	0	0	1500	150	0	0
65	180	9	0	0	200	0	20	0
64	500	50	0	0	750	0	75	0
63	500	50	0	0	750	0	75	0
62	500	50	0	0	750	0	75	0
61	500	50	0	0	750	0	75	0
60	500	50	0	0	500	0	30	0
59	500	50	0	0	500	0	50	0
58	500	50	0	0	500	0	50	0
57	500	50	0	0	300	0	50	0
56	500	0	50	3	500	0	50	3
55	500	0	50	10	500	0	50	10
54	500	0	10	10	500	0	50	10
53	500	0	50	10	500	0	50	10
52	500	0	50	10	500	0	50	10
51	500	0	50	10	500	0	50	20
50	500	0	50	10	500	0	50	30
49	500	0	50	10	500	0	50	20
48	0	0	0	0	0	0	0	0

The parameter $\langle Q \times r^2 \rangle_i$ is the mean extinction coefficient (with an accuracy up to factor π) for mode i, normalized to one particle per volume element

$$\langle Q \times r^2 \rangle_i = \frac{1}{n_i(z)} \times \int_0^\infty n_i(z, r) \times Q_i \left(\frac{2 \times \pi \times r}{\lambda} \right) \times r^2 \times \mathrm{d}r, \tag{5}$$

where $Q_i \left(\frac{2 \times \pi \times r}{\lambda} \right)$ is the extinction efficiency factor, λ — wave length.

The parameter $\langle Q \times r^2 \rangle_i$ used in the calculation of synthetic spectra is given in Tab. 4 for wave numbers of insignificant influence of gaseous absorption.

For all modes 75% water soluble sulphuric acid particles are assumed. The data on the real and imaginary refraction coefficient are taken from [16, 17]. Formula (1) neglects scattering processes, i.e. the single scattering albedo of particles is $\omega_0 = 0$. The parameter ω_0 is given for one of the modes in Tab. 4. ω_0 is in general sufficiently small justifying the neglect of multiple scattering processes in first approximation, though

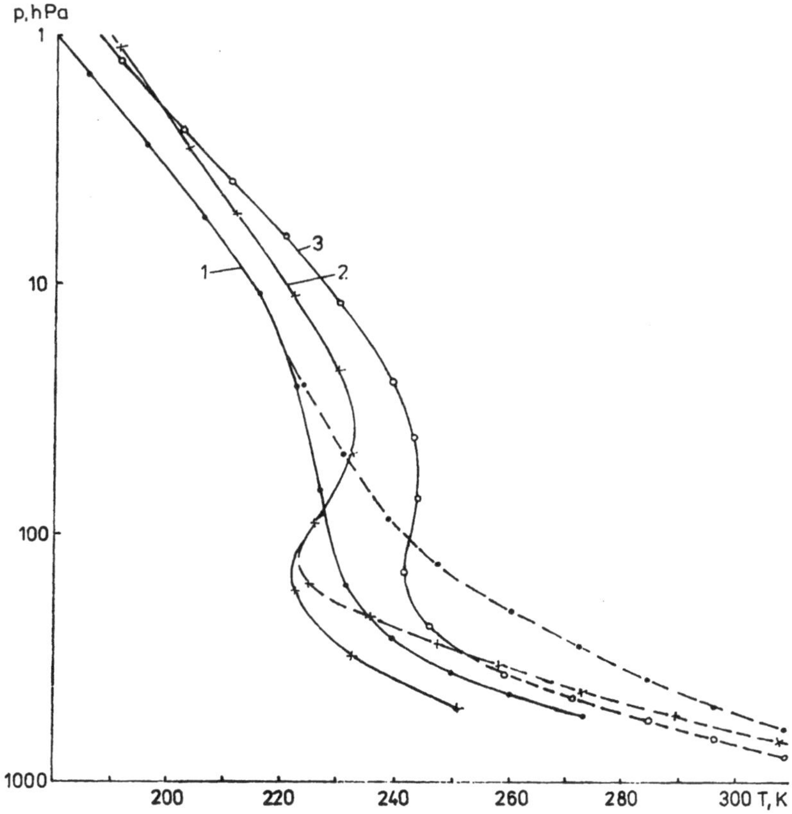

Fig. 3a

Fig. 3. Temperature profiles for groups of spectra on October 14, 1983 obtained directly by retrieval from averaged spectra (full lines) and corrected with model temperature profiles (dotted) (Tab. 1). Fig. 3a):
1 — groups 1 and 7, 2 — groups 2, 3 — group 3; (Fig. 3b): 4 — group 4, 5 — group 5, 6 — group 6

for more exact synthetic spectra scattering should be considered (see chapter 4). In the present stage this has not been done.

Fig. 1 shows a synthetic spectrum, calculated for latitudes $\varphi < 60°$ on the base of an a-priori model for 14 wave numbers outside the gaseous absorption bands. The mean temperature profile for group 1, shown in Fig. 3, was used. There is a good coincidence of the synthetic and measured spectrum for the narrow range 1000—1300 cm^{-1}, but outside its boundaries the synthetic spectrum has considerably higher brightness temperatures. The temperature difference is very great in the region $\nu < 550$ cm^{-1} where the clouds have maximum transmission (see Tab. 4). Obviously, the a-priori model has to be corrected for at altitudes $\lesssim 70$ km. This correction can be carried out either for the total number of particles or for parameters of the particle size distribution function or for both.

We varied the boundary altitude and the number density of modes 2 and 2′. The minima of mode 2 were introduced at 60 to 65 km altitude, and the maximum of mode 3 at 50 km altitude. Direct measurements indicate the existence of these peculiarities. The density ratio of particles of modes 1 and 2 (or 1 and 2′) was chosen to 10. The most important variation which is necessary for the correspondence of synthetic and measured spectrum is the increase of the number of modes 2 and 2′ particles above 60 km altitude. We

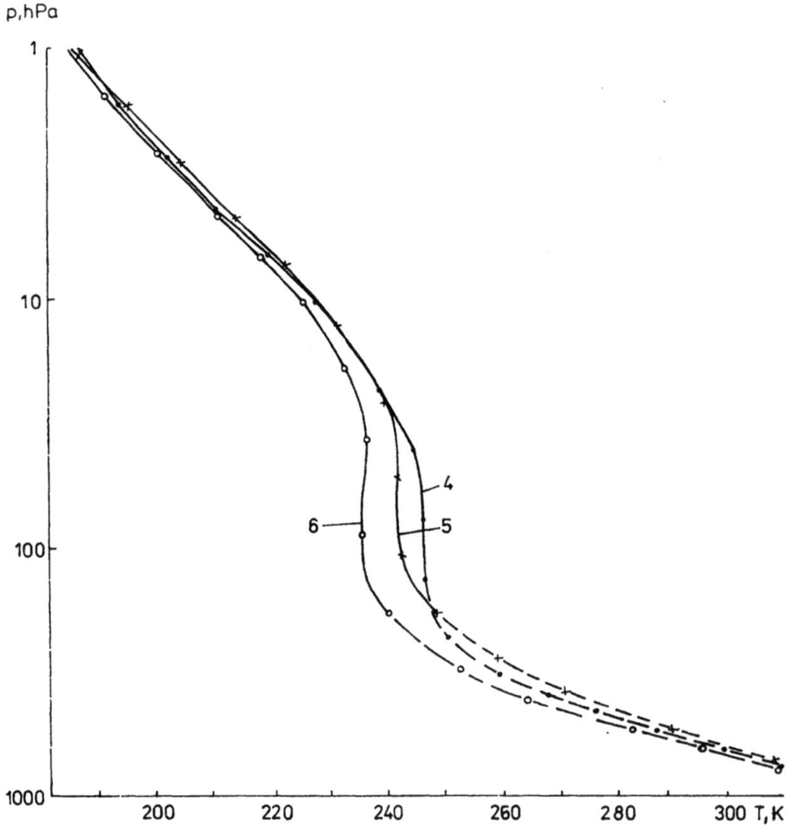

Fig. 3b

changed mode 2 into 2′ in the altidude interval 57—65 km, otherwise we had to increase the particle number of mode 2 by several times.

The right part of Tab. 5, the modified a-priori model, which gives exact (with an error $\Delta T \leqq 1$ K) correspondence of calculated and measured values T_B for two wave numbers 365 and 1152 cm^{-1} and also for nearly the whole spectrum too, among them in the spectral region of 900 cm^{-1}, belonging to H_2SO_4.

Obviously, the good correspondence of the synthetic and the measured spectrum can be considered an essential new argument in favour of the fact that sulphuric acid is a main component of the cloud layer and its supposed characteristics are approximately correct. Simultaneously, the model represented in the right part of Tab. 5 cannot be considered unequivocal. It is a certain realistic approximation which gives a satisfactory synthetic spectrum and does, at the same time, not contradict the data of direct measurements and Earth based observations. The direct measurements show somewhat smaller particle numbers for mode 2 [6] at 57—65 km altitude, but this difference is hardly significant as only a single measurement is available as reference and as the existing difference can be compensated by changing the size distribution appropriately. Nephelometric measurements [13] estimated particle number densities of 200—350 cm^{-3} in 58—62 km altitude with $r_{eff} = 1.0$—1.35 for a sufficiently narrow size distribution.

In Fig. 4, the altitude dependence of the extinction coefficient and the optical depth of the modified model is given for several wave numbers showing that the optical depth $\tau = 1$ at 365 cm^{-1} corresponds to 57 km altitude, and $\tau = 2$ to about 54 km altitude. In this way the outgoing infrared spectral radiance, originates in different cloud layers, including the 'middle' cloud layer (52—57 km).

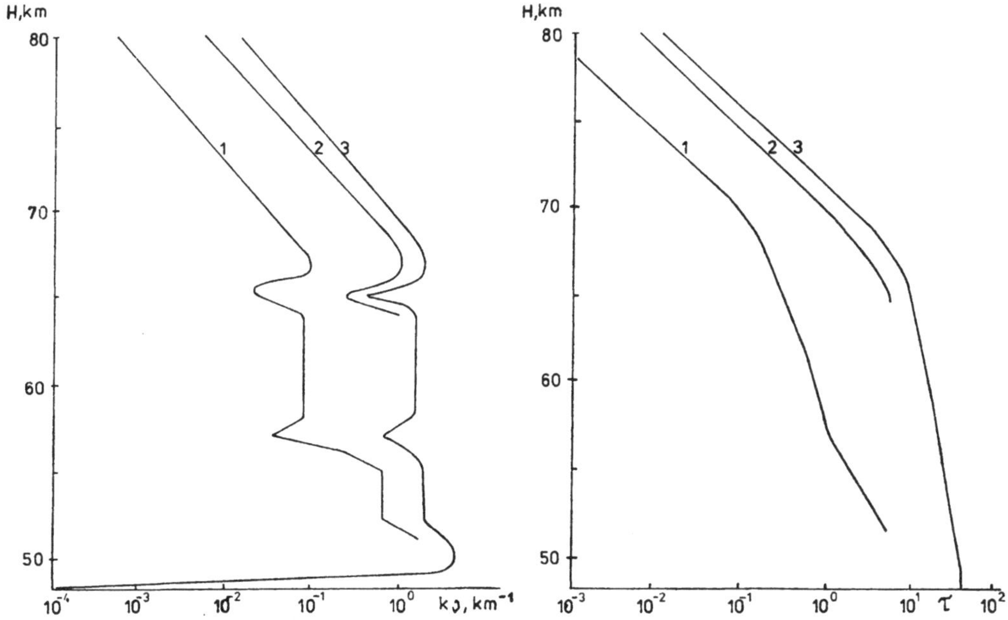

Fig. 4. Extinction coefficient and optical depth in dependence on altitude for the modified a-priori model. 1 — 365 cm^{-1}, 2 — 1152 cm^{-1}, 3 — 18000 cm^{-1} (visible region of the spectrum)

We state that the spectra at low latitudes ($\varphi < 60°$) are identical on the night and day side of the planet (see Fig. 1). This is valid for all orbits demonstrating the scarcely differing cloud structure in a great altitude interval from 54 to 75 km during day and night in low latitudes. According to measurements of Venera 9 and Venera 10 significant diurnal variations of the cloud structure at low latitudes were supposed [18, 19]. Either this conclusion was not correct, or the cloud structure varies in low latitudes with a great characteristic time constant.

3. Simplified models

It is obvious that the modified a-priori model can be used at low latitudes, but at $\varphi > 60°$ it requires fundamental modifications. Because of no direct measurements at that latitudes there is hardly a reason for applying such a complicated model there. Since the physical quantity which directly determines the density of the outgoing radiation at a given wave number is the vertical structure of the extinction coefficient, it is purposeful to look for simplified versions of the vertical structure using Fig. 5. The simpliest possibilities are:

A. At each given wave number the extinction coefficient remains constant below a certain altitude H_0 which is common for every wave number. The extinction coefficients at different wave numbers are determined from Tab. 4 (modes 2 and 1 are taken into account). Characteristic parameters of the model are the extinction coefficient k_ν and the altitude H_0.

The altitude of optical depth $\tau = 1$ at wave number ν equals

$$H_1(\nu) = H_0 - (k_\nu)^{-1}. \tag{6}$$

B. At each given wave number the optical depth and the extinction coefficient vary exponentially with a certain scale height. The extinction coefficients at different wave numbers are also given as in variant A. Characteristic parameters are the scale height (uniform for all wave numbers) and the altitude, corresponding to optical depth $\tau = 1$ for one of the wave numbers. The optical depth of the clouds at wave number ν and altitude H are connected by formula

$$\tau_\nu = e^{[H_1(\nu) - H]/H_a}, \tag{7}$$

where H_a — aerosol scale height, the extinction coefficient equals

$$k_\nu = \frac{1}{H_a} e^{[H_1(\nu) - H]/H_a}, \tag{8}$$

$$k_\nu = k_\nu^1 + k_\nu^2, \tag{9}$$

where k_ν^2 and k_ν^1 are extinction coefficients of modes 2 and 1. It is assumed that the relation of particle number densities of these modes is constant.

The particle number density of mode 2 at the altitude H equals

$$N_2 = \frac{k_\nu^2}{\pi \times \langle Q \times r^2 \rangle_\nu} \tag{10}$$

in model A. It does not depend on the altitude lower than level H_0. In model B it varies with altitude according to

$$N(H) = \frac{e^{[H_1(\nu) - H]/H_a}}{H_a \times \pi \times \langle Q \times r^2 \rangle_\nu}. \tag{11}$$

As a whole models of type A correspond better to the profiles given in Fig. 4, and they give on the average better correspondence of synthetic and observed spectra than B as shown by numerical experiments. The B type models lead to somewhat greater particle number densities in the middle cloud layer. Nevertheless, since we know nothing about the cloud structure at latitudes $\varphi > 65°$, there is no reason to renounce completely models of this type, at least for the description of the upper cloud layer. In Tab. 6 and in Figs. 5 and 6 evaluations of cloud parameters are given obtained by method A. Tab. 7 contains the corresponding parameters for method B. On principle, a combination of both variants (constant extinction coefficient below a certain altitude and an exponential decreasing above) can be applied resulting in a three parameter model which is more complicated for retrievalal gorithms. We do not exclude the possibility of its application in future.

In case of a temperature profile with inversion ambiguity is possible, approving 2 or 3 different models (both of type A and type B) with different parameters. Group 2 serves as an example. It has been supposed that the ambiguity can be removed by means of determination of synthetic spectra in the CO_2 absorption range, but this has not been done until now. In [20] results of similar determinations for the short-wave wing of the 15 μm CO_2 band (see Fig. 3 in this paper) are given, showing that similar spectra can be calculated with different models for group 2.

The fundamental argument in favour of the selection of a lower position of the upper boundary (64.5 km for type A model) is the great depth of the SO_2 band, at 1360 and 519 cm^{-1} in this spectral group, though it cannot be considered decisive since local-temporal variations of the SO_2 content are not impossible.

Table 6

Model of constant extinction coefficient

No. of groups Oct. 14, 1983	latitude	altitude of upper boundary km	extinction coefficient km^{-1}		number of particles mode 2**), cm^{-3}
			1152 cm^{-1}	365 cm^{-1}	
1	34–50°	72.0	0.91	0.073	110
2*)	63–70	64.5	2.5	0.20	370
3	79–85	70.0	2.5	0.20	370
4	86–87	65.0	4.0	0.32	595
5	85–81	60.0	5.0	0.4	745
6	78–71	72.5	0.83	0.067	125
7	59–50	72.0	0.91	0.073	110

*) with ambiguity — three different solutions for altitude of upper boundary and number density of mode 2 are possible (see text and Figs. 5, 6)

**) ratio of number density of mode 1 and 2 particles was chosen 10

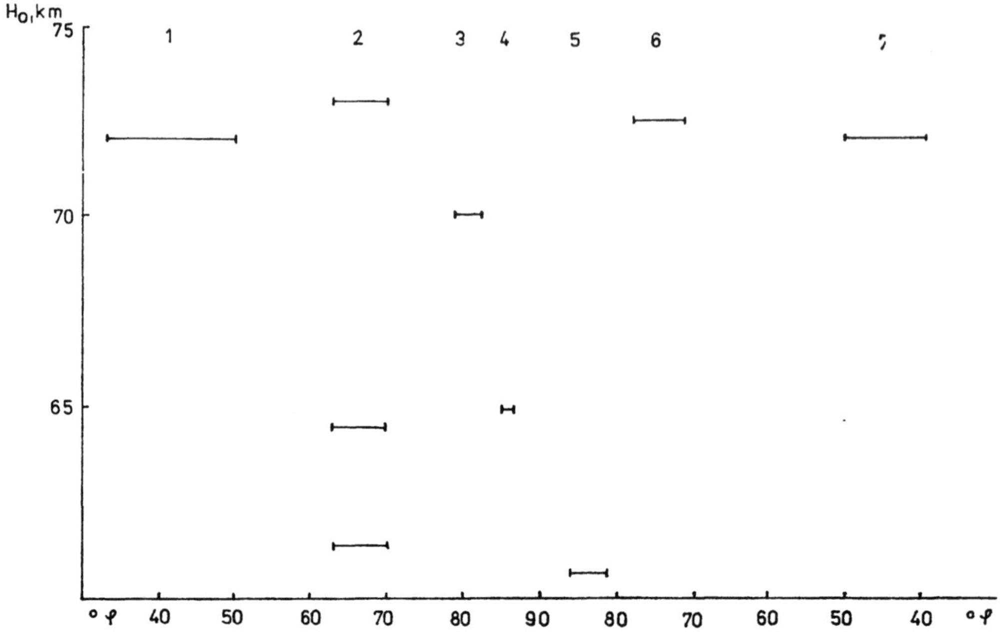

Fig. 5. Altitude of the upper cloud boundary in dependence on latitude for models with constant particle concentration (Tab. 6). The numbers correspond to the group numbers (Tab. 1). For group 2 models are given with $H_0 = 73$ km and $N = 180\ cm^{-3}$ and $H_0 = 61$ km and $N = 615\ cm^{-3}$, which are missing in Tab. 6

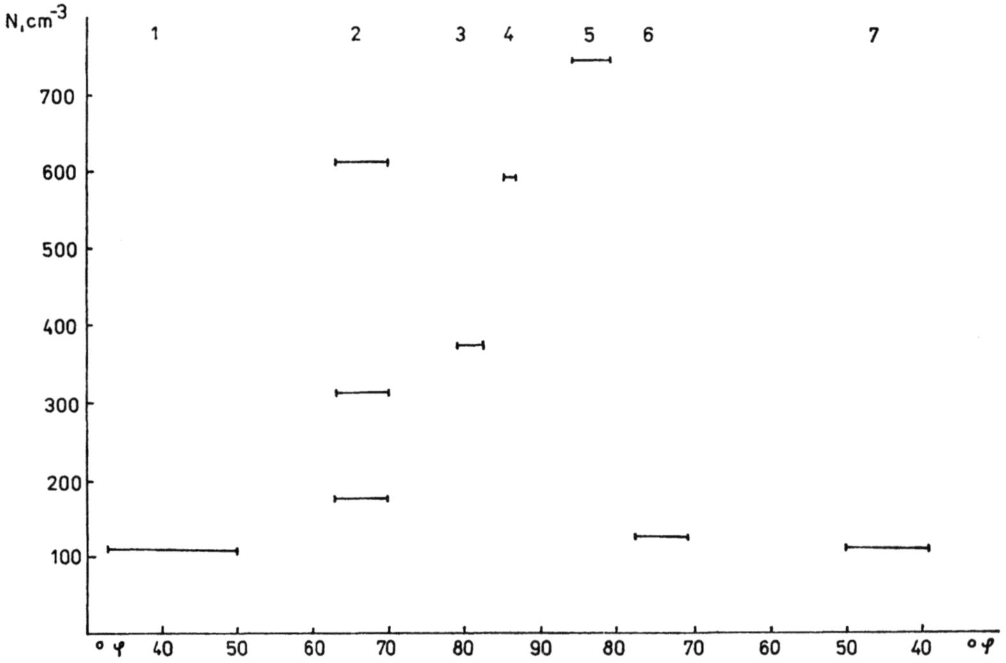

Fig. 6. Number density of particles in dependence on latitude for the same models as in Fig. 5

Table 7

No. of groups	latitude	altitude corresponding to $\tau = 1$, km		aerosol scale height	number density of mod 2 particles corresponding to $\tau = 1$**)	
		1152 cm^{-1}	365 cm^{-1}		1152 cm^{-1}	365 cm^{-1}
1	34–50	68.5	58.5	3.8	31	385
2*)	63–70	69.0	58.4	4.2	28	350
		62.0	58.3	1.4	82	1150
3*)	79–85	69.0	66.0	1.1	135	2060
		67.0	59.4	3.0	50	630
4	86–87	65.0	59.9	2.0	74	950
5	85–81	60.0	57.7	1.1	135	1720
6	78–71	67.0	55.6	4.5	33	415
7	59–50	68.5	58.8	3.8	31	385

*) with temperature inversion
**) ratio of number density of mode 1 and 2 particles was chosen 30

Additional arguments in favour of a lower cloud top instead of the highest position of the upper boundary (72 km) for the night part of the orbit on October 14, are the much lower cloud tops in high than in middle latitudes, and the substantially larger opacity of polar cloud near the upper boundary (see Figs. 5, 6).

On the day side of the orbit on October 14, the difference between high and low latitudes is smaller despite the occurrence of the so-called 'hot spot' at 78–82° latitudes, where the cloud top has a minimum and the number concentration a maximum (see Figs. 5, 6).

At high latitudes, there is obviously a belt with lower cloud tops and higher particle concentration. This belt is obviously not closed, and its boundaries are not strictly fixed, they vary [15].

As already mentioned [1–3], the 'hot spot' at latitudes 78–82° is possibly one of the poles of the 'dipole' found 5 years before our measurements in the Pioneer Venus infrared experiment [15]. The 'hot spot' sometimes occurs at the increasing part of the orbit (to the pole), sometimes on the decreasing one, and sometimes two weak spots occur. The spectral groups No. 5 on October 14 and also No. 5 on October 23 are in correspondence with the well expressed event of the occurrence of a 'hot spot'. Here, the upper boundary of the clouds is about 10 km lower than at latitudes $\varphi < 60°$ explainable by the existence of downward movements at altitudes of about 60 km in that region of the planet.

4. Unsolved problems

The method of determining the vertical extinction coefficient profile based on measurements of the outgoing radiance density for different viewing angles (limb darkening), is well-known. Measurements of Venera 9 and Venera 10 [18, 19] have been analyzed in this way, though it is clear now that a too wide spectral band (8–40 μm) reduces considerably the value of such an analysis, since in equatorial regions the level $\tau = 1$ is situated about 15 km lower at $\lambda > 18$ μm than at $\lambda < 12$ μm. Though in our case the dependence $k(H)$ can be inverted in a wide altitude range on principle using different wave numbers difficulties, connected with the geometry of the experiment as a wide field of view of the instrument and a limited accuracy in the determination of the limb position. In spite of these difficulties limb darkening analysis of Venera 15 measurements is intended in future.

It has to be stated once more that temperature retrieval and cloud layer parameter estimation must be solved in a self-consistent manner. So far such a self-consistent solution has not been obtained, and the quantitative data given here have preliminary character. A good correspondence of synthetic spectra and measured ones would be impossible if the hypothesis of sulphuric acid cloud particles (at least of mode 2) were not true. The assumed H_2SO_4 concentration was 75%. It is useful to analyse the admissible range of sulphur and concentration consistent with the measured spectra and to look for additional restrictions which can be imposed on the composition of cloud particles. At this stage the correspondence of synthetic and measured spectra is entirely satisfactory though some discrepancies remain at about 500 and 1500 cm^{-1}.

Their origin can be connected with peculiarities of the composition, altitude distribution, errors in laboratory data or in the assumed distribution according to dimensions. Additionally, the approximation of a purely radiating medium is not completely exact. The radiance density of a homogeneous semi-infinite medium with scattering is in case of a spherical indicatrix

$$I = I_0 \times \sqrt{1 - \omega_0} \times H(\omega_0), \tag{12}$$

where I_0 — radiance density without consideration of scattering, ω_0 — single scattering albedo, $H(\omega_0)$ — H-function. At small viewing angles ($z_0 = 0$) and for single scattering albedo $\omega_0 < 0.4$ the H-function is approximated by

$$H = 1 + 0.4 \times \omega_0 . \tag{13}$$

The radiance density correction with respect to scattering can reach 10% in this case at some wave numbers (e.g. for $\nu = 844$ cm^{-1}). The simplest method of scattering correction is the application of a correction coefficient, calculated for a homogeneous medium. Surely, in future scattering must be included in the interpretation of synthetic spectra. Up to now only one orbit has equantitatively been investigated concerning aerosol parameters. Visual inspection of the graphs of all orbits gives no hint of any distinct variations, indicating qualitatively similar conditions in the cloud layer, though all orbits must be analyzed with a standard procedure, perhaps on the base of that one probably presented here.

Conclusions

The following conclusions can be stated:

1. The spectra obtained during the 'scanning' of the planet along the meridian can be divided into several groups. Their peculiarities are at least in part be connected with the peculiarities of the cloud layer structure in different regions of the planet.
2. Measured and synthetic spectra calculated on the assumption of sulphuric acid cloud particles agree well and confirm, thus, this hypothesis.

3. At low latitudes ($\varphi < 60°$) the cloud transmission in the long-wave part of the spectrum ($\nu < 500\ cm^{-1}$) is so great that a significant part of the outgoing radiation is generated in the middle cloud layer ($53 < H < 58$ km).
4. The spectra measured at low latitudes agree with synthetic spectra calculated under the assumption of an a-priori model of particle size distribution and number density concentration which is with some modifications in accordance with in situ and remote Earth based measurements.
5. Spectra obtained in all latitudes agree reasonably well with synthetic spectra calculated under the assumption of simple cloud models:
   Model A with constant particle concentration within the cloud layer below a certain upper boundary H_0 and no patricles above H_0;
   Model B with exponential decrease of particle concentration on altitude.
   Models of type A show a somewhat better agreement of calculated and measured spectra.
6. The cloud structure at latitudes $\varphi < 60°$ is homogeneous on the day and night side of the planet for the orbit on October 14. First glance qualitative spectral analysis of all other measuring cycles indicates this statement as a general one.
7. The cloud characteristics are fundamentally different at 64—78° latitudes on October 14, before the transition to the polar region (night) and after it (day). The upper boundary of clouds is lower and more distinctly accentuated at night than during day when the cloud characteristics resemble the low latitude ones. The difference is not constant and not connected with local solar time according to qualitative spectral analysis of other cycles; there is obviously at these latitudes a region extended in longitude (but not closed) with increased particle concentration near the upper cloud boundary.
8. The cloud top is lower within the hot spot situated at 78—82° latitudes than in all other regions of the planet.

References

[1] Oertel, D., Moroz, V. I., Linkin, V. M. et al.: "Venera-15" i "Venera-16": pervye rezul'taty ehksperimenta po infrakrasnojj spektrometrii. Pis'ma v Astron. Zh. **10** (1984) 4, 243.

[2] Oertel, D., Moroz, V. I., Spänkuch, D. et al.: Infrared spectrometry of Venus from "Venera-15" and "Venera-16". Adv. Space Res. **5** (1985) 9, 25—36.

[3] Oertel, D., Moroz, V. I., Nopirakowski, J. et al.: Infrared experiment aboard the automatic interplanetary stations Venera 15 and Venera 16. 1. Methodics and first results. Veröffentlichungen des Forschungsbereichs Geo- und Kosmoswissenschaften, Heft 18, 1990.

[4] Spänkuch, D., Zasova, L. V., Schäfer, K. et al.: Infrared experiment aboard the automatic interplanetary stations Venera 15 and Venera 16. 2. Preliminary results of temperature profile retrieval. Veröffentlichungen des Forschungsbereichs Geo- und Kosmoswissenschaften, Heft 18, 1990.

[5] Kerzhanovich, V. V., Marov, M. Ya., Moroz, V. I.: Proposals for VIRA (Venus International References Atmosphere): dynamics and structure below 100 km. Preprint IKI AN SSSR Pr-831 (1983).

[6] Knollenberg, R. G., Hunten, D. M.: The microphysics of the clouds of Venus: results of the Pioneer Venus particle size spectrometer experiment. J. Geophys. Res. **85** (1980) A 13, 8039.

[7] POLLACK, J. B., TOON, O. B., WHITTEN, R. C. et al.: Distribution and source of the UV absorption in Venus atmosphere. J. Geophys. Res. **85** (1980) A 13, 8141.

[8] ESPOSITO, L. W., KNOLLENBERG, R. G., MAROV, M. YA.: The clouds and hazes of Venus. In: Venus/ed. D. M. Hunten, L. Colin, T. M. Donahue, and V. I. Moroz. The University of Arizona Press, Tucson, 1983.

[9] MOROZ, V. I.: The atmosphere of Venus. Space Sci. Rev. **29** (1981) 1, 1.

[10] RAGENT, B., ESPOSITO, L. W., TOMASKO, M. G. et al.: Particulate matter in the Venus atmosphere. Report submitted on COSPAR, Graz, 1984.

[11] HANSEN, J. E., HOVENJER, J. W.: Interpretation of the polarization of Venus. J. Atmos. Sci. **31** (1974), 1137.

[12] YOUNG, A. T.: Are the clouds of Venus sulfuric acid? Icarus **18** (1973) 564.

[13] MAROV, M. YA., LYSTSEV, V. E., LEBEDEV, V. N. et al.: The structure and microphysical properties of the Venus clouds: Venera-9, 10 and 11 data. Icarus **44** (1980) 3, 608.

[14] MARTONCHIK, J. V.: Sulfuric acid cloud interpretation of the infrared spectrum of Venus. Astrophys. J. **193** (1974), 495.

[15] TAYLOR, F. W., BEER, R., CHAHINE, M. T. et al.: Structure and meteorology of the middle atmosphere of Venus: infrared remote sensing from the Pioneer Orbiter. J. Geophys. Res. **85** (1980) A 13, 7963.

[16] PALMER, K. F., WILLIAMS, D.: Optical constants of sulfuric acid: application to the clouds of Venus? Applied Optics **14** (1975) 1, 208.

[17] JONES, A. D.: Optical constants of sulfuric acid in the far infrared. J. Q. Spec. Rad. Trans. **16** (1976), 1017.

[18] KSANFOMALITI, L. V., DEDOVA, E. V., OBUKHOVA, L. F. et al.: Infrakrasnoe izluchenie oblakov Venery. Kosmich. Issled. **14** (1975), 5, 768.

[19] KSANFOMALITI, L. V.: Venera 9 and 10: Thermal radiometry. Icarus **41** (1980) 1, 36.

[20] MOROZ, V. I., ZASOVA, L. V.: Infrakrasnoe izluchenie Venery: priblizheniye metody raschjota spektra v polosakh pogloshhenija atmosfernykh gazov. Kosmich. Issled. **23** (1985) 2, 259.

4. Infrared experiment aboard the automatic interplanetary stations Venera 15 and Venera 16

4. Preliminary results of spectral analysis in the region of H_2O and SO_2 absorption bands

V. I. Moroz, W. Döhler, E. A. Ustinov, K. Schäfer, L. V. Zasova, D. Spänkuch, A. V. D'jachkov, R. Dubois, V. M. Linkin, D. Oertel, V. V. Kerzhanovich, J. Nopirakowski, I. A. Macygorin, H. Becker-Ross, A. A. Shurupov, W. Stadthaus, A. N. Lipatov

Abstract

First results of Venusian spectra interpretation of the Venera 15 infrared experiment are presented. A great number of absorption details are seen, belonging to the rotation band but also to the fundamental rotational-vibrational $\nu_2 H_2O$ band. Two SO_2 absorption bands were identified (ν_2 at about 519 cm^{-1} band and ν_3 at about 1360 cm^{-1}). The H_2O and SO_2 mixing ratios are 30 ppm at 58 km and 2 ppm at 66 km, respectively, with an uncertainty of factor 3 according to preliminary estimates. These figures are based on midlatitude spectra of October 14, 1983.

In the infrared experiment of Venera 15 and Venera 16 about 1500 spectra of the planet were obtained in the range from 250 to 1600 cm^{-1} [1—2]. These spectra contain many absorption details which belong mainly to CO_2, to a less degree to H_2O, and to some degree to H_2SO_4 (in the aerosol phase) and SO_2. The CO_2, H_2O and H_2SO_4 bands were identified at once, but SO_2 bands are only sporadically seen in the spectra. In this paper arguments are presented which confirm the identification of SO_2, and examples are given of first quantitative estimates of H_2O and SO_2 mixing ratios (contents) based on the analysis of a small number of spectra.

Equivalent absorption width and the form of absorption features do not only depend on the absorber content but also on the vertical temperature profile and on the cloud extinction coefficient. In case of a monotonous temperature decrease with height the absorption feature is the stronger the deeper the level of optical depth $\tau = 1$ is situated in that part of the spectrum. If in case of a temperature inversion the level $\tau = 1$ is near to its lower boundary, the absorption feature can be absent for any absorber content or can even be transformed into an emission feature. The gaseous absorption visible in the measured spectrum increases with the dropping of the cloud top (cloud is equivalent to a decrease of the aerosol extinction coefficient) for one and the same mixing ratio. Therefore, before any quantitative determination of absorber mixing ratios can be get information on atmospheric temperature profiles [3] and cloud extinction coefficients [4], must be derived.

1. Qualitative comparison of Earth and Venus spectra in the 280—1600 cm^{-1} range

It is interesting to compare the spectra of the outgoing radiation of Earth and Venus. In Fig. 1, one Earth spectrum at cloudless conditions, measured by an instrument of the same type from the satellite Meteor 28 in 1977 [5] is shown together with one Venusian spectrum. Both instruments (constructed and manufactured in the GDR) have similar characteristics in spectral resolution and sensitivity. The equivalent absorption widths of the 15 μm CO_2 band in both planetary spectra differ not very much, regardless of the fact that CO_2 is the main component (about 97%) in the Venus atmosphere, but only a minor constituent (volume mixing ratio $F_{CO_2} \approx 330$ ppm) in the Earth atmosphere. The reason is evident: On Venus, the spectrum within this band is formed above the 100 hPa pressure level because of the presence of a cloud layer. The H_2O spectral components are much stronger in the Earth spectrum than in the Venusian one. Therefore, one can draw the qualitative conclusion that there is much less water above the Venusian clouds than in the troposphere of the Earth. This conclusion is not new but we will see below that the quantitative analysis of the Venusian spectrum at wave numbers smaller than 500 cm^{-1} allows for the first time the estimation of the H_2O content at 50—60 km altitude, higher than in situ measurements made so far, and lower than from retrievals using near infrared spectra measured from Earth.

Another important difference between both spectra is the presence of a strong ozone band in the Earth spectrum and its absence in the Venusian one. And vice versa, there are components in the Venusian spectrum absent in the spectrum of Earth: these are the bands of H_2SO_4 (at about 500 and 1000 cm^{-1}) and of SO_2 (1360 and 519 cm^{-1}). The

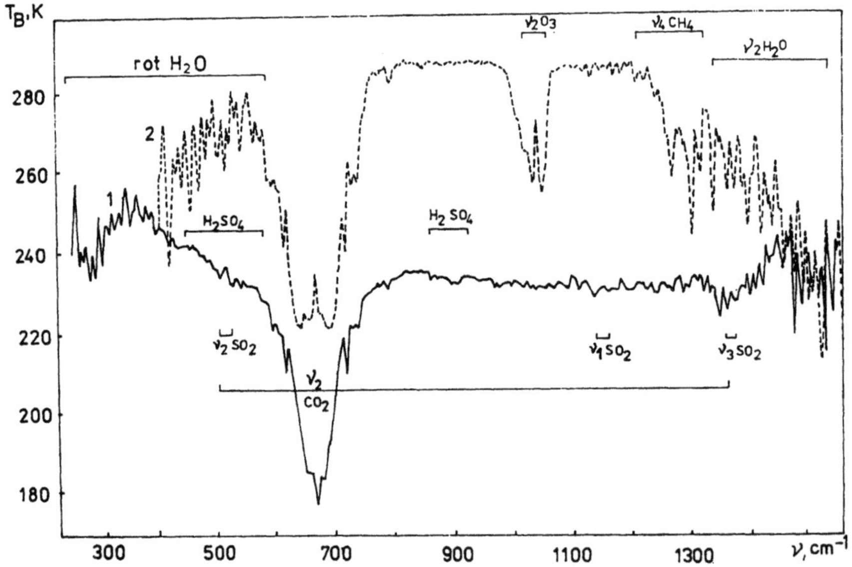

Fig. 1. Venusian infrared radiance spectra (1) and Earth infrared radiance spectra (2), obtained by means of infrared Fourier spectrometers. Spectrum 1 was measured aboard Venera 15 on November 17, 1983, in low latitudes, spectrum 2 aboard Meteor 28, on July 5, 1977, at night, above South France. At the abscissa — wave number ν, at the ordinate — brightness temperature T_B

ν_4CH_4 band (1305 cm^{-1}) is visible in the Earth spectrum, but is missing in the Venusian spectrum. The spectral components belonging to H_2O and SO_2 are for the first time detected in the infrared spectrum of the Venusian thermal radiation though they have been observed in the Venusian reflection spectra from Earth repeatedly [6—11], and sometimes with higher resolution [8]. This is explained by the fact that the Earth atmosphere is nearly or completely non-transparent in the corresponding parts of the spectrum, and even the use of instruments on high-altitude aeroplanes [9—11] did not prove a sufficiently effective mean.

In the H_2O rotation band, measurements had been carried out by the infrared radiometer of the Pioneer Venus satellite [12], however, with a very wide filter (182—286 cm^{-1}) which did not permit the separation of the water vapour contribution with sufficient confidence only.

2. The radiance spectrum for the a-priori model of vertical H_2O and SO_2 distributions

In the preparatory phase of the experiment (1979—1982) calculations of the expected Venusian radiance spectrum were made for a a-priori model of the vertical H_2O and SO_2 distributions [13] based on data published until 1979. This model is shown in Fig. 2.

The sources of information used can be divided into 2 groups — in situ measurements aboard Venera 11, 12 and the Pioneer-Venus large probe and remote measurements from Earth and from the Pioneer satellite. The first group gives evaluations of the mixing ratio only below 50—55 km, the second one on the layer above 65—70 km.

The remote spectroscopic measurements were carried out in the near infrared (H_2O) or in the near ultraviolet (SO_2) spectral range, and their quantitative interpretation is ambiguous since the model of simple reflection was used in most cases. These models

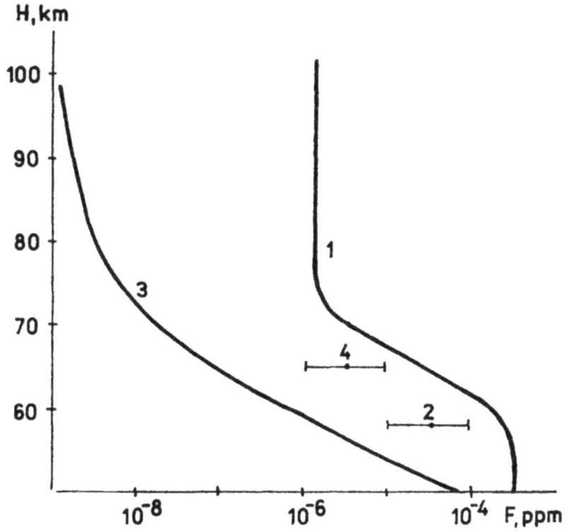

Fig. 2. A-priori models of vertical distributions of H_2O (1) and SO_2 (3) for the calculation of synthetic spectra before the mission. Given are, too, the estimates (2 — H_2O, 4 — SO_2) based on the analysis of the spectra of low latitudes of October 14, 1983. Abscissa — volume mixing ratio *F*, ordinate — altitude *H*

suppose clouds as a non-transparent surface for radiation with the absorbing gas as a whole above that surface. In reality, it is necessary to consider the radiation of a diffuse medium where aerosol particles and gas molecules are mixed. Such a model has used in our calculations.

The calculation of synthetic spectra needs cloud as well as temperature profile models. The corresponding a-priori models are described in [3, 4].

The transmittance calculation for gases was carried out line-by-line using the AFGRL catalogue of spectral lines [14] for a set of atmospheric levels p_i with a spectral resolution as well as a wave number step of 0.1 cm^{-1}.

The quasimonochromatic transmittances $t_{g\nu_i}$ were than convoluted with the spectral response function $f(\nu - \nu')$

$$t'_{g\nu_i} = \int_{-\infty}^{\infty} t_{g\nu_i'} \times f(\nu - \nu') \times d\nu' .$$

The transmission functions for the cloud model $t_{a\nu_i}$ were determined separately.

The total transmittance is

$$t_{\nu_i} = t'_{g\nu_i} \times t_{a\nu_i} . \tag{1}$$

The spectral response function of the Fourier spectrometer with apodization was represented by the function

$$f(\nu - \nu') = \frac{N}{\sigma} \times \frac{\sin\left[N \times \frac{\pi}{\sigma} \times (\nu - \nu')\right]}{N \times \frac{\pi}{\sigma} \times (\nu - \nu')} \times \left[0.54 + 0.46 \times \frac{\frac{N^2}{\sigma^2} \times (\nu - \nu')^2}{1 - \frac{N^2}{\sigma^2} \times (\nu - \nu')^2}\right], \tag{2}$$

with $\sigma = 3556$. For onboard processing of interferograms the number N of sampling points was 900, in case of processing the spectra on Earth $N = 1422$. The width of the spectral response function (at level 0.5) was 7.5 cm^{-1} and 5 cm^{-1}, respectively.

The outgoing radiance is

$$I_\nu = \int_{-\infty}^{\infty} B[T(p)]\, K_\nu(p, T)\, d \lg p , \tag{3}$$

where T — temperature, p — pressure, $B_\nu(T)$ — Planck function, $K_\nu(p, T)$ — weighting function. K_ν is determined by the expression

$$K_\nu(p, T) = -\frac{\partial t_\nu}{\partial \lg p} , \tag{4}$$

where t_ν — transmission function, calculated according to formula (1).

In Fig. 3 (full line) a part of a synthetic spectrum is shown, calculated by the above described method with a resolution of 5 cm^{-1}. The mean temprature model (see [3]) was used. Figure 3 contains, too, corresponding parts of two measured spectra with the same spectral resolution. The spectral components caused by H_2O are deeper in the synthetic spectra than in the measured ones. We note that the synthetic spectra were calculated with the passive integration only to the 600 hPa level which is completely sufficient for the region $\nu > 700\ cm^{-1}$ but somewhat insufficient for the region $\nu < 500\ cm^{-1}$ since the clouds are much more transparent here [4]. If integration is made to deeper levels, the calculated absorption features of H_2O would become still deeper.

Analogous calculation for SO_2 showed a reverse situation where even the strongest bands of this molecule at 1360 and 519 cm^{-1} are not visible in the synthetic spectra, of the a-priori models, but had been identified in measured spectra (see below).

3. Identification of SO_2 bands

The calculation of the synthetic spectrum showed that SO_2 bands can not be observed if the a-priori model is correct. Therefore the first detection of an absorption near 1360 cm^{-1} caused some doubts to identify this absorption with the $\nu_3 SO_2$ band. Additionally, a weak isotopic SO_2 band is situated exactly here, and a CO_2 band, too, induced by pressure. Further calculations, however, showed that the equivalent absorption width near 1360 cm^{-1} is in many cases somewhat too great for the last two bands.

In spectra obtained in the warm "polar region" [4, 15] the brightness temperature in

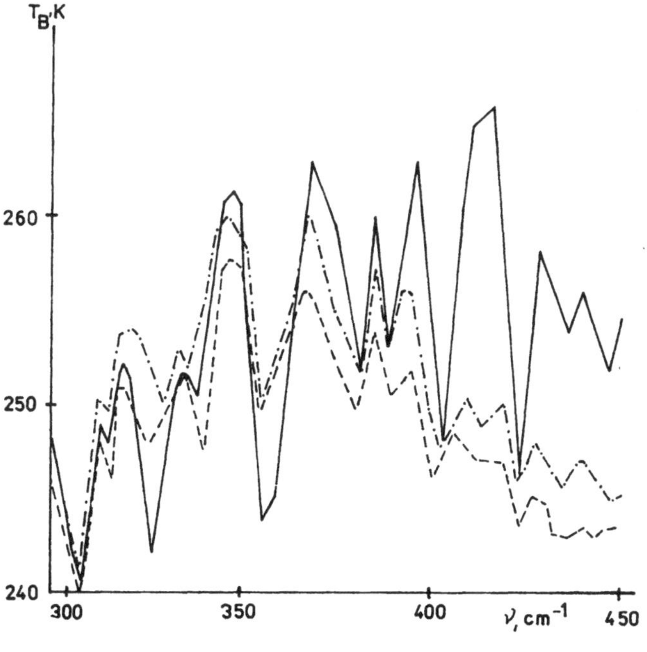

Fig. 3. Spectral range 300–450 cm^{-1}. Synthetic spectrum — full line, calculated with a-priori models of the vertical profile of H_2O (Fig. 2), and of the temperature profile and clouds (described in the text). The line marked with strokes and the line marked with strokes and dots — two individual spectra (middle latitudes). Spectral resolution of all three spectra — 5 cm^{-1}

the spectral range from 850 to 1400 cm^{-1} remains nearly constant, only with one exception — the component at 1360 cm^{-1}. The reason for this constant brightness temperature spectrum is the nearly isothermic temperature profile in the region of the upper cloud layer and the high concentration of aerosol particles. It is evident that if stronger CO_2 bands are missing they must be missed in this range, too.

Finally, in several spectra, especially with increased resolution (on Earth processing of interferograms — see [2]), a component was observed near 520 cm^{-1}, the position and form of which corresponds to the $\nu_2 SO_2$ band. There are also difficulties connected with the strength of CO_2 bands, and they are also overwhelmed due to the existence of smooth high latitude spectra with many CO_2 absorption bands not seen there. Tracks of the $\nu_1 SO_2$ band near 1150 cm^{-1} are also found.

Thus, the qualitative identification of SO_2 can certainly accepted but the quantitative analysis will be more complicated than for H_2O.

4. H_2O mixing ratio in middle latitudes

In the thermal radiance spectrum of Venus there are two spectral ranges identified with absorptions of water vapour. One of them is in the range $250 < \nu < 450$ cm^{-1}. From Fig. 4 it is evident that the spectral components which can be observed in this range, correspond with respect to their position to the strongest lines or groups of lines of the H_2O rotational band. In this part of the spectrum there is a great signal-to-noise-ratio. Spectral signatures in the 1400—1600 cm^{-1} range, belong mainly to the rotational vibrational $\nu_2 H_2O$ band. Here, the spectral signatures in the individual spectra can not

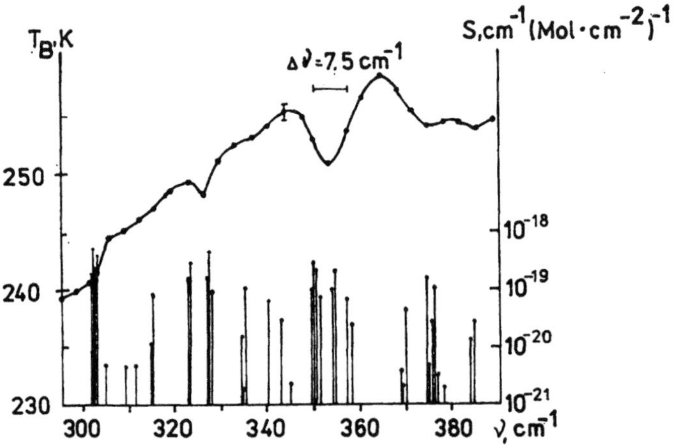

Fig. 4. Spectral range 295—350 cm^{-1} of onboard processing of interferograms (see [2])
Full line — average of 5 spectra, obtained on October 14 in middle latitudes $34 < \varphi < 50°$. The points denote wave numbers for which radiances in the spectrum were determined in the calculation of spectra from interferograms (by means of onboard electronic data processing).
Vertical line above — typical quantity of the mean quadratic deviation, horizontal one — width of the instrument function (7.5 cm^{-1} — spectra obtained in aboard interferogram processing).
The vertical line below shows the position of the lines of the rotational H_2O band, their length is proportional to the logarithm of the integral absorption coefficient in the given line (scale right).

clearly be identified (signal in the order of noise) though in the averaged spectra they can be identified.

In [4] it was shown that the level $\tau = 1$ in the spectral range $300 < \nu < 450$ cm^{-1} for aerosol extinction is situated at 57—58 km altitude. Since the components belonging to H_2O are relatively weak it is evident that just this altitude range makes the greatest contribution to their generation. The ν_2H_2O band is generated at significantly higher altitudes, the level $\tau = 1$ is at about 65 km altitude in middle latitudes. Therefore, in future it should be possible to determine the H_2O mixing ratio profile at least at two points quite different in altitude to determine from our measurements.

Examples of averaged spectra for different latitude zones and for different features of the cloud layer (low latitudes, cold collar, warm 'polar cap') are given in [4]. There it was shown that in these zones temperature profile as well as cloud structure are quite different from each other [1, 4]. On principle, H_2O profile can also be different.

To obtain quantitative evaluations of H_2O the following procedure was applied. For a group of similar spectra representing the above mentioned characteristic regions the altitude dependence of temperature was retrieved (profile $T(h)$). Then, a cloud model was searched for, which together with the retrieved temperature profile $T(h)$ gave a synthetic spectrum coinciding with the measured one outside the gaseous absorption bands. A set of H_2O mixing ratio models is assumed, and using the profiles $T(h)$ and the aerosol model, a set of synthetic spectra is calculated for spectral regions with water vapour absorption bands. The synthetic spectra can be compared with the measured ones — directly or after the construction of growth curves (see below). Mixing ratios with the best correspondence of synthetic and measured spectra are accepted as a reasonable approximation.

The greatest difficulty of this procedure is the 'exact' (line-by-line) determination of the gaseous transmission functions for different vertical distributions of the temperature and water vapour. To avoid the unacceptally high computer time an approximative method for the calculation of gaseous transmittances is applied [16]. This method is based on a certain line profile approximation, on the consideration of a sufficiently limited number of the strongest lines only and on the introduction of several other simplifications. Instead of transmittances quasi monochromatic absorption coefficients and quasi monochromatic radiances are calculated with a wave number step of about 0.1 of the width of the spectral response function. The convolution of these quasimonochromatic radiances with the spectral response function is carried out in the last stage. In [16] results of line-by-line calculated transmittances and transmittances calculated by this approximative method are compared, showing the efficiency and reliability of the proposed method for the analysis of our spectra in spectral regions of H_2O absorption lines. A modification of this procedure is described in [16], too, allowing the calculation of the outgoing radiance in the range of CO_2 and SO_2 bands correspondingly.

Selecting a sufficiently representative spectral absorption range, it is possible to determine its equivalent width as a function of the mixing ratio at the given altitude for a set of synthetic spectra. We will call this dependence growth curve. Each growth curve can be applied only for that cloud model and for that temperature profile that were used in its determination.

In the following we describe some examples obtained in the process of analyzing one of these spectral groups — these are the "low latitude" spectra of October 14, 1983. The main task of this analysis was intended as a sensitivity study regarding the choice of a cloud layer model and the different extrapolation of the temperature profile. Of course, temperature profiles retrieved from those spectra [3] were used. Nevertheless, in the altitude of the cloud layer the extrapolation by means of models is necessary in the present stage of investigation since in the retrieval of $T(h)$ the aerosol extinction has not been considered up to now.

In Fig. 5 the profile $T(h)$ is given which we used with two variants of extrapolation. Variant (b) is the main one, variant (a) the additional one.

In Tab. 1 the parameters of several aerosol models are given used for the calculation of synthetic spectra. Three of these models were obtained for middle latitudes [4]. All models were selected in such a way that the spectrum outside the gaseous absorption bands coincides with the measured one with an accuracy of 1—2 K.

For the calculation of synthetic spectra two hypotheses were applied regarding the vertical distribution of the H_2O mixing ratio in the vicinity of the level $\tau = 1$:

1) the mixing ratio F does not depend on altitude

$$F = F_0 = \text{const}, \tag{5}$$

2) the mixing ratio decreases exponentially with altitude

$$F = F_0 \times \exp\left[(H_0 - H)/H_1\right], \tag{6}$$

where $H_0 = 58$ km — altitude, corresponding to the level $\tau = 1$ for aerosol extinction (at 350 cm^{-1}), H_1 — parameter, characterizing the scale height of the absorber. In our experiments H_1 was taken as 2.5 km.

In Fig. 6, a set of synthetic spectra for the cloud model No. 1b (see Tab. 1) and of the exponential profile of H_2O is given as example. Moreover, there is an averaged spectrum

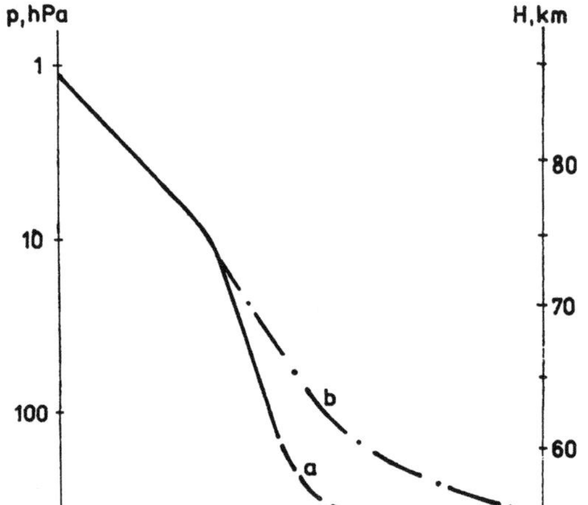

Fig. 5. Temperature profile, retrieved with consideration of aerosol influence in the 15 µm CO_2 band from the group of spectra obtained in October 14 in middle latitudes (full line), a and b — two different interpolations of this profile using model [22] (variant for middle latitudes)

Table 1

Cloud models, used for the calculation of synthetic spectra

No. of model*)	Model type	Particle concentration in dependence on altitude**)
1b	all 3 modes (concretized a-priori model)	see [4]
2b	mode 1 and 2, constant concentration	below 72 km $N_2 = 110$; $N_1/N_2 = 10$ above 72 km $N_1 = 0$; $N_2 = 0$
3b	mode 1 and 2, exponential altitude distribution	$N_2 = 31 \times \exp[(68.5 - H)/3.85]$; $N_1/N_2 = 10$
3a	for temperature profile a	$N_2 = 34 \times \exp[(65 - H)/3.50]$; $N_1/N_2 = 30$

*) models designated by a and b correspond with temperature profile *a* and *b* from Fig. 5
**) N_1, N_2 — number density of mode 1 and 2 particles, cm^{-3}

for low latitudes (orbit on October 14, 1983). Another example (for a constant mixing ratio) is given in [16].

The measured spectrum in Figs. 4 and 6 was obtained with a resolution of 7.5 cm^{-1}, therefore the depth of the absorption components (ΔT_B) is smaller than in Fig. 3. The synthetic spectra given in Fig. 6 are also calculated with a resolution of 7.5 cm^{-1}. The difference in the 'incline' of the synthetic spectra and the measured one arises because in the determination one and the same aerosol extinction coefficient was used for the whole range (equals the extinction coefficient for $\nu = 350\ cm^{-1}$).

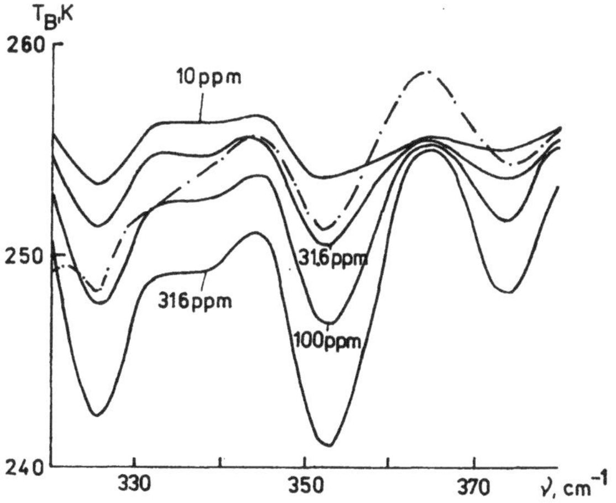

Fig. 6. Synthetic spectra of Venusian radiation in the area of the rotational H_2O band, calculated for the model with exponential altitude dependence of the mixing ratio (scale height 2.5 km) and with the concretized a-priori cloud model (see Tab. 1) for different values of the mixing ratio *F*. Line marked by strokes and dots — measured spectrum (average for 34—50° latitudes, orbit of October 14, 1983)

This incline scarcely influences the equivalent absorption width

$$W = \int \frac{I_{\nu_0} - I_\nu}{I_{\nu_0}} \, d\nu ,$$

where I_ν — radiance density with presence of H_2O, I_{ν_0} — radiance density without H_2O (level of the 'continuous spectrum'). In the measured spectrum for the absorption near $\nu = 352 \text{ cm}^{-1}$

$$W_{352} = 0.5 \pm 0.1 \text{ cm}^{-1}.$$

The main error source of the equivalent absorption width is the indefiniteness in the definition of the level of the continuous spectrum. It can be described statically, and the error of $\pm 0.1 \text{ cm}^{-1}$ is an approximate estimation.

In Tab. 2 and Fig. 7 theoretic growth curves $W_{352}(F)$ calculated after the synthetic spectra are given. From these data the sensitivity of the mixing ratio to the choice of the cloud models to the temperature profile and also to the character of the vertical cloud profile is obvious. The most realistic combination of temperature and aerosol profile models is 1b (see Tab. 1), but it is obvious that even in this case the uncertainty in estimating F (at an altitude of 58 km) is within one order of magnitude dependent on the assumption on the vertical H_2O profile. If the mixing ratio is assumed constant with height then $F_0 \approx 10$ ppm, if it is fast decreasing (according to the law given by formula (5)), then $F_0 \approx 100$ ppm. The preliminary estimation is, therefore,

$$F_{H_2O} \approx 30 \text{ ppm}$$

at 58 km altitude with the note that the possible deviation of this estimation from the real magnitude can reach 3 times or on third.

In Fig. 8 results of H_2O estimates are given, obtained in different experiments indicating that the supposition on a fast decrease of F_{H_2O} in the altitude range 55—60 km seems to be more realistic than the hypothesis of a constant mixing ratio with height. The conditions of equilibrium of the partial pressure of H_2O with the presence of sul-

Table 2

Equivalent absorption with of synthetic spectra near 352 cm^{-1}, calculated for different cloud models and different temperature profile extrapolations

Signatures of Fig. 7	1 b 2	1 b 1	2 b 1	3 b 1	3 a 1
temperature profile	b	b	b	b	a
cloud model (see Tab. 1)	1 b	1 b	2 b	3 b	3 a
altitude dependence of F_{H_2O}*)	exp	const	const	const	const
mixing ratio**):					
10^{-6}					0.22
10^{-5}	0.30	0.42	0.33	0.16	0.61
$3.2. \times 10^{-5}$	0.39	0.71	0.50	0.35	
10^{-4}	0.60	1.11	0.71	0.62	1.35
$3.2. \times 10^{-4}$	0.76				

*) const — altitude independent, exp — exponential distribution (see formula (5, 6))
**) at 58 km altitude for exponential distribution

phuric acid droplets — the mode 2 particles, obviously, consist of — also require a fast decrease of F_{H_2O} in this altitude interval [17].

Fig. 8 demonstrates that the assessment of water vapour mixing ratios at 55—60 km altitude is an actual task since significant differences exist between the results of different measurements of F_{H_2O} at 45—55 km altitude and those from higher altitudes where lines of the H_2O rotational band are generated. One group of experiments supplies here large mixing ratios (1000—2000 ppm), the other one significantly smaller ones (200 to 400 ppm). Further analysis of our data using also other parts of the spectrum, especially of the 6.3 μm band is reason for some hope to find the vertical H_2O profile in the upper part of the cloud layer independently and to contribute to the classification of the present contradiction. Further important questions which have to be discussed in the future concern the local-temporal variations of the H_2O mixing ratio. Up to now we can only state that the H_2O content on the day- and on the night-side of the planet in middle latitudes is of the same order of magnitude according to the data of the orbit on October 14. This is confirmed by equal equivalent absorption widths for the corresponding spectral groups, with identical temperature and cloud structure profiles, too [3, 4].

5. SO_2 mixing ratio

An estimation of the SO_2 content was made for the same middle-latitude-spectra. The synthetic spectra were calculated for this reason in the region 1200—1500 cm^{-1} with

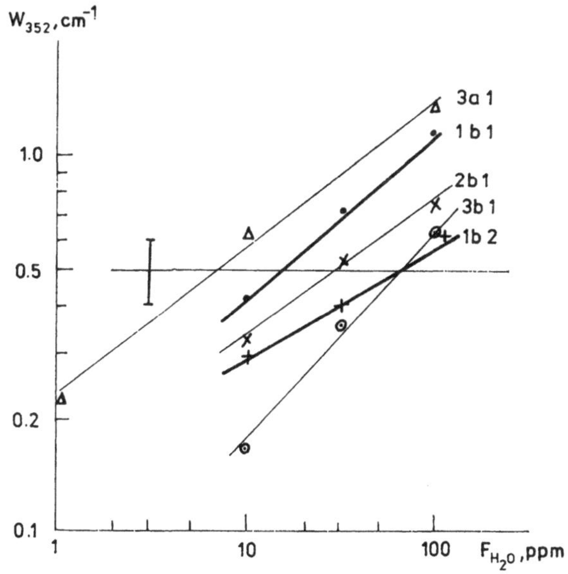

Fig. 7. Growth curves of equivalent absorption width at $\nu \approx 352$ cm^{-1}. The first number and letter correspond to the model number according to Tab. 1, the last number indicates if the mixing ratio is assumed as independent from altitude (1), or varying with height (2) (see formula (6). Abscissa — H_2O mixing ratio (in case of variation the magnitude at 58 km altitude is taken), ordinate — equivalent absorption width. Line parallel to the abscissa — measured equivalent absorption width. the vertical line characterizes the measurement error

consideration of the CO_2 absorption (1385 cm^{-1} band). The growth curves for the 1360 cm^{-1} SO_2 band were then obtained. The calculations are simplified by the fact that the distances between the rotational lines of SO_2 are very small (0.03 cm^{-1}) so that for pressures $p > 20$ hPa the approximation of strong superimposing lines [18] can be applied. The absorption coefficients measured by Moskalenko et al. [18] were used. The calculation of synthetic spectra is described in more detail in [16]. The same hypotheses about temperature profile and clouds for H_2O were applied as well as the same assumption on the altitude profile of the SO_2 mixing ratio, namely:

1) the mixing ratio is altitude-independent, 2) the mixing ratio decreases with altitude; in this case formula (6) was used, too, but with $H_1 = 65$ km, since for the given region of the spectrum the level $\tau = 1$ lies approximately at this altitude.

The first hypothesis gives the mixing ratio $F \approx 0.7$ ppm, the second one -- one order of magnitude larger. The preliminary estimation is

$$F_{SO_2} \approx 2 \text{ ppm},$$

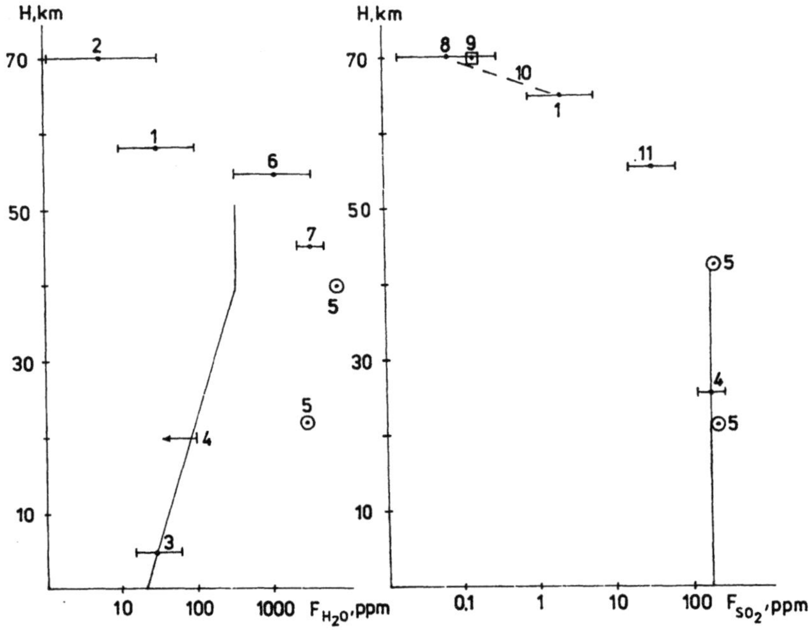

Fig. 8. Vertical profiles of H_2O mixing ratio (left) and SO_2 mixing ratio (right) in the atmosphere of Venus according to data of various experiments

1 — present paper, 2 — Earth based observations of H_2O absorption in the 0.8 μm spectral region [23], 3 — spectrophotometry aboard Venera 11, 12, 13, 14 [24, 25], 4 — gas chromatography aboard Venera 12 [26], 5 — gas chromatography aboard Pioneer Probe [27], 6 — gas chromatography aboard Venera 13 [28], 7 — electrolytic sensor aboard Venera 13, 14 [29], 8 — Earth based observations of SO_2 absorption bands in the 0.3 μm spectral range [19], 9 — measurements of SO_2 bands in the 0.2—0.3 μm spectral range aboard Pioneer Vernus Orbiter [20], 10 — interpretation of Earth based measurements by Barker in paper [21], 11 — photometry in the 0.36 μm spectral range aboard Venera 14 [30]

its accuracy is characterized by a factor of 3. This value of F_{SO_2} is significantly larger than the assessment based on measurements in the ultraviolet region (Earth based measurements [19] and aboard the Pioneer Venus Orbiter [20]) and it coincides qualitatively with the consequence of a sharp decrease of F_{SO_2} with increasing altitude in the 64—68 km altitude range (see [21]) from Earth based observation. A common analysis of both kinds of data is necessary for a better assessment of the vertical SO_2 profile.

It is possible that there are local-temporal variations of SO_2. The ν_2 519 cm^{-1} SO_2 band is seen quite well in the spectra obtained at 50—70° latitudes including the cold collar, it is weaker towards lower latitudes and it is not visible in equatorial latitudes as well as in polar latitudes. The quantitative analysis of these effects will be the subject of one of the following publications.

Conclusions

1. A great number of absorption features were identified, belonging to the H_2O rotational band, but also to be fundamental rotational-vibrational $\nu_2 H_2O$ band.
2. Two SO_2 absorption bands were identified (ν_2 at about 519 cm^{-1} and ν_3 at about 1360 cm^{-1}).
3. The quantitative estimation of H_2O and SO_2 depends in critical manner on the choice of temperature profiles and cloud models. Though information on both quantities be derived directly from the spectra measured, the remaining uncertainties distinctly influence the quantitative analysis.
4. The H_2O mixing ratio at 50 km altitude is estimated to 30 ppm with an accuracy up to a factor of 3.
5. The SO_2 mixing ratio at 66 km altitude is estimated to 2 ppm with an accuracy up to the same factor as for H_2O. Both mixing ratio assessments were obtained on the basis of analyses of spectra recorded on October 14, 1983 in middle latitudes.

References

[1] Oertel, D., Moroz, V. I., Linkin, V. M. et al.: "Venera-15" i "Venera-16": pervye rezul'taty ehksperimenta po infrakrasnojj spektrometrii. Pis'ma v Astron. Zh. **10** (1984) 4, 243.

[2] Oertel, D., Moroz, V. I., Nopirakowski, J. et al.: Infrared experiment aboard the automatic interplanetary stations Venera 15 and Venera 16. 1. Methodics and first results. Veröffentlichungen des Forschungsbereichs Geo- und Kosmoswissenschaften, Heft 18, 1990.

[3] Spänkuch, D., Zasova, L. V., Schäfer, K. et al.: Infrared experiment aboard the automatic interplanetary stations Venera 15 and Venera 16. 2. Preliminary results of temperature profile retrieval. Veröffentlichungen des Forschungsbereichs Geo- und Kosmoswissenschaften, Heft 18, 1990.

[4] Zasova, L. V., Spänkuch, D., Moroz, V. I. et al.: Infrared experiment aboard the automatic interplanetary stations Venera 15 and Venera 16. 3. Some conclusions on the structure of clouds, based on the analysis of spectra. Veröffentlichungen des Forschungsbereichs Geo- und Kosmoswissenschaften, Heft 18, 1990.

[5] Spänkuch, D.: Arbeiten des Meteorologischen Dienstes der DDR auf dem Gebiet der indirekten Sondierung. Zeitschrift für Meteorologie **30** (1980), 205.

[6] Sinton, W. M. and Strong, J.: Radiometric observations of Venus. Astrophys. J. **131** (1960) 2, 470.
[7] Moroz, V. I., Davydov, V. D., Zhegulev, V. S.: Fotometricheskie i spektroskopicheskie nabljudenija planet v diapazone 8–14 mkm. Astron. Zh. **46** (1969) 2, 136.
[8] Kunde, V. G., Hanel, R. A. and Herath, L. W.: High spectral resolution ground-based observations of Venus in the 450–1250 cm^{-1} region. Icarus **32** (1977) 2, 210.
[9] Reed, R. A., Forrest, W. J., Houck, J. R. and Pollack, J. B.: Venus: the 17 to 38 micron spectrum. Icarus **33** (1978), 554.
[10] Aumann, H. H. and Orton, G. S.: The 12–20 mkm spectrum of Venus: implications for temperature and cloud structure. Icarus **38** (1979) 2, 251.
[11] Aumann, H. H., Martonchik, J. V. and Orten, G. S.: Airborne spectroscopy and spacecraft radiometry of Venus in the far infrared. Icarus **49** (1982) 2, 227.
[12] Taylor, F. W., Beer, R., Chahine, M. T. et al.: Structure and meteorology of the middle atmosphere of Venus: infrared remote sensing from the Pioneer Orbiter. J. Geophys. Res. **85** (1980) A 13, 7963.
[13] Döhler, W., Schäfer, K., Spänkuch, D.: Line by line gaseous transmission for Venus standard atmosphere in the 250 cm^{-1} to 1400 cm^{-1} region. Report for VIRA workshop, Hamburg, 1983.
[14] McClatchey, R. A., Benedict, W. S., Clough, S. A. et al.: AFCRL atmospheric absorption line parameters compilation. AFCRL-TR-73-0096. L. G. Hanscom field, Bedford, Massachussets, 1973.
[15] Linkin, V. M., Schäfer, K., Macygorin, I. A. et al.: Infrared experiment aboard the automatic interplanetary stations Venera 15 and Venera 16. 5. Preliminary results of the analysis of fields of brightness temperatures and thermal fluxes. Veröffentlichungen des Forschungsbereichs Geo- und Kosmoswissenschaften, Heft 18, 1990.
[16] Moroz, V. I., Zasova, L. V.: Infrakrasnoe izluchenie Venery: priblizhennye metody raschjota spektra v polosakh pogloshhenija atmosfernykh gazov. Kosmich. Issled. **23** (1985) 2, 259.
[17] Knollenberg, R. G., Hunten, D. M.: The microphysics of the clouds of Venus: results of the Pioneer Venus particle size spectrometer experiment. J. Geophys. Res. **85** (1980) A 13, 8039.
[18] Moskalenko, N. I., Terzi, V. F., Parshin, S. N. et al.: Issledovanija spektra pogloshhenija sernistogo gaza v IK-oblasti. Izv. AN SSSR, fiz. atm. i okeana **14** (1978) 12, 1275.
[19] Barker, E. S.: Detection of SO_2 in the UV spectrum of Venus. Geophys. Res. Lett. **6** (1979) 2, 117.
[20] Esposito, L. W., Winnick, J. R., Stewart, A. I.: Sulfur dioxide in the Venus atmosphere: distribution and implication. Geophys. Res. Lett. **6** (1979) 7, 601.
[21] Zasova, L. V., Krasnopolsky, V. A., Moroz, V. I.: Vertical distribution of SO_2 in upper cloud layer of Venus and origin of UV absorption. Adv. Space Res. **1** (1981), 31.
[22] Kerzhanovich, V. V., Marov, M. Ya., Moroz, V. I.: Proposals for VIRA (Venus International Reference Atmosphere): dynamics and structure below 100 km. Preprint IKI Pr-831 (1983).
[23] Barker, E. S.: Observations of Venus water vapor over the disk of Venus: the 1972–74 data using the H_2O lines at 8197 and 8176 Å. Icarus **25** (1975) 2, 268.
[24] Moroz, V. I., Parfent'ev, N. A., San'ko, I. F.: Spektrofotometricheskijj ehksperiment na spuskaemykh apparatakh "Venera-11" i "Venera-12". 2. Analiz spectral'nykh dannykh "Venera 11" metodom slozhenija sloev. Kosmich. Issled. **17** (1979) 5, 727.
[25] Moroz, V. I., Moshkin, B. E., Ehkonomov, A. I. et al.: Spektrofotometricheskijj ehksperiment na spuskaemykh apparatakh "Venera-13" i "Venera-14". Predvaritel'nye rezul'taty analiza spektrov v oblasti polos pogloshhenija H_2O. Kosmich. Issled. **21** (1983) 2, 254.
[26] Gel'man, B. G., Zolotukhin, V. G., Lamonov, N. I. et al.: Analiz khimicheskogo sostava atmosfery Venery na AMS "Venera-12" gazovym khromatografom. Kosmich. Issled. **17** (1979) 5, 708.

[27] Oyama, V. I., Carle, G. C., Woeller, F. et al.: Pioneer Venus gas chromatography of the lower atmosphere of Venus. J. Geophys. Res. **85** (1980) A 13, 7891.

[28] Mukhin, L. M., Gel'man, B. G., Lamonov, N. I. et al.: Gazokhromatograficheskijj analiz khimicheskogo sostava atmosfery Venery na spuskaemykh apparatakh AMS "Venera-13" i Venera-14". Kosmich. Issled. **21** (1983) 2, 225.

[29] Surkov, Ju. A., Ivanova, F. V., Pudov, A. N. et al.: Sodershanie vodjanogo para v atmosfere Venery po dannym AMS "Venera-13" i "Venera-14". Kosmich. Issled. **21** (1982) 2, 231.

[30] Ehkonomov, A. I., Moshkin, V. E., Moroz, V. I. et al.: Ehksperiment po UF-fotometrii na spuskaemykh apparatakh "Venera-13" i "Venera-14". Kosmich. Issled. **21** (1983) 2, 254.

5. Infrared experiment aboard the automatic interplanetary stations Venera 15 and Venera 16

5. Preliminary results of the analysis of fields of brightness temperatures and thermal fluxes

V. M. LINKIN, K. SCHÄFER, I. A. MACYGORIN, W. DÖHLER, V. I. MOROZ, R. DUBOIS, A. V. D'JACHKOV, D. SPÄNKUCH, V. V. KERZHANOVICH, D. OERTEL, L. V. ZASOVA, J. NOPIRAKOWSKI, E. A. USTINOV, H. BECKER-ROSS, A. N. LIPATOV, W. STADTHAUS, A. A. SHURUPOV

Abstract

The measured data obtained by Venera 15 in the infrared spectrometric experiment are in this paper represented in the following manner: 1. latidudinal dependence of brightness temperature (along the route) for selected wave numbers; 2. temporal (and solar time) dependence of brightness temperature for selected latitudes; 3., 4. analogous dependences for the integral of the spectrum ('flux'). A preliminary analysis of this material shows the following: 1. on the northern hemisphere there are three characteristic regions — low latitudes (40—60°), where brightness temperatures and thermal fluxes are constant, the 'cold collar' (60—75°, varying with latitude) with great variations, and a warm 'polar cap' ($\varphi > 75°$) with 'hot spots'; 2. the thermal asymmetry at latitudes $\varphi < 60°$ is missing; 3. the hot spots are possibly twofold; 4. the cold collar does not rotate together with the atmosphere.

In [1—3] a preliminary analysis of the results of the infrared experiment aboard Venera 15 is given, based on spectral analysis. The same data can be presented in another way — as meridional profiles of the brightness temperature at selected wave numbers for each measuring cycle or as zonal means, if results of different cycles are used. Moreover, it is interesting to investigate the latitudinal and local-solar-time-dependence of the thermal flux, i.e. of the integral of the spectrum.

An example of a meridional profile was given in [4, 5] for the orbit on October 14, 1983. In the following additional examples for several other orbits, are given as well as zonal means (brightness temperatures and thermal fluxes in fixed latitude belts in dependence on time).

1. Additional information on the measurement geometry

The main characteristics of the geometry of the experiment are given in [4, 5] where a diagram can be found showing the positions where measurements were made: they are situated along the arc of a big circle which begins near the equator (at first limb viewing), proceeds in 2.7° distance from the pole (here, the viewing direction is near to the nadir) and interrupts in the equatorial zone at the other side of the pole (second limb viewing) characterizing the so-called typical measuring cycles. Aboard the interplanetary station Venera 15 47 of such cycles were made in the time interval from October 12 to December 14, 1983. Additionally, two orbits with a quite different geometry more suitably for low latitudes were arranged by 'turning' the pointing axis of the interferometer parallel to the equatorial plane in one cycle and with an angle to it (for cutting of the southern hemisphere) in another one. Thus, the regions of the southern hemisphere and of the equator could be investigated at more favourable conditions, but the spatial resolution was worse than in the typical cycles because of the greater distance namely about 60 km with nadir pointing and about 900 km in the other cycle.

In the following the part of the typical cycle before the passing of the nearest point to the pole is called the 'ascending' route and the residual part the 'descending' one. In mid-October, when the first measurements were made, the ascending route was on the night side approximately 30° from the morning terminator, and the descending part on the day side — 30° in longitude from the evening terminator. The route of the November 5 cycle nearly went along the terminator, and on the following dates the ascending part of the route was already situated on the day side of the planet and the descending one on the night side. The last cycle on December 14 had the angle of 65° longitude between the route and the terminator.

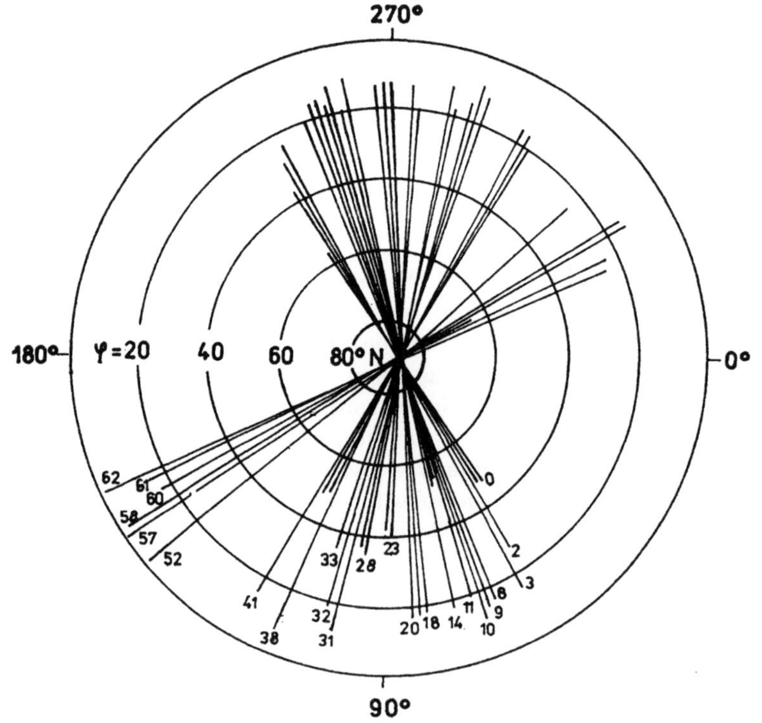

Fig. 1. Position of the measuring routes of typical measuring cycles of the automatic interplanetary station Venera 15 in coordinates of latitude and local solar time. Flight days (the cycle of October 12, 1983 corresponds to zero days) are marked at the beginning of the ascending route parts

Fig. 1 shows the diagram with the routes of typical cycles for different dates. The cycles are not complete in some cases because of different technical limitations. The measuring period of a typical cycle took about 20 minutes from limb to limb.

In that time, the cloud components in middle latitudes are displaced by about 120 km (if a speed of 100 m/s is supposed). The duration of solar days on Venus is about 116 Earth days. Conse quently, it can be considered that the information on the meridional profile of brightness temperatures and fluxes obtained from measurements during one cycle is practically that of one moment — the Venusian atmosphere scarcely varies during the measuring period. Therefore, the representation of the data as meridional profiles is an absolutely unequivocal information source on such effects as the existence (or absence) of the diurnal asymmetry of the latitudinal dependence of atmospheric characteristics in a given time moment.

The matter is worse with the temporal dependence of any measured parameter. If it remains constant it can be concluded that it depends neither on time nor on longitude or solar time (in that interval for which data are available) and that effects connected with the cloud movement are not observed. If variations exist as in many cases they can hardly be interpreted correctly without additional hypotheses. We will take, as always, only two situations: the absence of 'dynamics', i.e. of the variations, or its presence. Future analysis as, e.g. statistical analysis of quasiperiodic processes will, hopefully, improve the situation.

In the following we will denote the temporal dependence of any parameter at a given latitude as a 'zonal' profile, knowing that this profile can include the dependence on longitude as well as on solar time.

2. Meridional and zonal profiles of brightness temperatures

That part of the spectrum investigated in our experiment (from 250 to 1600 cm^{-1}) can roughly be divided into two parts: the short-wave ($\nu < 667$ cm^{-1}) and the long-wave part ($\nu > 667$ cm^{-1}). The latitudinal dependence of the brightness temperatures outside the 15 μm (667 cm^{-1}) gaseous absorption band is different for both parts of the spectrum. The physical reason for this fact is obvious: the extinction coefficient of the upper Venusian cloud layer is one order of magnitude smaller in the long-wave part of the spectrum than in the short-wave one [2]. Thus, radiation at $\nu = 365$ cm^{-1} is originated in deeper levels (for middle latitudes approximately 10 km deeper [2]) than in the short-wave part of the spectrum.

In the following brightness temperatures are given for the wave numbers 1152, 365, 872, 667 and 626 cm^{-1}. The choice of the first two is connected with the fact that the cloud extinction coefficient has a maximum at 1152 cm^{-1} and a minimum at 365 cm^{-1}. Wave number 872 cm^{-1} has been chosen since it nearly coincides with one of the channels of the infrared radiometer aboard Pioneer Venus Orbiter [6] and the comparison of the results is interesting.

Wave numbers 1152, 365 and 872 cm^{-1} are situated outside the gaseous absorption band. For comparison, two channels inside the $\nu_2 CO_2$ absorption band — 667 cm^{-1} (band centre) and 614 cm^{-1} (wing) were added.

In Figs. 2—5 the meridional profiles of thermal flux, brightness temperature and brightness temperature difference $\Delta T_B = T_B\,(365) - T_B\,(1152)$ for 4 measuring cycles are given.

Here and in the following we will denote the brightness temperatures for wave number ν by $T_B(\nu)$. The results are given without any corrections, e.g. it had not been tried to consider the "limb-darkening".

The zonal profiles of Fig. 6 are the average of all measurements (see Fig. 1), within a given latitude belt.

The behaviour of the brightness temperature profiles in Figs. 2—6 is characterized by the following fundamental peculiarities.

1. The meridional profiles T_B (1152) and T_B (365) are nearly parallel with great differences of about 20—30 K for latitudes $\varphi < \varphi_0$ with $40° < \varphi_0 < 50°$. There, typical brightness temperatures are $220 < T_B$ (1152) < 230 K and $240 < T_B$ (365) < 260 K. The meridional profile T_B (872) proceeds approximately 10 K higher than the profile T_B (1152). In the cycle with "turn" (Fig. 5) where the equatorial region was observed nearly in nadir geometry the difference $\Delta T_B = T$ (365) $- T$ (1152) is about 40 K with T_B (1152) $= 232$ K and T_B (365) $= 274$ K. The brightness temperatures decrease with in-

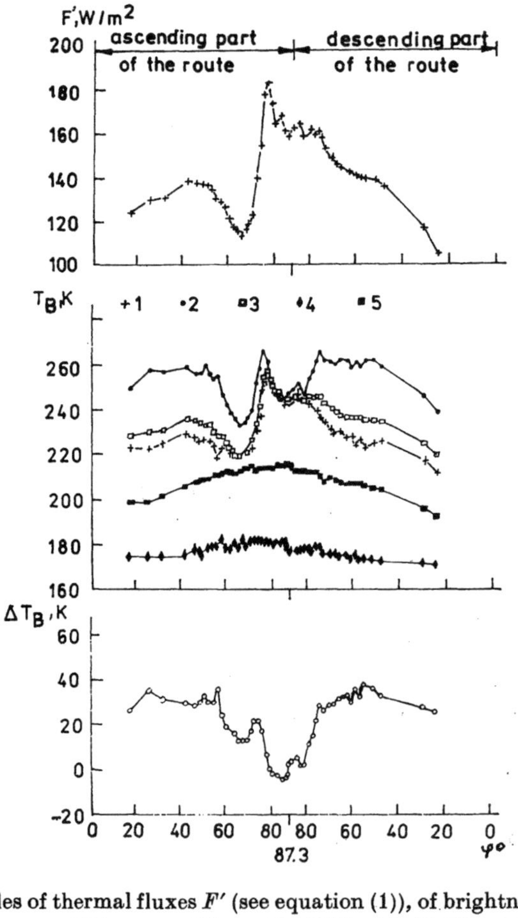

Fig. 2. Meridional profiles of thermal fluxes F' (see equation (1)), of brightness temperatures $T_B(\nu)$, and of brightness temperature differences $\Delta T_B = T_B$ (365)$-T_B$ (1152) of the orbit of October 15, 1983

1 — T_B (1152), 2 — T_B (365), 3 — T_B (872), 4 — T_B (667), 5 — T_B (626)

creasing latitude (obviously, because of the effect of limb darkening) for low and mid-latitudes ($\varphi < \varphi_0$). Periodic structures are determinable for $\nu = 365$ cm^{-1}. The temperatures are for the ascending and descending part of the route identical with an accuracy up to some degrees, though within these few degrees a certain asymmetry is being observed between ascending and descending part. The maximum temperature variations are 7—9 K in these latitudes for two months.

2. There is a relatively narrow latitude belt northern of φ_0 with an extension in latitude from 10 to 30° where both brightness temperatures, T_B (1152) and T_B (365), have a minimum. The center of the minimum varies from 62 to 74° latitude. This minimum is observed, as a rule, only at one side of the pole, but sometimes also at both sides. In the last case the depths of the minima are not identical. The deeper one is on the ascending part of the route (area of morning terminator). In this latitude belt the magnitude ΔT_B is only about 15 K and, hence, considerably smaller than at lower latitudes. The zonal profiles of T_B (1152) and of T_B (365) have significant variations there (at 70° latitude) within the limits 210 to 240 K, for T_B (1152) and 0 to 37 K for ΔT_B.

The upper limit corresponds to the case if the cold region is not passed through by the measurement route. The comparison of meridional and zonal profiles shows that the cold zone has a significant longitudinal extension though it is not closed. It is expressed

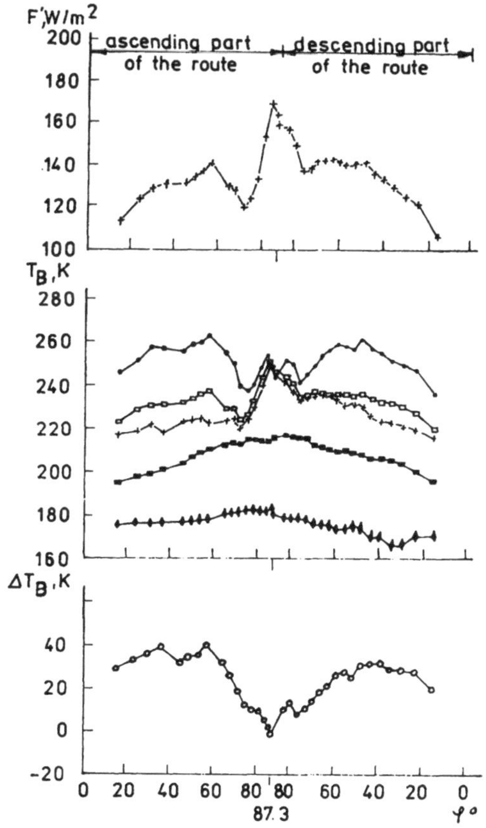

Fig. 3. The same as Fig. 2, of the orbit of October 21, 1983

more intense in the environment of the morning terminator. As agreed, in the following we will call this area "cold collar".

The cold collar is nearly always present on the ascending part of the route and as good as never observed on the descending one. This is evidence of the fact that such a cold feature does not rotate together with the atmosphere (the rotation period of the atmosphere is in this region, obviously, about 4 days [7]). This is possibly connected with the position of this feature relative to the sun (diurnal wave). There is some hint of a decrease of the minimum temperature towards the end of the observation period.

The analysis of one of the typical spectra of this region shows a temperature inversion there [1], and though the upper cloud boundary is obviously situated deeper than in equatorial latitudes [2] it is colder than for low-latitude clouds. The inversion can also be seen by comparing the meridional profiles T_B (872) and T_B (586) (edge of the gaseous absorption band) in the cold collar area with T_B (586) $>$ T_B (872). The meridional T_B profiles for wave numbers within the ν_2CO_2 absorption band have no minimum in this zone even for wave numbers near to the edge of this band (T_B (586)). This is evidence that the 'cooling' is localized in the altitude range of the cloud top.

3. North of the cold collar the curves T_B (1152) and T_B (365) proceed upward. This increase of the brightness temperature begins in different cycles not at the same lati-

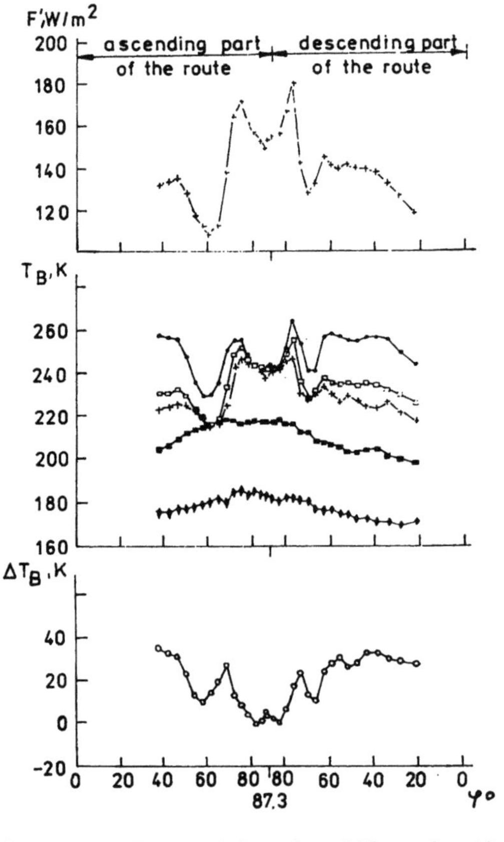

Fig. 4. The same as Fig. 2, of the orbit of November 10, 1983

tude, generally, higher than 70°, but sometimes at 65° already. Within this latitude range the 'polar cap'' is situated where the brightness temperatures T_B (1152) and T_B (365) are nearly identical: T_B (1152) $\approx$ T_B (365). Their absolute values are 240–245 K every where, with the exception of one or two 'hot spots' (see below). The total width of the polar cap varies from 25 to 40° in latitude. It is characteristic that the meridional profiles T_B (1152) have the maximum in the polar cap. However, the meridional profiles T_B (365) outside the hot spots show still lower tremperatures here than in latitudes $30° < \varphi < 50°$.

4. Inside the "polar cap" there is a relatively small 'hot spot' with a width of 5–7° in latitude (level 0.5 from peak T_B value) in most cases. The meridional profile shows, as a rule, one such spot (Fig. 3). In rare cases they are two (Fig. 4), and there are relatively many cycles where marked spots cannot be identified at all. T_B (365) has a well marked spot in the centre, with 254–266 K, where as distinct from the residual part of polar cap, the temperature difference has a significant value $10 < \Delta T_B < 15$ K, and the difference of the temperature in the polar cap from the mean one is extremely significant — from 14 to 18 K. The maximum amplitude of the T_B variations in the polar cap is not strictly constant. It is within the limits of 75 to 85° latitude (for the middle of the spot).

Another important peculiarity of the hot spots is the great equivalent width of the CO_2 bands 960 and 1060 cm^{-1} [4, 5]. The spectral analysis of a hot spot in the cycle on October 14 [2] showed that here the upper boundary of the clouds is situated at 60 km altitude (i.e. 12 km deeper than in low latitudes).

The zonal profile T_B and ΔT_B at 80° latitude shows a rather significant dynamics (Fig. 6). There is no doubt that the main reason for the variations is the passing of the

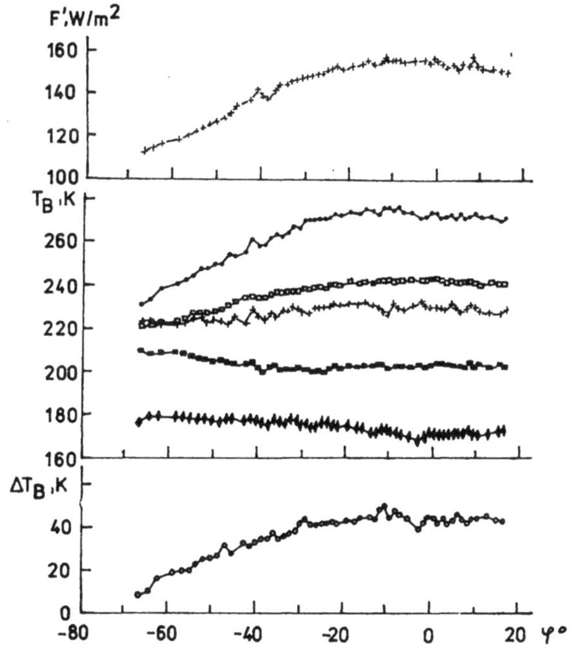

Fig. 5. The same as Fig. 2, of the orbit of November 11, 1983, when the viewing direction of the instrument was orientated nearly parallel to the equatorial plane

hot spots by the route, but also the change of the structure of the polar cap itself along the route. The latter is possibly connected with the rotation of the atmosphere. It is difficult to determine the rotation period according to our data. However, the T_B profiles to be observed in single cases turn out to be similar in successive cycles. For instance, the profiles on October 14 and 15 in the polar region turn out to be approximately symmetric relative to the pole. If this is really the passing of one and the same structure and the same hot spot, the rotation period can amount to 2 days (which is close to the Pioneer Venus data).

In the polar region often components with wave character with an amplitude of 3—4 K to 10 K and a wave length of 8—10° are seen especially for T_B (365), i.e. at lower altitudes.

5. The curves T_B (667) and T_B (614) are smooth, the brightness temperatures systematically increase towards the pole. This reflects the properties of the temperature fields in the strato-mesosphere discussed in a special paper.

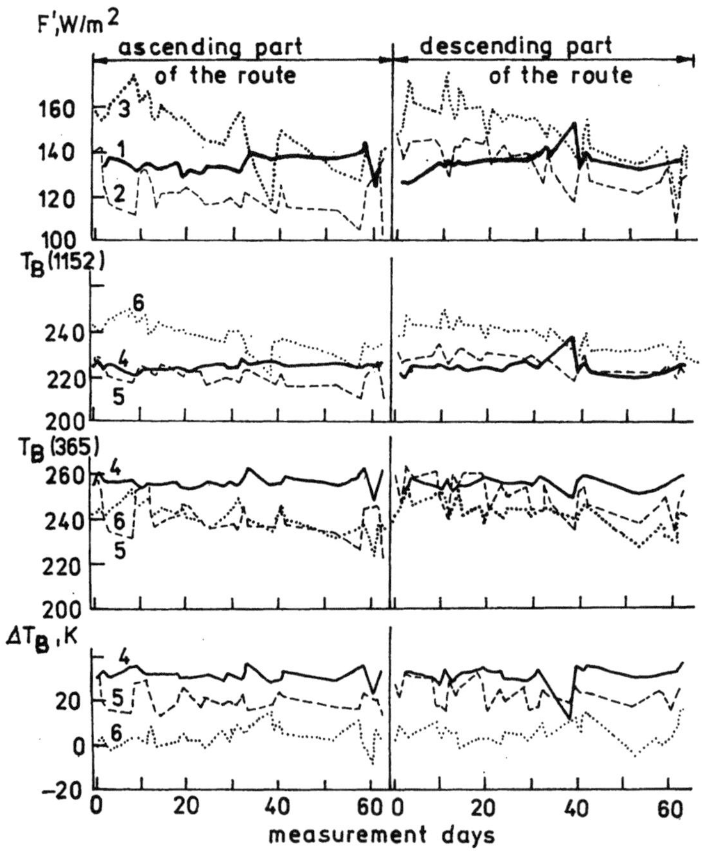

Fig. 6. Zonal profiles (averaged in the given latitudinal belts) F', T_B, and ΔT_B

$1 - \varphi = 40 \pm 5°$; $2 - \varphi = 65 \pm 5°$; $3 - \varphi = 85 \pm 5°$; $4 - \varphi = 40 \pm 10°$; $5 - \varphi = 65 \pm 5°$; $6 - \varphi = 85 \pm 5°$

6. In the latitude belt $\pm 20°$ from the equator the brightness temperatures are identical on the northern and southern hemisphere and they are varying as good as symmetrically.

3. Thermal fluxes

In Figs. 2—6, also the quantity

$$F' = \pi \times \int_{274.3}^{1656.5} I(\nu) \times \mathrm{d}\nu \quad (1)$$

is given for different latitudes and different cycles.

Here, $I(\nu)$ is the measured radiance density. We call the magnitude F' thermal flux since it resembles the hemispheric thermal radiation flux of the unit area:

$$F = 2 \times \pi \times \int_0^1 \int_0^\infty I(\nu, \mu) \times \mu \times \mathrm{d}\mu \times \mathrm{d}\nu, \quad (2)$$

where μ — cosinus of the radiation zenith distance angle.

The difference between (1) and (2) is caused by two reasons. The first one is the 'cutting off' of the spectrum because of the limited spectral range of the instrument. The systematic difference between F' and F is estimated to about 10% by the cutting-off-effect. It is more difficult to estimate the correction connected with the angle dependence (limb darkening) $I(\nu, \mu)$. This can be done by two methods — application of limb darkening curves found by other measurements (Venera 9, 10 and Pioneer-Venus) or by calculating synthetic spectra based on cloud layer models, e.g. of that type discussed in [2].

In future both ways will be tried. Here we confine to the uncorrected data of the thermal flux — to the F' values.

The thermal flux data shown in Figs. 2—6 allow the separation of the same characteristic regions which had already been defined on the basis of brightness temperatures analysis done in the preceding chapter.

We will specify the main peculiarities of the behaviour of the thermal fluxes in these regions.

1. In low latitudes ($\varphi < 50°$) the thermal flux increases monotonously with latitude; variations on the ascending and descending part of the route are similar, but they have a small asymmetry. The diurnal variations (Earth days) are small (about 1%). At the end of the measuring period an increase of the flux at these latitudes by some per cent was observed.

2. The cold collar occurs in the thermal flux, too. The radiation minimum is less distinct on the ascending part of the route, too. The decrease of the thermal flux in comparison with the flux in 40° latitude amounts about 25% (for the area of the morning terminator). The asymmetry of the thermal flux in the cold collar on the ascending (areas adjacent to the morning terminator) and descending (areas adjacent to the evening terminator) parts of the route is strongly expressed. In single cycles (October 13 and 22) the cold collar is hardly be seen.

3. The polar region emits on the average 25% more energy than low latitudes. Thus it is justified to call the polar cap "warm". Hot spots are identifiable but strongly weaken-

ed; the absolute maximum of the measured thermal fluxes $F' = 180$ W/m² corresponds to them. It is interesting to note that the mean thermal flux at latitudes higher than 40° decreased significantly to the end of the measuring period (i.e. with approximation of the ascending part of the route to the subsolar point) whereas in low latitudes an increase could be stated.

4. Thermal fluxes in the equatorial region increase monotonously towards the equator where 160 W/m² are reached.

4. Comparison with results of other measurements

It is interesting to compare the results mentioned above with the data of other measurements, especially with Earth based observations, measurements of Venera 9, 10 and of the Pioneer Venus Orbiter.

The brightness temperatures in the transparency window 8—13 μm were measured from the Earth many times (see outline [7]). As a rule, values in the order of magnitude 228—230 K in the equatorial zone are obtained which is practically coinciding with our results.

Aboard Venera 9 and Venera 10 the measurements were carried out by means of a wide-band radiometer in the wave length range of 8—40 μm. Hereby, the spectral characteristics of the filter had a complicated form — it did not let pass the radiation in the range 12—18 μm. A direct comparison is scarcely informative, but it can be stated an approximate coincidence of brightness temperatures, calculated according to magnitudes F' in low latitudes with brightness temperatures which were observed of Venera 9, 10. However, there is a serious divergence. L. V. Ksanfomality [8, 9] interpreted the date of Venera 9 and Venera 10 as an evidence of the asymmetry of thermal fluxes on the night and on the day side: the night fluxes were systematically greater than the day ones. Analogous conclusions were drawn by some observers who had measured the infrared radiation of Venus from the Earth [7]. Our measurements show in a convincing manner that not only the fluxes but also the spectra of the thermal radiation are similar on the day and on the night side in the range of local solar time from 56 to 155°.

This does not exclude the presence of any anomalies at another local solar time. Moreover, it is not excluded that variations of cloud layer characteristics in middle latitudes exist with great characteristic periods (many months or years).

Pioneer Venus Orbiter radiometer measurements were carried out in some narrow spectral channels (600—750, 676—686, 722—736, 751—771, 867—883 cm⁻¹) and in one wide band one. The pericenter of the orbit was situated, as in our case, above the northern hemisphere. While passing along the route above the planet a scanning was carried out in angle and plane, approximately perpendicular to the orbit plane and in this way a two-dimensional distribution of radiances was obtained, covering the greater part of the northern hemisphere. The maps of brightness temperatures, obtained by Pioneer Venus in the 875 cm⁻¹ channel, show the same main areas as our measurements: the zone of low latitudes with nearly constant brightness temperatures (also no distinctive marks of thermal asymmetry), the cold collar and the warm polar cap. Inside the cap there are always observed two hot spots, symmetrically regarding the pole and situated at a latitude of about 80°. Their radius was about 5°.

At first sight it may seem that there is a great divergence to our data, which show the presence, as a rule, of one warm spot only. Nevertheless, there are no serious contradictions: Our routes did not pass exactly the pole, and therefore the crossing of two maxima of the dipole simultaneously is less probable than of one (see Fig. 7). On the basis of our data only the conclusion about the existence of the dipole would not have been possible, but it must be stated that our measurements do not contradict the presence of such a structure. Moreover, the fact that in case of two hot spots their strengths were weaker than of the single spot observed, can be considered as an argument in favour of a constant presence of a double structure (see Fig. 7).

Conclusions

1. On the northern hemisphere of the planet the brightness temperature field contains three characteristic regions:

   a) low latitudes (up to 40—60°), where the brightness temperatures and thermal fluxes are nearly constant;
   b) the cold collar (60—75°), obviously not closed in latitude with great variations of T_B and thermal fluxes;
   c) the "warm" polar cap, with maximum and relatively stable thermal fluxes. Hot spots as regions of limited spatial dimensions are located in the polar cap with absolute maximum thermal flux and brightness temperatures outside the ν_2CO_2 band system.

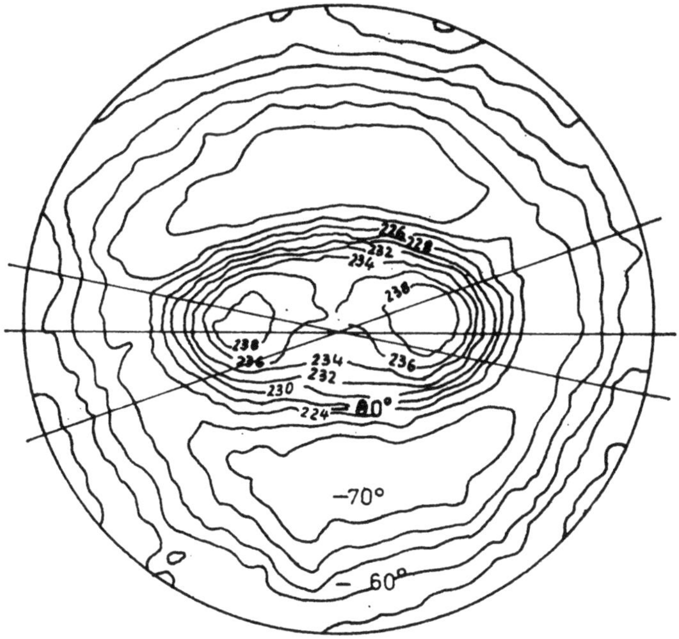

Fig. 7. Possible variants of crossing of the 'dipole' structure by the measuring route for a typical cycle

2. The thermal asymmetry (brightness temperature is greater on the day than on the night-side) is missing in low latitudes. This relates at least to the local solar time interval 56—155°.
3. Our observations are not able to give an independent confirmation of the double structure of the hot spots near the north pole ('dipole'), but they do not contradict the hypothesis that this structure as a whole has been alive during the 5 years from the first observation from the Pioneer Venus Orbiter.
4. The cold collar does not rotate together with the atmosphere but is located around the morning terminator.

References

[1] SPÄNKUCH, D., ZASOVA, L. V., SCHÄFER, K. et al.: Infrared experiment aboard the automatic interplanetary stations Venera-15 and Venera-16. 2. Preliminary results of temperature profile retrieval. Veröffentlichungen des Forschungsbereichs Geo- und Kosmoswissenschaften, Heft 18, 1990.

[2] ZASOVA, L. V., SPÄNKUCH, D., MOROZ, V. I. et al.: Infrared experiment aboard the automatic interplanetary stations Venera 15 and Venera 16. 3. Some conclusions on the structure of clouds, based on the analysis of spectra. Veröffentlichungen des Forschungsbereichs Geo- und Kosmoswissenschaften, Heft 18, 1990.

[3] MOROZ, V. I., DÖHLER, W., USTIONOV, E. A. et al.: Infrared experiment aboard the automatic interplanetary stations Venera 15 and Venera 16. 4. Preliminary results of spectral analysis in the region of H_2O and SO_2 absorption bands. Veröffentlichungen des Forschungsbereichs Geo- und Kosmoswissenschaften, Heft 18, 1990.

[4] OERTEL, D., MOROZ, V. I., NOPIRAKOWSKI, J. et al.: Infrared experiment aboard the automatic interplanetary stations Venera 15 and Venera 16. 1. Methodics and first results. Veröffentlichungen des Forschungsbereichs Geo- und Kosmoswissenschaften, Heft 18, 1990.

[5] OERTEL, D., MOROZ, V. I., LINKIN, V. M. et al.: "Venera 15" i "Venera 16".: pervye rezul'taty ehksperimenta po infrakasnojj spektrometrii. Pis'ma v Astron. Zhurn. **10** (1984) 4, 243.

[6] TAYLOR, F. W., BEER, R., CHAHINE, M. T. et al.: Structure and meteorology of the middle atmosphere of Vernus: infrared remote sensing from the Pioneer orbiter. J. Geophys. Res. **85** (1980) A 13, 7963.

[7] KERZHANOVICH, V. V., MAROV, M. YA., MOROZ, V. I.: Proposals for VIRA (Venus International Reference Atmosphere): dynamics and structure below 100 km. Preprint IKI Pr-831 (1983).

[8] KSANFOMALITI, L. V., DEDOVA, E. V., OBUKHOVA, L. F. et al.: Infrakrasnoe izluchenie oblakov Venery. Kosmich. Issled. **14** (1976) 5, 768.

[9] KSANFOMALITI, L. V.: Venera 9 and 10: Thermal radiometry. Icarus **41** (1980) 1, 36.

6. The Venusian Polar Dipole

F. W. TAYLOR

Abstract

The Venusian polar dipole is one of the more remarkable phenomena in the Solar System to have been discovered and observed during the recent programmes of planetary exploration led by the USA and the USSR. The dipole consists of a giant vortex of surprisingly complex structure and behaviour located in the middle atmosphere at the north pole of the planet, with a similar feature presumed to exist at the south pole also. In this paper, the phenomenon is described and characterised and a qualitative attempt is made to account for its presence. However, although nearly a decade has passed since the northern dipole was discovered, there is still no detailed theory which models its appearance and behaviour satisfactorily. A new and fairly simple experiment is proposed which, if mounted on a future space mission, might supply the missing information to rectify the situation.

Introduction

The Venusian Polar Dipole was discovered in December 1978 by the Pioneer Venus Orbiter (PVO) Infrared Radiometer (OIR) as a very bright feature observed in successive scans of the planet at a wavelength of 11.5 μm (Taylor, McCleese and Diner, 1979). The authors believed at first that the infrared feature originated in a single large depression in the clouds, caused by the downward flow of air over a wide area around the pole, as part of the general circulation pattern of the middle atmosphere. The cloud tops on Venus lie at a pressure of some 50 hPa, about 65 km above the surface of the planet.

A middle atmosphere circulation in which each hemisphere contained a single large Hadley-type cell had been inferred from television images of Venus obtained by Mariner 10 (Suomi and Limaye, 1978). This, if real, together with the rapid zonal circulation of the atmosphere at these levels, also inferred from television images, was expected to give rise to a giant vortex over each pole. In the centre of the vortex, it was argued, the rapidly rotating, descending air would depress the level of the clouds to altidudes where the prevailing temperature was higher. This then gives rise to enhanced infrared emission, particularly at wavelengths such as those between 8 and 14 μm where the main gaseous constituent of the Venusian atmosphere, carbon dioxide, has a high transmission.

The principal surprise from the Pioneer measurements was not the existence of anomalously high levels of infrared emission from the polar regions, which supports the interpretation outlined above, but its detailed character, which is proving extremely difficult to understand even after years of study.

The OIR data was obtained as line scans across the planet, using the rotation of the PVO spacecraft to scan, in a small field of view giving a resolution on the planet of a few km at periapsis and a few tens of km near the N pole. (Beause of the viewing geometry, the south pole was not observed at all by the OIR). These scans could be built up into images, a set of which has been published in the papers by Taylor et al. (1980) and Taylor, Hunten and Ksanfomality (1982). During a 72 day period, the OIR obtained images of the dipole nearly every day, with every fourth day in the high data rate 'imaging' mode. These show quite clearly that there are *two* bright features, falling either side of the pole, and rotating rapidly around it, leading to the name 'polar dipole'.

The appearance shown in this diagram (Fig. 1) is a synthesis of the very variable structure exhibited by the dipole during the relatively short time it was observed. Most of the time, the bright regions had a 'chevron' shape; and often, bright filaments connected the two chevrons, usually crosswise as shown in the diagram.

This odd shape is made more complex by the obscure relationship which the dipole bears to the *polar collar*, a dark (cold) feature which surrounds the dipole at latitudes from 60 to 75 deg N approximately. Unlike the dipole, the collar does not rotate rapidly and is mainly a temperature, rather than a cloud morphology, feature (Schofield and Taylor, 1983). Also unlike the dipole, which is obviously a feature with a mainly wavenumber two character, the collar is primarily wave one, as can be seen from its crescent-shaped appearance in the images (see figs. in Taylor et al., 1980; c.f. Fig. 1).

The rotation of the dipole has been analysed by Schofield and Diner (1983). They measured the rotation of the feature between successive images independently, in order to estimate reliably the measurement uncertainty. They found that the motion had three components which exceeded the error; a steady rotation with a period of 2.945 $\pm$ 0.003 days (retrograde); a secular decrease in period with a deceleration of 0.21 $\pm$ 0.05 deg day^{-1}, and a sinusoidal component with a peak amplitude of 0.39 $\pm$ 0.06 and a period of about 40 days. No attempt has been made to explain this behaviour, and indeed the interpretation of terms of this kind from a relatively short data set requires extreme caution. For example, Schofield and Diner point out that the quadratic term in their fit would imply, if taken literally, that the rotation of the dipole would cease altogether in 627 $\pm$ 151 days and then reverse. This seems unlikely and, until further data is obtained, the most that probably can be said is that the motion of the dipole is basically retrograde rotation with a period of about 3 days, and that other components of motion are present which are extremely complex.

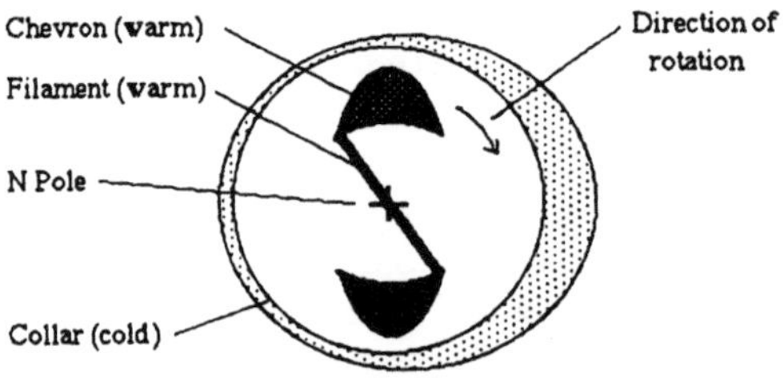

Fig. 1. Diagram of the main characteristics of the dipole. The diameter of the middle of the collar is about 5000 km, and the brightness temperature contrast (at. 11.5 μm) between the hottest part of the chevron and the coldest part of the collar is about 45 K

In addition to its variable rotation rate, the dipole also appears to nutate or in some sense vary the position of its axis of rotation with respect to that of the planet. On at least one occasion (plate 3a in Taylor et al., 1980) the apparent centre of the dipole was observed to be displaced by about 5° or about 500 km from the pole.

Elson (1982, and personal communication, 1987) has had some success in finding normal modes which exist under conditions approximating the polar Venusian atmosphere and in which wavenumber 2 is dominant. Studies of this kind raise the question of whether the dipole is a thermal feature (due to temperature contrasts on a constant pressure surface) or a cloud morphology feature (due to cloud height contrasts within a temperature field where all constant pressure surfaces are at least approximately at the same temperature). It was not possible to differentiate using PVO/OIR data alone but, by measuring the equivalent widths of carbon dioxide bands, the Venera 15/16 spectrometers were able to show that the latter effect predominates. Zasova et al. (1985) found that the hottest parts of the dipole were some 10 km lower than their surroundings.

It is also worth noting here that, although the viewing geometry from the Earth is very unfavourable (remembering that the axial tilt of Venus is only about 2°), the polar collar has been observed from the Earth on several occasions (eg. Diner et al., 1976), and part of the dipole was seen at least once (D. J. Diner, personal communication). The significance of this is twofold: it confirms that the dipole, and especially the collar, is a long-lived feature and not just a freak of the period observed by Pioneer (and Venera), and it also strongly suggests that dipoles exist at *both* poles since a southern collar has been observed on more than one occasion. The collars as seen from Earth are as often absent as present, but this is expicable in terms of their crescent-shaped, wave-one appearance with most of the amplitude on one or other side of the disc.

An Heuristic Model of the Dipole

What in general terms, is the physical structure and likely cause of the dipole feature? The answer to that question is almost as far from our grasp now as it was the day the feature was first revealed. However, it is always possible to speculate in qualitative terms in ways that account for what is seen without attempting to show in detail the cause or even that such structures are stable or viable. Such an *heuristic* model is shown in Fig. 2.

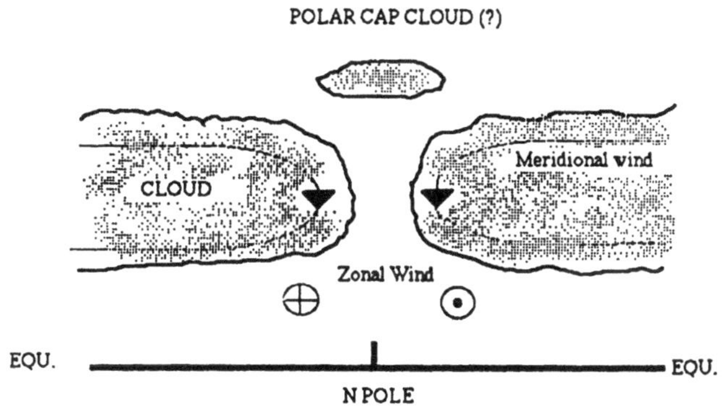

Fig. 2. Heuristic model of the cloud structure over the North pole of Venus

The idea of the polar vortex is difficult to give up because the evidence for the slow poleward propagation of rapidly zonally-rotating parcels of air is so strong. Also, at least one of the OIR 12.5 μm images (plate 5 in Taylor et al., 1980) shows spiral streaks in the outer edge of the collar which may or may not be related to similar classes of feature in the ultraviolet images, but which are strongly suggestive of a vortex. The collar itself is the manifestation in the temperature field of a jet stream at latitudes between about 50 to 60° N. This separates the low-latitude from the high-latitude regimes. South of the jet, large amounts of mass and angular momentum are being transported polewards at the cloud-top level; inside the collar this mass and momentum is propagated downwards inside the dipole. Perhaps the rotation rate of the eye of the vortex, and its shape, represent the most stable configuration for conserving mass and angular momentum during this process.

No conceivable dynamical process produces chevron-shaped descending regions and it seems more likely that the region of maximum subsidence is in fact elliptical, consisting of a symmetrical vortex structure with a superimposed disturbance of predominantly wavenumber 2. A roughly circular patch of relatively stagnant cloud may overly this, forming a polar cap such as is common on other planets. (The cap cannot be more than a few km higher than the mean cloud top height, however, otherwise it would have been resolved by the other channels of the OIR, which were designed for vertical sounding). The chevrons are then the tips of the elliptical vortex, which extend out beyond the polar cap clouds (Fig. 2).

The linear filaments which are sometimes seen connecting the opposite cusps of the chevrons are much harder to explain. They must be features of the cap cloud, somehow induced by the rotating clouds beneath. It is virtually impossible, even with complete license, to begin to speculate in any detail as to what mechanism could give rise to such a curious effect. Nor does our model give any hint as to why the axis of rotation of the dipole apparently strays some 500 km from the pole in less than a day and returns just as quickly. The apparently systematic fluctuations in the rotation rate are equally mysterious.

Further Observations

Some things which are quite clear are that the dipole is a major solar system phenomenon, connected with the general circulation of Venus' atmosphere, and one which we are ill equipped to explain. Like the Jovian Great Red Spot, the absence of viable theories which can be tested, or in this case any theory at all, leaves us uncomfortably in doubt

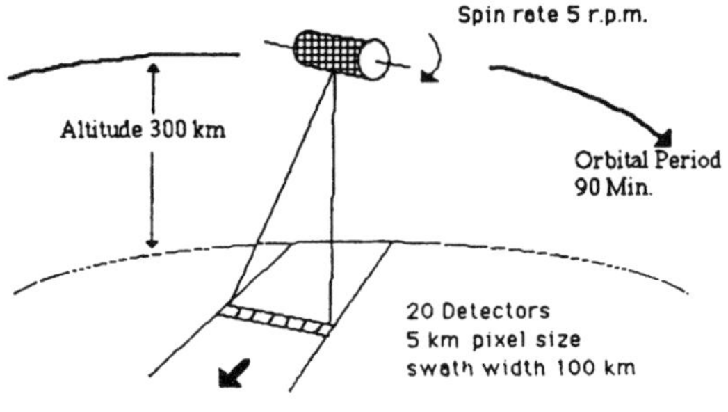

Fig. 3. Concept of a simple instrument to obtain motion pictures of the dipole in the thermal infrared

as to our basic ability to understand even gross features of planetary atmospheric circulations. Furthermore, enough time has elapsed to suggest that there is simply not enough information on which to base an hypothesis, and that new studies are needed. In particular, infrared images with better resolution and covergage than PVO are our best hope of making the vital breakthrough. Such data would not be difficult to acquire. The following concept was derived within the framework of a possible British, later European-led mission to Venus with a simple and inexpensive spacecraft. Neither of these flight opportunities are very likely to materialise in the foreseeable future, however.

The Venus Atmosphere Dynamics Imaging Radiometer (VADIR) is considerably less complicated than the PVO-OIR, having only two spectral channels, identical to OIR channels 4 and 5 (764 ± 7 cm^{-1} and 870 ± 10 cm^{-1} respectively). The concept (Fig. 3) is of a spinning spacecraft in a near-circular orbit about 300 km above the cloud-tops, with the spin axis parallel to the surface of the planet and to the orbital path. For a spin rate of 5 r.p.m., the sub-spacecraft point advances by nearly 100 km each spin. A linear array of 20 detectors, each sized for a 5×5 km^2 footprint on the planet, would then provide slightly overlapping coverage from each swath. In this way, a detailed image of the dipole could be built up each orbit, or approximately fifty times each 72 hr dipole rotation period. From a simple calculation, or by scaling the OIR performance, it is straightforward to show that the signal to noise ratio in each picture element (pixel) would be around 100 to 1. From such images, motion pictures could be constructed and used to study in detail the structure and variability of the dipole. The Pioneer pictures, which over the pole were at much lower spatial and temporal resolution, nevertheless reveal sufficient detail in the dipole structure that we can be sure there will be many small-scale cloud features which could be tracked to determine cloud top wind fields within the vortex. In this way, it is to be hoped that the true nature of this complex and fascinating phenomenon, a major puzzle in the field of planetary atmospheric dynamics, will eventually be revealed.

References

DINER, D. J., WESTPHAL, J. A., and SCHLOERB, F. P. Infrared Imaging of Venus: 8–14 micrometers. Icarus **27** (1976), 191–196.

ELSON, L. S., Wave instability in the polar region of Venus. J. Atmos. Sci., **39** (1982), 2356–2362.

SUOMI, V. E. and LIMAYE, S. S., Science, **201** (1978), 1009–1011.

TAYLOR, F. W., MCCLEESE, D. J., and DINER, D. J., Polar clearing in the Venus clouds observed from the Pioneer Venus Orbiter. Nature, **279** (1979) 5714, 613–614.

TAYLOR, F. W., BEER, R., CHAHINE, M. T., DINER, D. J., ELSON, L. S., HASKINS, R. D., MCCLEESE, D. J., MARTONCHIK, J. V., REICHLEY, P. E., BRADLEY, S. P., DELDERFIELD, J., SCHOFIELD, J. T., FARMER, C. B., FROIDEVAUX, L., LEUNG, J., COFFEY, M. T., and GILLE, J. C., Structure and meteorology of the middle atmosphere of Venus: infrared remote sounding from the Pioneer Orbiter. J. Geophys. Res. **85** (1980) A 13, 7963–8006.

TAYLOR, F. W., HUNTEN, D. M., and KSANFOMALITY, L. V., The thermal balance of the middle and upper atmosphere of Venus. In: Venus, eds. D. M. HUNTEN, L. COLIN, T. DONAHUE, and V. MOROZ, University of Arizona Press, 1982.

ZASOVA, L. V., SPÄNKUCH, D., MOROZ, V. I. et al.: Infrared experiment aboard the automatic interplanetary stations Venera 15 and Venera 16. 3. Some conclusions on the structure of clouds, based on the analysis of spectra. Veröffentlichungen des Forschungsbereichs Geo- und Kosmoswissenschaften, Heft 18, 1990.

7. Thermal structure of the middle atmosphere of Venus from Venera 15 data

R. Dubois, L. V. Zasova, D. Spänkuch, V. I. Moroz, K. Schäfer, E. A. Ustinov, D. Oertel, V. M. Linkin, W. Döhler, J. Güldner

Abstract

A Fourier spectrometer aboard Venera 15 measured the outgoing spectral radiance of Venus from 250 to 1600 cm^{-1} (6 to 40 μm) in the nothern hemisphere during 40 polar orbits from 12 October to 14 December 1983. The spectra differ significantly in magnitude of brightness temperature and its spectral dependence. Different Venusian regions are characterized by typical, significantly different radiance and brightness temperature spectra. At least five groups can be identified, one typical for equatorial regions, one for mid-latitudes, two for the subpolar region (cold collar, warm spots), and one for the polar region.

The 15-μm-CO_2 fundamental absorption band was used for temperature retrievals in the altitude range between 60 and 95 km. Supplementary numbers in the range 365 cm^{-1} to 1313 cm^{-1} were added to infer simultaneously the opacity profile of the upper haze- and -cloud region.

The temperature profiles for the five typical spectra are presented and the models of the middle atmosphere of Venus (VIRA 1985, Adv. Space Res. Vol. 5, No. 11, pp. 3—58) are compared with the temperature profiles derived from the Fourier spectrometer soundings of Venera-15 for the same latitude belts.

Furtheron, the dependence of these temperature profiles on the local time of day in a solar-fixed coordinate system is discussed with graphical plots of temperature contours in the middle atmosphere in latitude-pressure coordinates including derived optical depths for the upper haze- and -cloud region.

Introduction

Many measurements using remote sensing methods and atmospheric probes have been made of the atmospheric temperature structure and cloud morphology of Venus since the early years of exploration of Venus by spacecraft in the sixties. A number of models representing the structure of the Venusian atmosphere have been proposed. The early temperature-height (pressure) models were reasonably accurate for the deep atmosphere, but have become increasingly uncertain for cloud levels and above.

From a thorough comparison and synthesis of data from Venera landers, Pioneer Venus Probes, the Orbiter Infrared Radiometer (OIR) on the Pioneer Venus Orbiter and the radio occultations (ORO) the uncertainties were found to be smaller for altitudes up to 100 km in the Venus International Reference Atmosphere (VIRA) [1]. Due to variations of up to 10 K the definition of data sets for the middle atmosphere was the most difficult problem. During the OIR experiment as many as more the half a million soundings of the middle atmosphere were made with most of the data taken in the Venusian northern hemisphere. This OIR data set [2] of the temperature structure of the Venusian middle atmosphere has been used as the basis of comparison for the other data, because it represents with measurements during 72 days in late 1978 and early 1979 some time average. However, with reference to [1] the variability about the mean state of the VIRA are probably up to about 5 K below 80 km altitude and no more than 10 K between 80 and 100 km.

The uncertainties in the temperature sounding of the Venera-15 Infrared Fourier Spectrometer (VIFS) are rather small and the altitudinal resolution is acceptable (see Table 1 and 3) providing the means to verify the typical thermal structure in the Venusian middle atmosphere and the VIRA temperature models [1], [2] from approximately 60 km in the upper cloud region up to 100 km. The representativeness of the zonal mean temperature structure is, however, restricted by the limited sounding coverage of the Venusian northern hemisphere (40 polar orbits from 12 October to 14 December 1985 only).

Typical emission spectra for different Venusian regions

A first-look inspection of the measurements of the Venera-15 Fourier spectrometer shows that the spectra (from 250 to 1600 cm^{-1}) differ significantly in the magnitude of the brightness temperature (T_B) and its spectral dependence. Emission spectra are determined by the thermal and cloud structure and by the chemical composition of the emitting atmosphere. Changes in these parameters affect the shape of the spectra. Therefore, the emission outside the 668 cm^{-1}-CO_2 band system can be grouped into five classes representative of different regions of Venus (see. Fig. 1) with the following characteristics:

1. Equatorial spectrum:
   Latitudinal range from 0.5° N to 10° N, sharp decrease of T_B outside absorption bands with increasing ν, differences $\sim$40 K between $\nu = 365$ cm^{-1} and $\nu = 1153$ cm^{-1}, maximum T_B of $\sim$275 K at 365 cm^{-1}, T_B at 668 cm^{-1} 5—10 K colder than for other classes.
2. Mid-latitude spectrum:
   Latitudinal range from 36° N to 71° N, similar to equatorial spectra with about 10 K lower T_B outside the 668 cm^{-1} absorption band, T_B at 668 cm^{-1} is 5 K warmer than in equatorial regions.
3. Cold subpolar spectrum (polar collar):
   Latitudinal range from 66° N to 73° N, lowest T_B of all groups outside the 668 cm^{-1} CO_2 band with $\sim$220 K between 800 and 1000 cm^{-1} and 230 K at 365 cm^{-1}.
4. Polar spectrum:
   Latitudinal range from 85° N to 87° N, $T_B \sim 242 \pm 1$ K nearly constant outside the 668 cm^{-1} band.
5. Hot polar spectrum (dipole):
   Latitudinal range 77° N to 80° N, warmest T_B ($\sim$250 K) in comparison with other groups for $\nu > 800$ cm^{-1}, T_B nearly constant outside absorption bands similar to polar spectrum (No. 4) but with $\sim$8 K higher temperatures.

The maximum brightness temperature difference is about 45 K at 365 cm^{-1} between the cold subpolar and the equatorial spectrum. The brightness temperature contrast between the cold polar collar spectrum and the hot polar spectrum is about 20 K at 365 cm^{-1}, about 30 K at 872 cm^{-1} and 1153 cm^{-1}.

The VIFS radiance spectra and some aspects of their interpretation are dicussed in detail by Moroz et al. [3].

The temperature retrieval from the Fourier spectrometer soundings

The temperature profile and the cloud model which best fit to the emission spectra measurements are retrieved using a first-guess model and an iterative technique together with the relaxation method, similar to those of Chahine [4] and Smith [5]. All temperature profiles and cloud model parameters of the present paper are the result of a routine computation with a standard version of the retrieval program. Hence, the result is not always the optimum retrieved temperature profile and cloud model.

To compare the retrieved temperatures with VIRA [1] adequately the VIRA temperature models at 85° N, 75° N, 60° N, 45° N, 30° N latitude form the data set for the

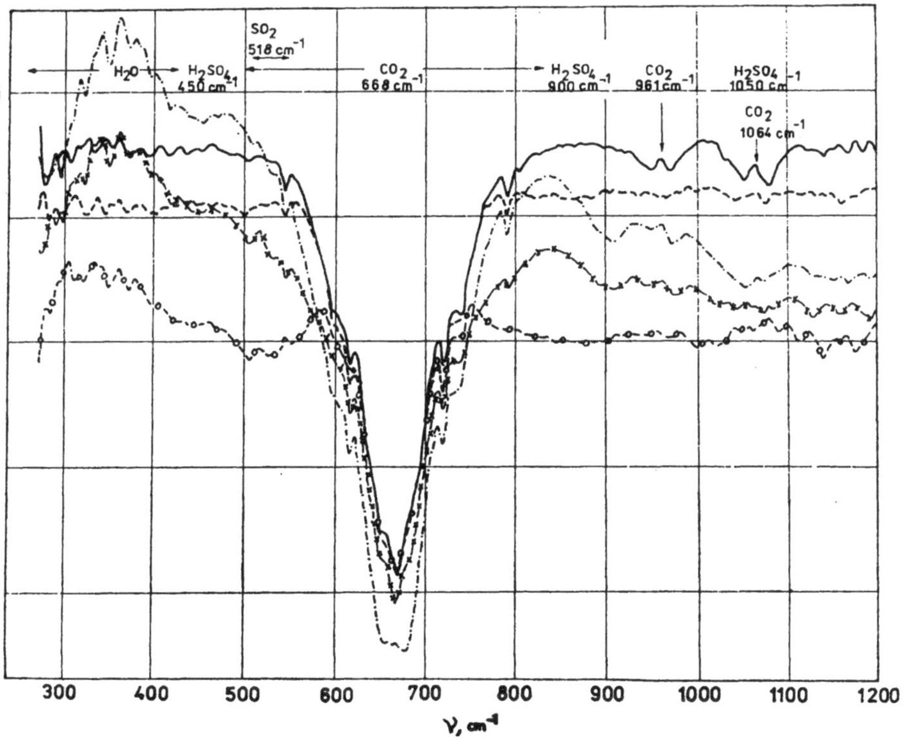

Fig. 1. Examples of the groups of brightness temperature (T_B) spectra typical of different Venusian regions selected from single different data of the Fourier spectrometer measurements •—•— Equatorial spectrum; ×—×— Mid-latitude spectrum; ○—○— Cold subpolar spectrum — — — Polar spectrum; ——— Hot polar spectrum

first guess. The program selects that initial temperature profile which is nearest in latitude to the measurement. The temperature is retrieved as a function of atmospheric pressure. The altitude required for cloud parameter estimations is derived from the hydrostatic equation taking the 0.5 hPa pressure level of the corresponding VIRA model as the reference because only there are very small pressure variations with height ($\Delta p_{max} \sim 0.05$ hPa for the same altitude in km above a 6052.0 ± 0.25 km radius reference surface of Venus). Gas constant, compressibility and acceleration due to gravity are the same as in VIRA [1]. The weighting functions of the 668 cm^{-1}-CO_2 fundamental absorption band were used for the retrieval of altitude-dependent temperature profiles. A vertical resolution of approximately 3 km was reached with ten weighting function maxima in the pressure range from 0.5 hPa up to approximately 200 hPa (Table 1).

The derived temperature profile is increasingly dependent on the inital temperature model at the boundaries of the altitude interval defined by the weighting functions (above 0.5 hPa and below 200 hPa) [6] which explains the differences in the corresponding figures in [22].

The CO_2 transmission function was calculated line by line [7] using the AFCRL spectral line catalogue [21]. The temperature dependence of the transmission function was taken into account by a fast correction procedure reducing the retrieval error by a maximum of about 3 K. The local zenith angle of the measured spectra was included within the retrieval program. The VIFS soundings require a zenith angle correction of more than 60° for low latitudes ($\varphi < 30°$) in the standard polar orientation. Hence, the retrievals are less accurate in equatorial regions.

Due to a changed orientation of the Venera spacecraft [8, 9] only a small zenith angle correction of 9° is needed for equatorial spectra ($\varphi \lesssim 5°$), on 11 November 1983.

The influence of aerosol at the edges of the 668 cm^{-1}-CO_2-absorption band grows with increasing pressure. Without taking into account this influence the retrieved temperatures of the upper cloud lyaer are systematically underestimated [6, 10]. This under-

Table 1

The weighting functions of the 668 cm^{-1}-CO_2 fundamental absorption band

long-wave wing wavenumber cm^{-1}	weighting function maxima hPa	short-wave wing wavenumber cm^{-1}	weighting function maxima hPa
555.63	200	677.17	1
562.67	150	691.06	2
572.99	100	701.48	5
586.88	45	711.90	10
600.77	20	732.73	16
614.66	12	753.57	45
632.02	5	770.93	90
642.44	2	781.35	150
656.33	1	802.18	200
666.75	0.5		

estimation can be 10 K and more below an altitude of 70 km (~30 hPa) (cf. Fig. 2). Therefore, reliable temperature estimates below 70 km are only feasible by appropriately taking into account the cloud influence. Simultaneous estimates of temperature and cloud parameters are the best way to reduce this error. Due to the wide spectral range covered by VIFS such an approach can be made with the VIFS measurements. Differences in the cloud extinction coefficient are up to one order of magnitude in the spectral interval from 300 to 1 200 cm^{-1}. Altogether, ten wavenumbers have been selected for the simultaneous determination of the temperature and the upper cloud structure (Table 2). The cloud data of the Pioneer Venus Mission [11, 12, 13], summarized by Pollack [14] in a three-mode cloud model was modified and taken as the first guess

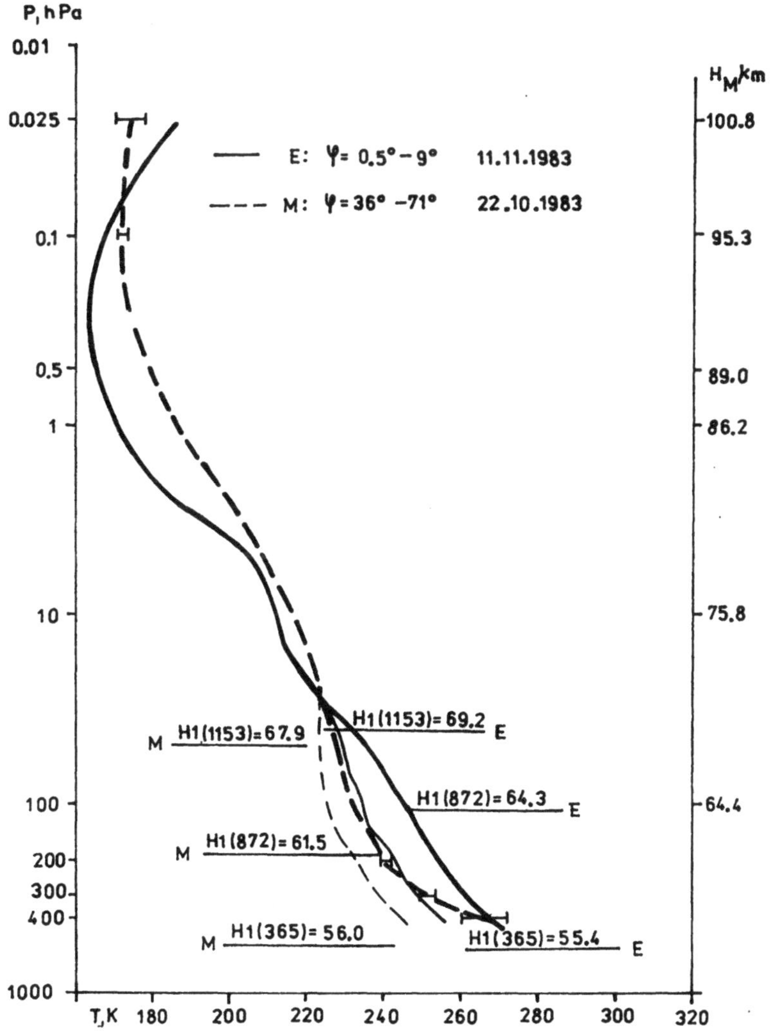

Fig. 2. Retrieved temperature profiles from typical equatorial (*E*) and mid-latitude (*M*) emission spectra. *H 1* indicates the altitude and corresponding pressure level of the cloud unit optical depth of the wavenumber labelled. Retrieval uncertainty is shown by bars

Table 2
Wavenumbers for the simultaneous determination of the temperature and the upper cloud structure

wavenumber, cm^{-1}	wavenumber, cm^{-1}
364.63	1000.12
406.30	1107.78
451.45	1152.92
871.64	1201.54
899.42	1312.66

for the upper haze-region and the upper cloud layer [10]. Problems remains, however, in the retrieval technique used, for the cloud parameters, in case of very strong temperature inversions and of isothermal layers. Hence, further improvements of the retrieval technique are necessary at high latitudes ($\varphi \geqq 60°$). Sulfur acid clouds have their maximum optical depth in the VIFS spectral region at 1152 cm^{-1} and their minimum at 365 cm^{-1} [3]. At both wavenumbers there is no gaseous absorption. Fig. 2 showing the retrieved temperature profiles contains also the pressure levels with a unit optical depth above those levels for both these wavenumbers and for 872 cm^{-1}. The value of the latter wave number permits a comparison with the result of the OIR 'cloud channel', which are, however, influenced by CO_2 absorption [6, 15].

The pressure level with unit optical depth above that level taken at 1152 cm^{-1} is a good estimate of the real cloud top temperature and, thus, the cloud top pressure. Additionally, the difference of the pressure levels with at the 1152 cm^{-1} and 365 cm^{-1} unit optical depth above contains information on cloud opacity.

It is difficult to give an analytical expression for the uncertainty in the retrieved temperatures and cloud properties in case of the relaxation technique used. The uncertainty given in Table 3 is the result of numerical simulations taking into account errors in measured radiances, calculated transmittances, in the limited vertical resolution due to the spread of the weighting functions, and in the influence of the first guess models. The temperature profiles are the average of the retrievals of the long-and short-wave wing of the 668 cm^{-1} CO_2-absorption band if the temperature difference between both retrievals is within the retrieval uncertainty. In some cases mainly at low latitudes ($\varphi \leqq 35°$), however, differences in the retrieved temperatures are considerably larger, probably caused by additional absorption in the long-wave wing so that the temperatures retrieved from the short-wave wing are more reliable.

Table 3
The uncertainty in the retrieved temperature profile (± K)

pressure	hPa	0.025	0.1	1	10	100	200	300	400
altitude	km	~101	~95	~86	~76	~64	~61	~60	~58
equatorial		3.2	1.5	1.1	1.0	2.3	1.6	2.2	3.5
mid-latitude		3.7	1.4	0.7	0.7	1.0	2.2	2.0	5.9
cold subpolar		4.2	1.5	0.6	1.1	4.0	2.7	3.5	3.4
hot polar		6.1	0.9	0.6	1.1	1.1	1.9	2.2	6.0
polar		6.1	1.1	0.7	1.6	2.3	2.9	1.9	3.7

The typical thermal structure of different Venusian regions

The five characteristic emission spectra (Fig. 1) are used to derive the typical thermal structure at different Venus latitudinal regions. Fig. 2 gives the temperature profile for low and middle latitudes as well as the pressure (altitude) values with spectral unit optical depths above the levels at the wave numbers labelled. Equatorial temperatures are lower than at mid-latitudes (and in polar regions (cf. Fig. 3)) in the altitudinal range from 72 km to 95 km (≙ 0.1 to 25 hPa) with differences up to 15 K around 1 hPa (~86 km). In contrast to higher latitudes we find at low latitudes a temperature inversion with a

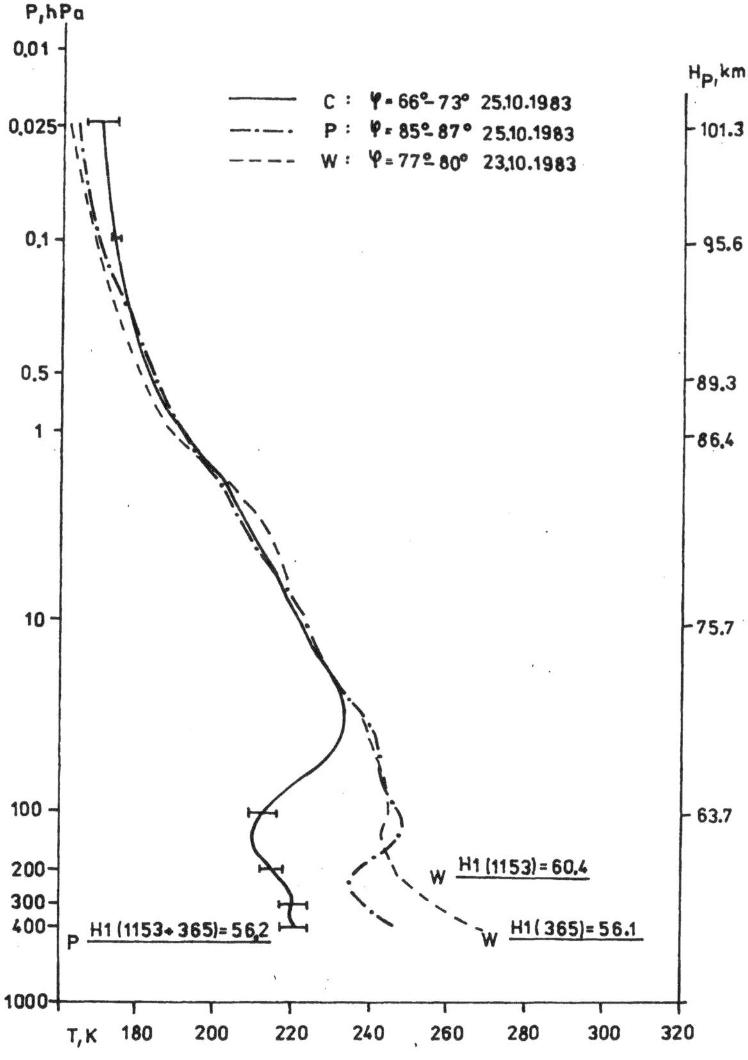

Fig. 3. Retrieved temperature profiles from typical cold subpolar, hot polar and polar emission spectra. Retrieval uncertainty is shown by bars

minimum temperature of about 164 K at 92 km. On the other hand, equatorial temperatures are the warmest around the 200 hPa level resulting in a less stable temperature stratification at lower than at higher latitudes in the 200 to 1 hPa range. Within the upper cloud region equatorial temperatures are up to 13 K warmer than at mid-latitudes as indicated by the colder midlatitude T_B spectra at the edges and outside the 668 cm^{-1} band. Unit cloud optical depth at 365 cm^{-1} is at the equator reached 2 km lower than at mid-latitudes. As the change in cloud top height given by the 1152 cm^{-1} channel is much less pronounced, only 0.2 km from low to middle latitudes, there is an increase in upper cloud opacity with increasing latitude.

Fig. 2 contains, also, temperature retrievals obtained by means of pure CO_2 transmittances to demonstrate the effect of an insufficient cloud estimation.

The polar temperature structure (Fig. 3) in the altitudinal range from 70 km to 85 km is for all three typical polar emission spectra approximately the same about 5 to 10 K higher than at midlatitudes. A strong inversion of about 24 K is the typical thermal feature of the cold subpolar (collar) region. The inversion lies between 30 hPa with a temperature maximum of 234 K and 150 hPa with a minimum of about 210 K. This represents the coldest thermal structure in the Venusian upper cloud layer below 67 km, which is in agreement with the lowest T_B of all typical spectra outside the 668 cm^{-1} band. The highest temperatures at 62 km are restrieved from the equatorial and the hot polar spectrum ('hot spots') with about 250 K. Hence, the greatest horizontal temperature difference in the middle atmosphere of Venus appears at that altitude with about 40 K. The temperature at the effective cloud top defined by unit cloud optical depth at 1152 cm^{-1} is lower and denser in the polar region than in other regions.

Comparison of temperature profiles from Venera-15 data with VIRA-models

The temperature structure presented here is retrieved from VIFS-measured radiances averaged in a solar-fixed coordinate system. This system is used because solar-fixed and latitude-dependent differences were observed to exceed daily and long-term radiance fluctuations at most longitudes, latitudes and altitudes in the Venusian middle atmosphere.

Taylor et al. [16] and Schofield et al. [17] showed that the global mean temperature field of the equatorial middle atmosphere varies longitudinally with a solar-fixed structure. At low latitudes wavenumber 1 is observed at the cloud top. A semi-diurnal thermal tide appears to be dominant with its maximum of $\Delta T \sim 8$ K at about 80 km altitude near sunrise and sunset. The equatorial temperature profile of the VIFS sounding on 11 November 1983 corresponds to the sunrise extreme of the semi-diurnal tide at Venusian local time, LT = 07.00 hrs.

The VIFS-temperature structure differs from the adequate OIR-profile (LT = 06.00 to 08.00 hrs.) [2] by not more than 1 to 3 K in the alttiude range of 60—78 km and 2 to 5 K between 79 and 92 km. Equal temperatures are derived at 0.1 hPa ($\sim$95 km).

For mid-latitudes and polar latitudes averaged temperature profiles are calculated from all available spectra of the 40 Venera-15 polar orbits for latitude intervals of $\pm 5°$. These averaged temperature profiles approximatelly represent the zonal mean temperature in spite of the absence of data during night and day (see Fig. 4). The variance of the Venusian middle atmosphere with local time can to some extent be extracted from

Figures 5 to 8. Additionally, a further subdivision was made using the solar zenith angle (θ) with the definition:

day: $\theta = 0° - 80°$,
night: $\theta = 100° - 180°$,
twilight: $\theta = 80° - 100°$.

Hence, the following six zones are defined (see Fig. 4):

'morning twilight' (MT), $\theta = 80° - 100°$
'evening twilight' (ET), $\theta = 80° - 100°$
'morning day' (MD), $\theta = 0° - 80°$
'evening day' (ED), $\theta = 0° - 80°$
'morning night' (MN), $\theta = 100° - 180°$
'evening night' (EN), $\theta = 100° - 180°$.

Representative mean profiles of these six zones can, however, not be obtained for all latitudinal belts as the number of measurements depends on latitude. There are more measurements at higher latitudes due to the Venera-15 orbit geometry.

The VIRA models [1] and the zonal mean Venera temperature profiles very well agree in the latitude ranges 30° and 45° (cf. Fig. 5 and 6). There are some differences between VIRA and our mean temperature profiles in the twilight regions at 30° latitude. A wave-like structure appears in the evening twilight temperature profile due to the insufficient number of measurements. The variability in the upper altitude range from 85 to 100 km is probably caused by the semidiurnal tide. This phenomenon is diminished at 45° latitude. The region represents the beginning of the deviation from the low-latitude temperature structure. The cold collar region is in the VIRA not wel

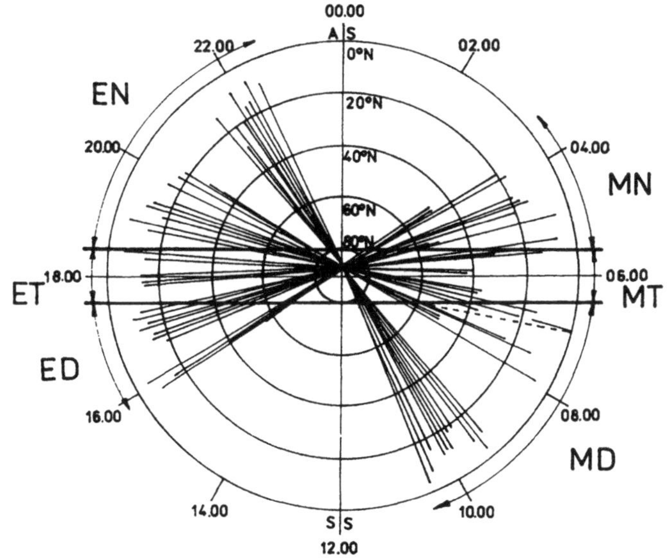

Fig. 4. Venera-15 orbits with VIFS data in solar fixed coordinates
AS: antisolar point; *SS*: subpolar point; *MT*: morning twilight; *ET*: evening twilight; *MD*: morning day; *ED*: evening day; *MN*: morning night; *EN*: evening night

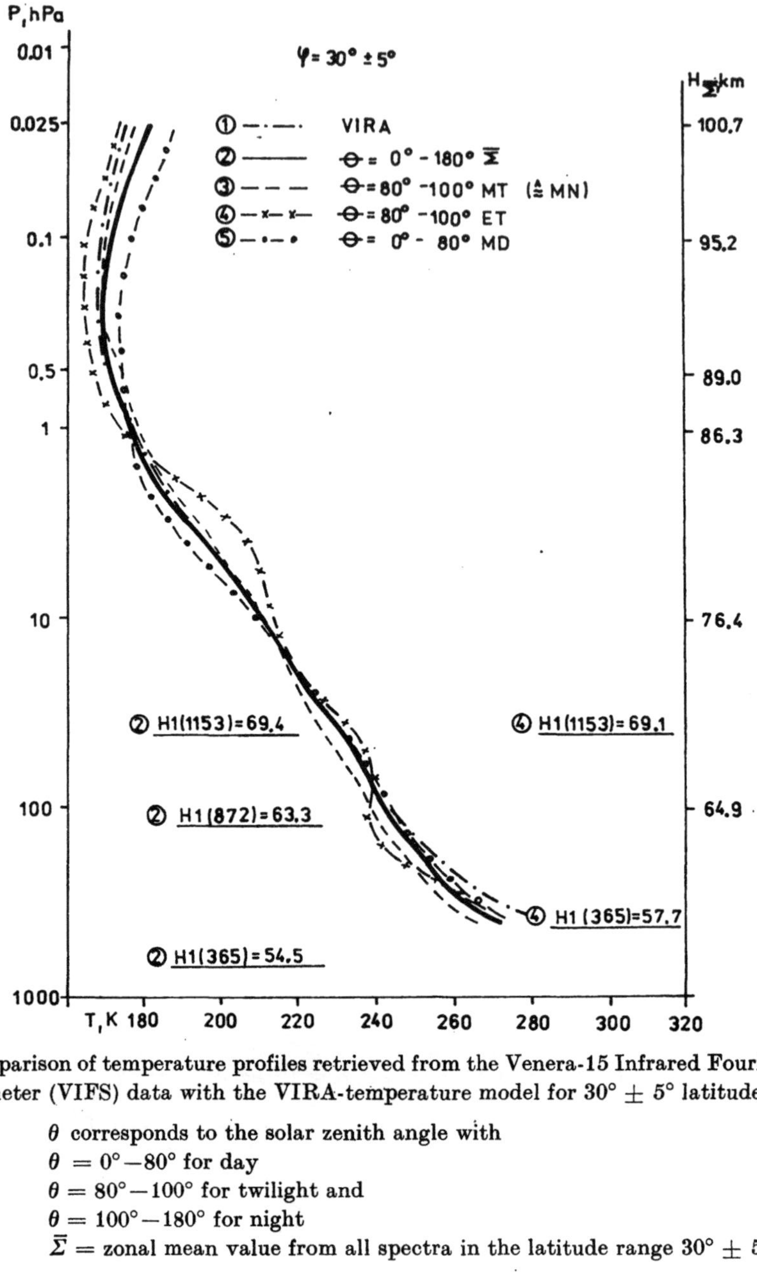

Fig. 5. Comparison of temperature profiles retrieved from the Venera-15 Infrared Fourier Spectrometer (VIFS) data with the VIRA-temperature model for 30° ± 5° latitude

θ corresponds to the solar zenith angle with
$\theta = 0°-80°$ for day
$\theta = 80°-100°$ for twilight and
$\theta = 100°-180°$ for night
$\bar{\Sigma}$ = zonal mean value from all spectra in the latitude range 30° ± 5°

The temperature profile in the morning twilight region (MT) is nearly the same as for the morning night region (MN)

2 ≙ Σ: LT = 03.43–08.26 hrs., number of days: 21, number of spectra: 51
and LT = 16.02–21.09 hrs., number of days: 23, number of spectra: 45
3 ≙ MT: LT = 05.20–05.34 hrs., number of days: 3, number of spectra: 6
4 ≙ ET: LT = 17.57–18.28 hrsg., number of days: 6, number of spectra: 12
5 ≙ MD: LT = 06.52–08.36 hrs., number of days: 13, number of spectra: 36

represented as the latitudes 60° and 75° do not meet the collar's core, bot only its edges. The VIFS-mean temperature structure at 60° latitude is identical with the VIRA model [1] in the pressure range from 0.02 to 7 hPa (i.e. from 100 to 77 km altitude). Below 77 km altitude the VIRA-profile is warmer (Fig. 7). The temperature gradient of the strong inversion in the VIFS-mean profile is greater from 40 to 100 hPa. Temperature variations with local time are, not surprisingly, greater at the altitude of the centre of the cold collar around 100 hPa (~64 km) and also at lower levels. The lowest tempera-

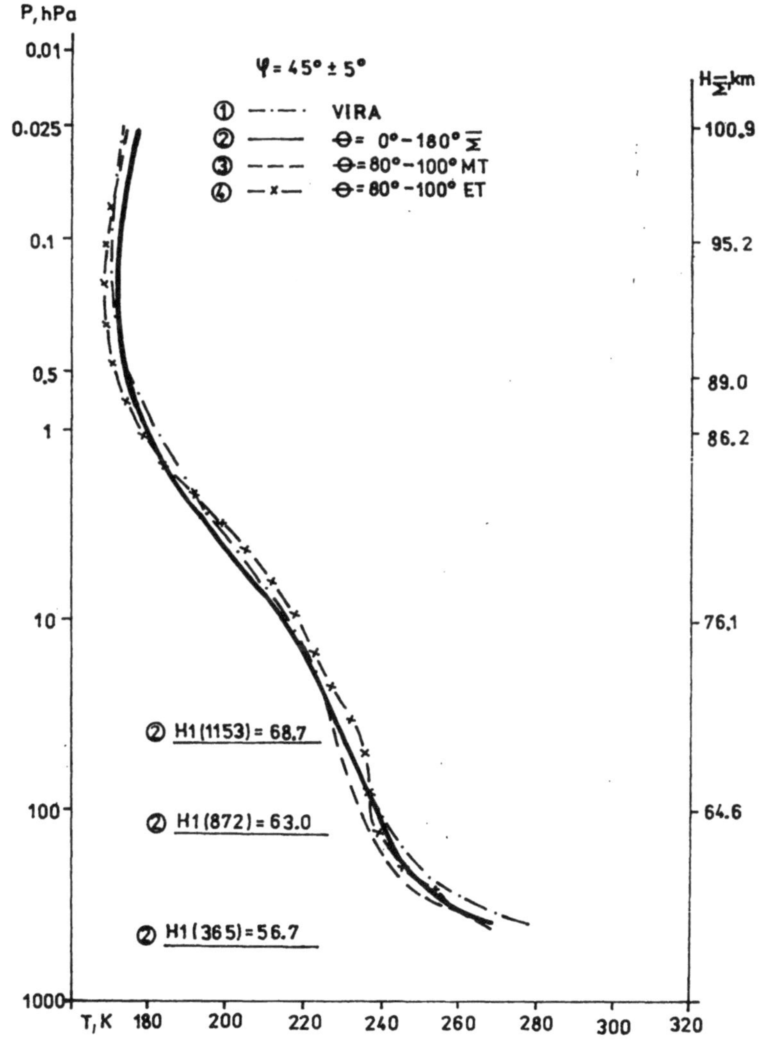

Fig. 6. Comparison of temperature profiles retrieved from the VIFS data with the VIRA temperature model for 45° ± 5° latitude

2 ≙ $\bar{\Sigma}$: LT = 03.21 – 10.11 hrs., number of days: 32, number of spectra: 88
and LT = 16.00 – 22.38 hrs., number of days: 23, number of spectra: 62
3 ≙ MT: LT = 05.38 – 06.20 hrs., number of days: 10, number of spectra: 23
4 ≙ ET: LT = 17.06 – 18.25 hrs., number of days: 6, number of spectra: 14

ture (about 210 K) is observed during morning-twilight and morning-day. During avening-day and evening-twilight, however, much warmer temperatures (230 K) occur et the same altitude, indicating a cooling by more than 20 K at the core-level from sunset to sunrise.

The mean temperatures of the evening-night zone with about 225 K and of the morning-night region with about 223 K confirm this cooling at this altitude. It is interesting

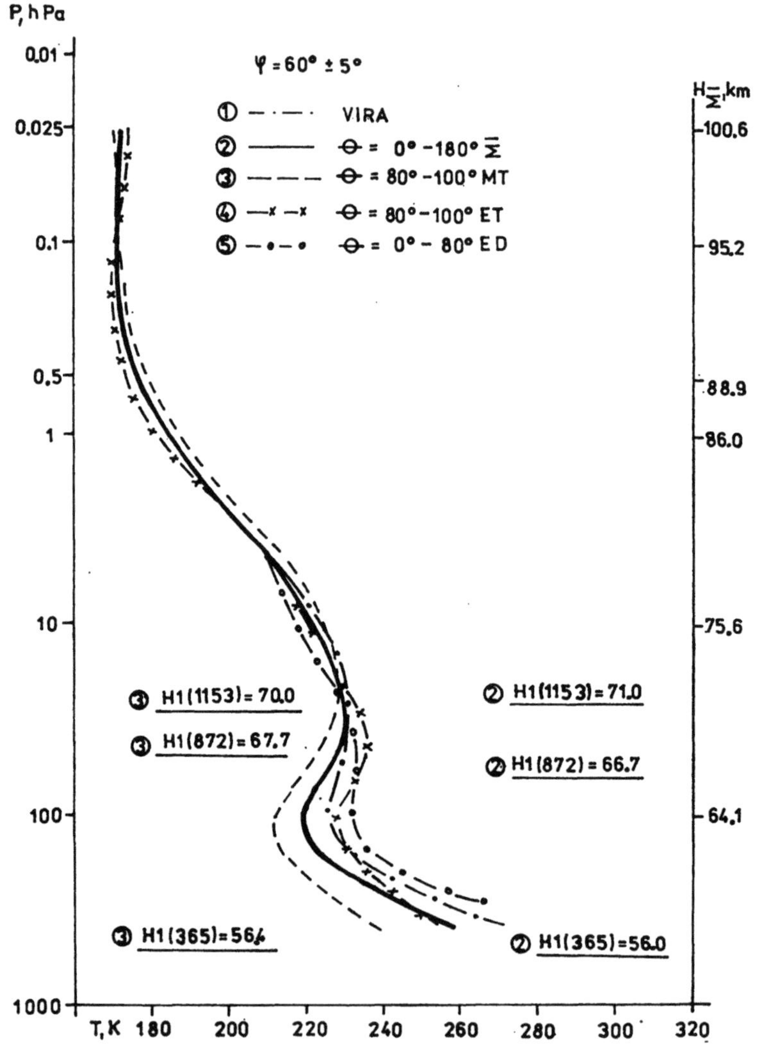

Fig. 7. Comparison of temperature profiles retrieved from the VIFS data with the VIRA temperature model for 60° ± 5° latitude

2 △ $\bar{\Sigma}$: LT = 02.23—08.29 hrs., number of days: 39, number of spectra: 133
and LT = 15.57—22.42 hrs., number of days: 22, number of spectra: 77
3 △ MT: LT = 04.43—07.38 hrs., number of days: 16, number of spectra: 39
4 △ ET: LT = 16.50—19.05 hrs., number of days: 9, number of spectra: 26
5 △ ED: LT = 15.57—16.00 hrs., number of days: 2, number of spectra: 9

to note that the same longitude dependence (in solar-fixed coordinates) is present in the OIR temperatures at 65° latitude (VIRA [2]) around the core level at 64—66 km with a minimum temperature of about 213 K in the morning (LT = 6.00—10.00 hrs.).

This cooling seems to be reduced to 10 K for the time interval from sunset with 232 K to sunrise with 223 K at 75° latitude around 130 hPa (~62.5 km). But on the whole,

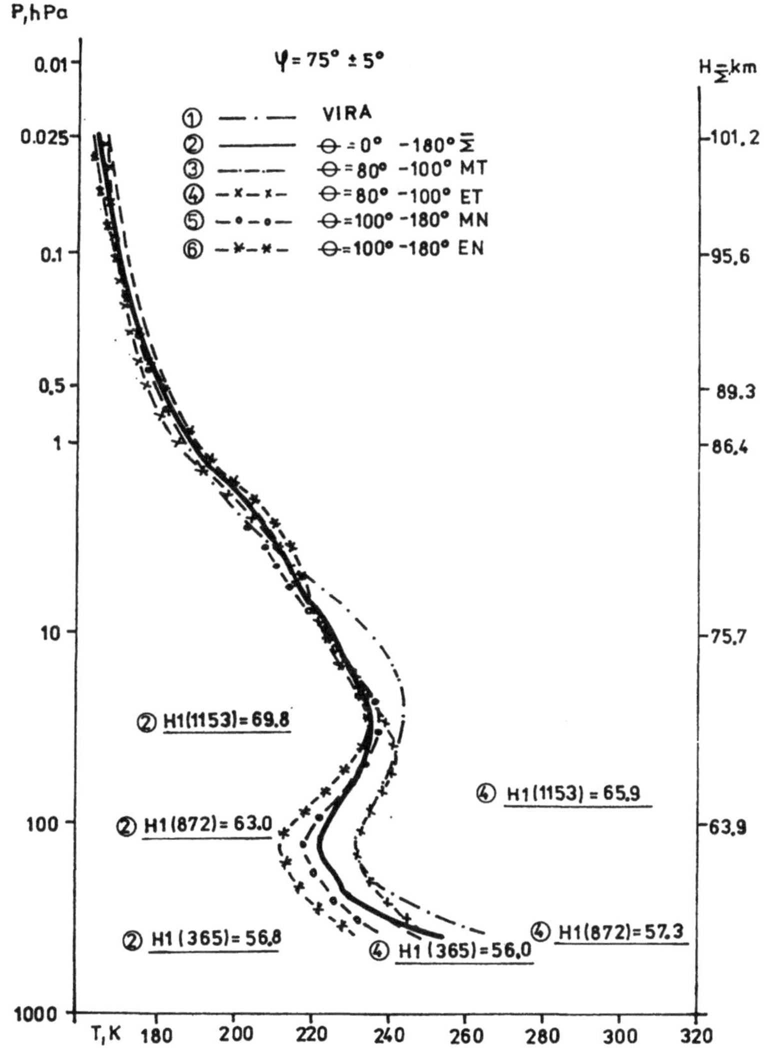

Fig. 8. Comparison of temperature profiles retrieved from the VIFS data with the VIRA temperature model 75° ± 5° latitude

2 ≙ $\bar{\Sigma}$: LT = 02.40—09.53 hrs., number of days: 39, number of spectra: 132 and LT = 16.42—23.21 hrs., number of days: 22, number of spectra: 87
3 ≙ MT: LT = 02.40—09.42 hrs., number of days: 35, number of spectra: 108
4 ≙ ET: LT = 16.27—21.12 hrs., number of days: 13, number of spectra: 46
5 ≙ MN: LT = 03.04—03.23 hrs., number of days: 4, number of spectra: 7
6 ≙ EN: LT = 20.15—23.21 hrs., number of days: 9, number of spectra: 41

there the temperature structure is more complex. The strong inversion in the evening-night zone (cf. Fig. 8) has a temperature minimum of about 213 K with up to 20 K lower temperatures than at sunset. The temperature profile during the morning-night shows an increased temperature of about 219 K at the core level of the cold collar with a further warming to about 223 K in the morning-twilight. The OIR-data set [2] shows, by and large, a very similar temperature structure at the core level at 75° latitude but with different absolute values. The OIR-temperatures decrease by about 5 K from sunset to midnight and increase up to sunrise by 12 K (at LT = 06.00 hrs.). However, during the morning the temperature drops dramatically by 17 K (maximum). This low temperature remains up to noon. A similar dramatic temperature decrease of about 20 K seems to be in the VIFS-data set for LT = 08.54—09.53 hrs. in the morning.

Although the temperatures retrieved from the OIR and VIFS measurements differ by more than 10 K at the corresponding location the similarity in the complex middle atmospheric temperature structure after 5 years is worthwhile mentioning. Before discussing the latitude ranges 75° and 85° of the Venusian middle atmosphere, it should be kept in mind that this is the latitude of the occurrence of the polar dipole. The dipole is a giant vortex of a surprisingly complex structure and behaviour and one of the more remarkable phenomena in the solar system hitherto discovered [18]. But there is still no detailed theory which models the dipole appearance and behaviour satisfactorily. The true nature of this complex phenomenon in the Venusian atmospheric dynamics has still not been revealed.

The VIFS-zonal mean temperature profiles for 75° and 85° latitude are up to 10 K lower than the corresponding VIRA model [1] below 80 km. The evening-twilight temperature structure agrees within the cloud layer in a pressure range of 40 to 200 hPa. The substantial systematic temperature difference in the polar temperature structure over an altitude range of over 20 km is a remarkable feature. The VIRA temperature model [1] was derived by preserving the temperature differences with latitude indicated by the OIR experiment — VIRA [2]. But two exceptions to this procedure were made — at latitude 75° and 85° — because the OIR and ORO temperature structures differ substantially with up to 8 K lower OIR-values between 60 and 70 km. The VIRA model [1] uses in this altitudinal range the ORO data. This choice was justified by a better altitudinal resolution of the ORO technique in this altitudinal range.

It is interesting to note that the OIR zonal mean temperature at $\varphi = 85°$ (VIRA [2]) agrees quite well with the VIFS zonal mean temperature profile. The temperature deviation is about 3 K at $\varphi = 75°$ and about 2 K at $\varphi = 85°$. For instance at $\varphi = 75°$ the temperatures derived from VIFS, OIR and ORO are at 70 km 236 K, 234 K and 244 K, respectively. The temperature inversion at 65° and 75° latitude in the VIFS profiles appears to be stronger than in the OIR-temperature structure caused by the better altitudinal resolution. VIFS soundings have smaller uncertainty in the altitude of the upper cloud layer in case of strong vertical temperature changes. Another comparison is feasible between the VIFS-, the ORO- and Venera Radio Occultation (VRO) data around 85° latitude. The VRO measurements were made during the first days of the Venera mission from 14 to 31 Oct. 1983 [20] and are presented as averages for individual selected latitudes between 81° and 89° with solar zenith angles between $\theta = 81°$—95°. The VRO temperatures are about the same or up to 4 K higher than the ORO values in the altitude range 60 to 76 km and about 5 K to 13 K higher than the VIFS temperatures in this altitude range. However, the VRO data show a remarkable time variability at higher altitudes, for instance of 21 K around the 75 km level for the same sounding location in a four-day interval. The temperature inversion derived from the VRO data amounts to 4 K to 17 K with a minimum temperature of about 236 K to 243 K around the 59 km level.

A temperature variability with a time scale of weeks is shown in Fig. 9 by two special selected temperature profiles derived from the VIFS data for the 85—88° latitude region. The outgoing infrared radiation flux measured in the spectral range of the VIFS in the latitudinal range 85° + 5° reaches a maximum of 175 W/m² × sr on 20 October 1983 and decreases approximately continously by nearly 30% to about 132 W/m² × sr

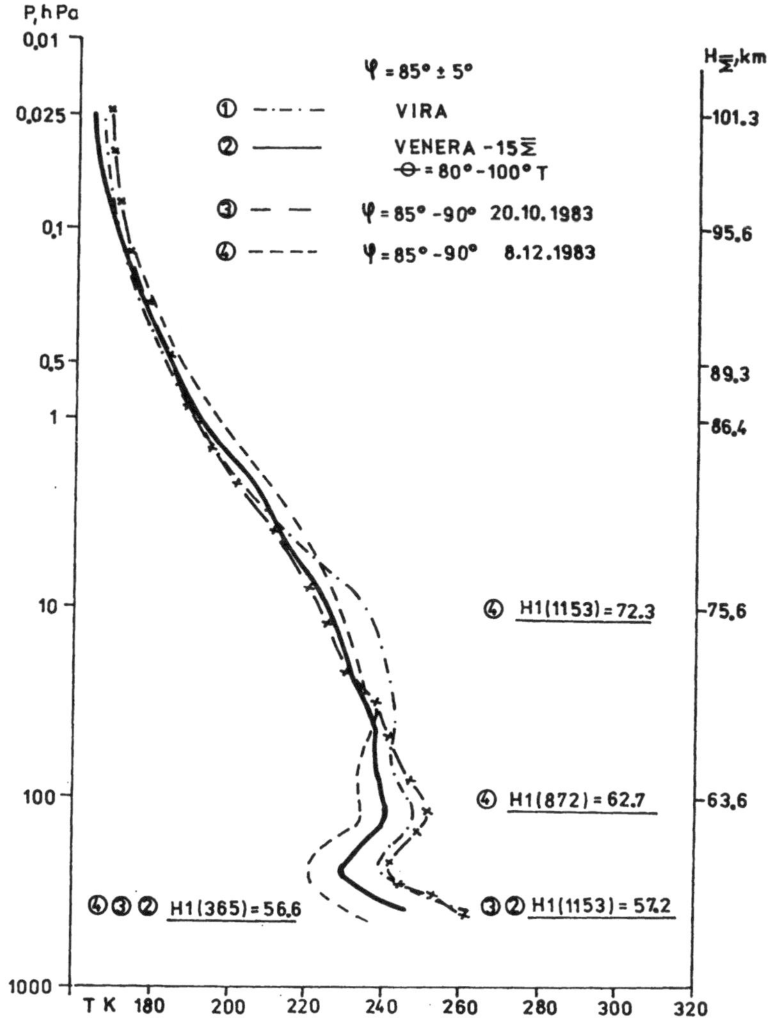

Fig. 9. Comparison of temperature profiles retrieved from the VIFS data with the VIRA temperature model 85° ± 5° latitude. Correction: Signature *3* —×—

Σ: LT = 00.00—12.00 hrs., number of days: 36, number of spectra 165 and LT = 12.00—24.00 hrs., number of days: 24, number of spectra 93 on 8 December 1983 (cf. Fig. 13) [19]. The temperature profile retrieved from the minimum emission spectrum agrees much more with the VIRA [1] polar model in its altitude dependence but is up to 8 to 15 K lower between 10 and 200 hPa (≙75—60 km). The temperature structure of the maximum emission spectrum is different. It is up to 5 K lower than from the minimum emission spectrum between 86 and 76 km (1—10 hPa) and higher than the VIRA model by about 3 K below 65 km (≙80 hPa).

The Venusian mean thermal structure and its variability

We summarize our findings about the temperature structure of the Venusian middle atmosphere and its variability in Figures 10 to 13. A survey of the above discussed temperature deviations between the VIFS-zonal mean values and the VIRA [1] models for the five latitude ranges 30° N, 45° N, 60° N, 75° N and 85° N is shown in the upper part of Fig. 10. We find greater temperature differences in the polar region caused by higher VIRA [1] temperatures below 80 km. The time variability is demonstrated in the lower part of Fig. 10 by temperature differences between the VIFS-zonal mean and zonal means for limited solar zenith angles and local time intervals. At 85° latitude mean tem-

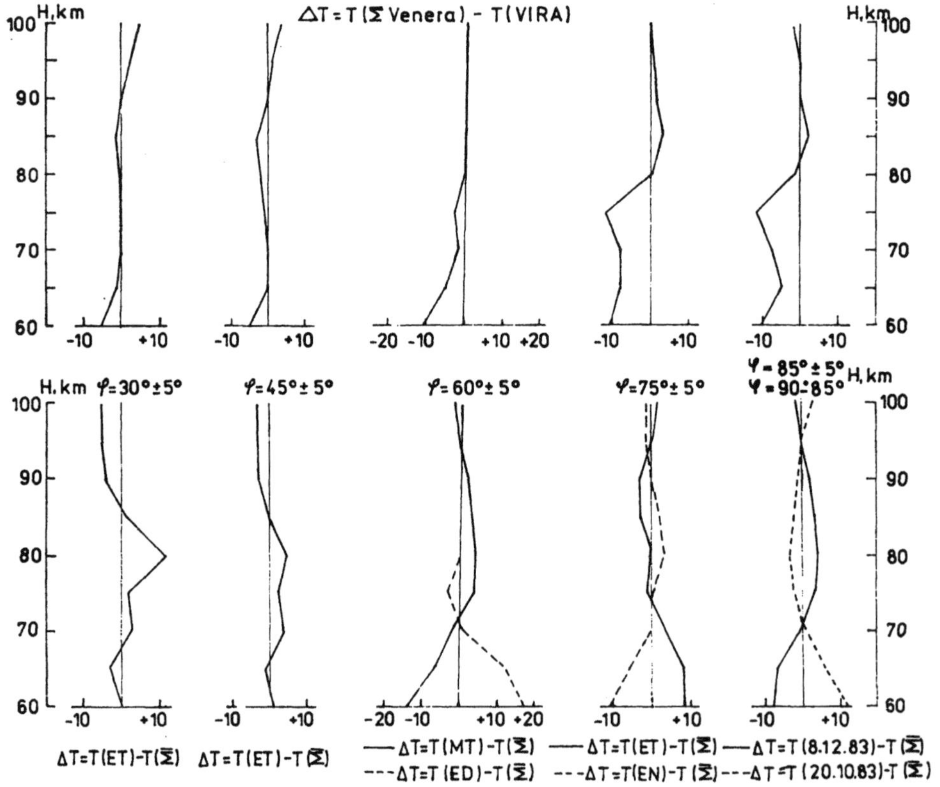

Fig. 10. Temperature difference between the VIFS zonal mean values ($\bar{\Sigma}$) and the VIRA model for five latitude ranges and temperature differences from the VIFS zonal mean for regions with a certain solar zenith angle and local time — cf. Fig. 4

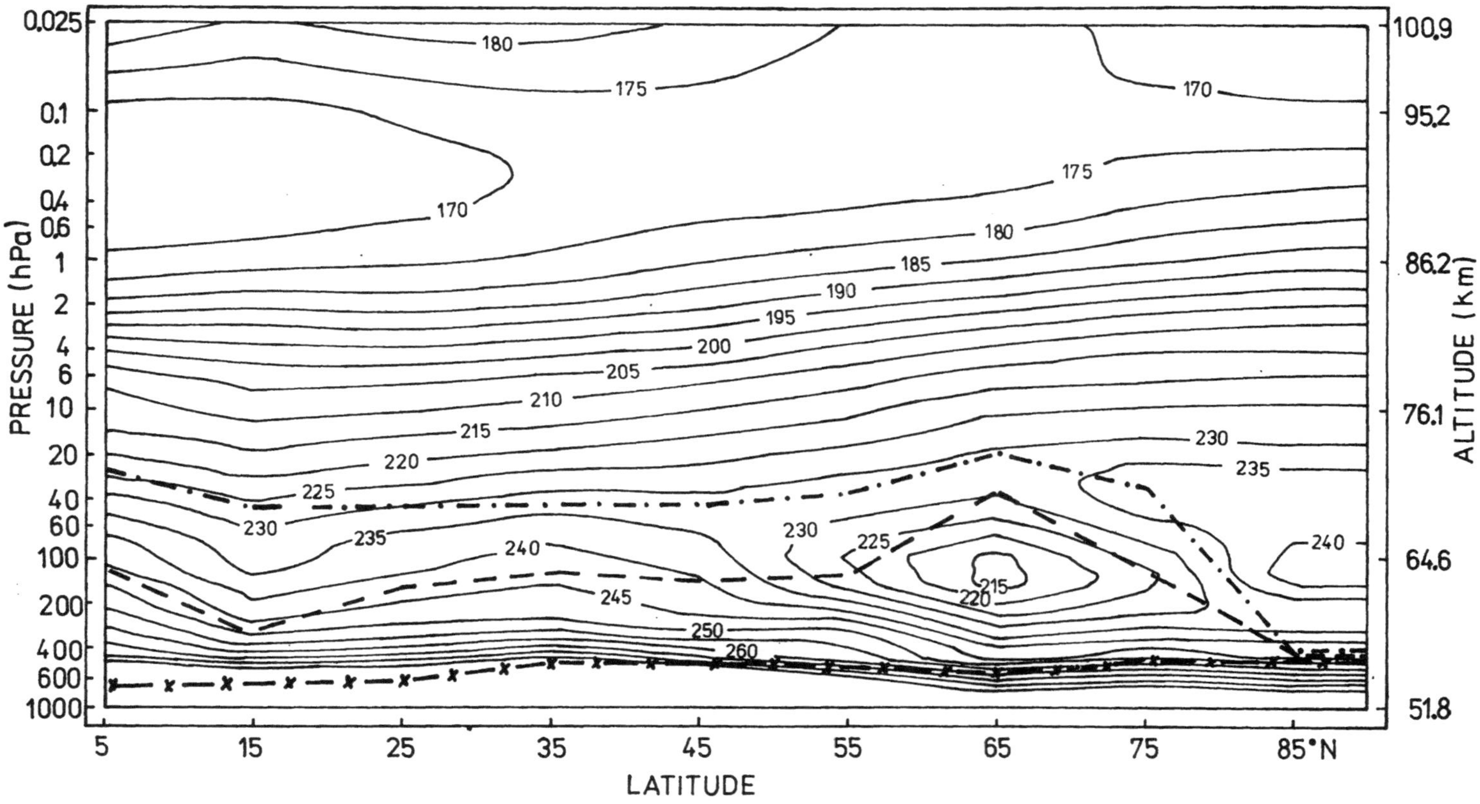

Fig. 11. The retrieved zonal-mean temperature field and cloud structure from VIFS data. Cloud unit optical depth is indicated for the wave numbers 1153 cm^{-1}, 872 cm^{-1}, and 365 cm^{-1}. The altitude H is valid for 45° latitude and for other regions as an approximation

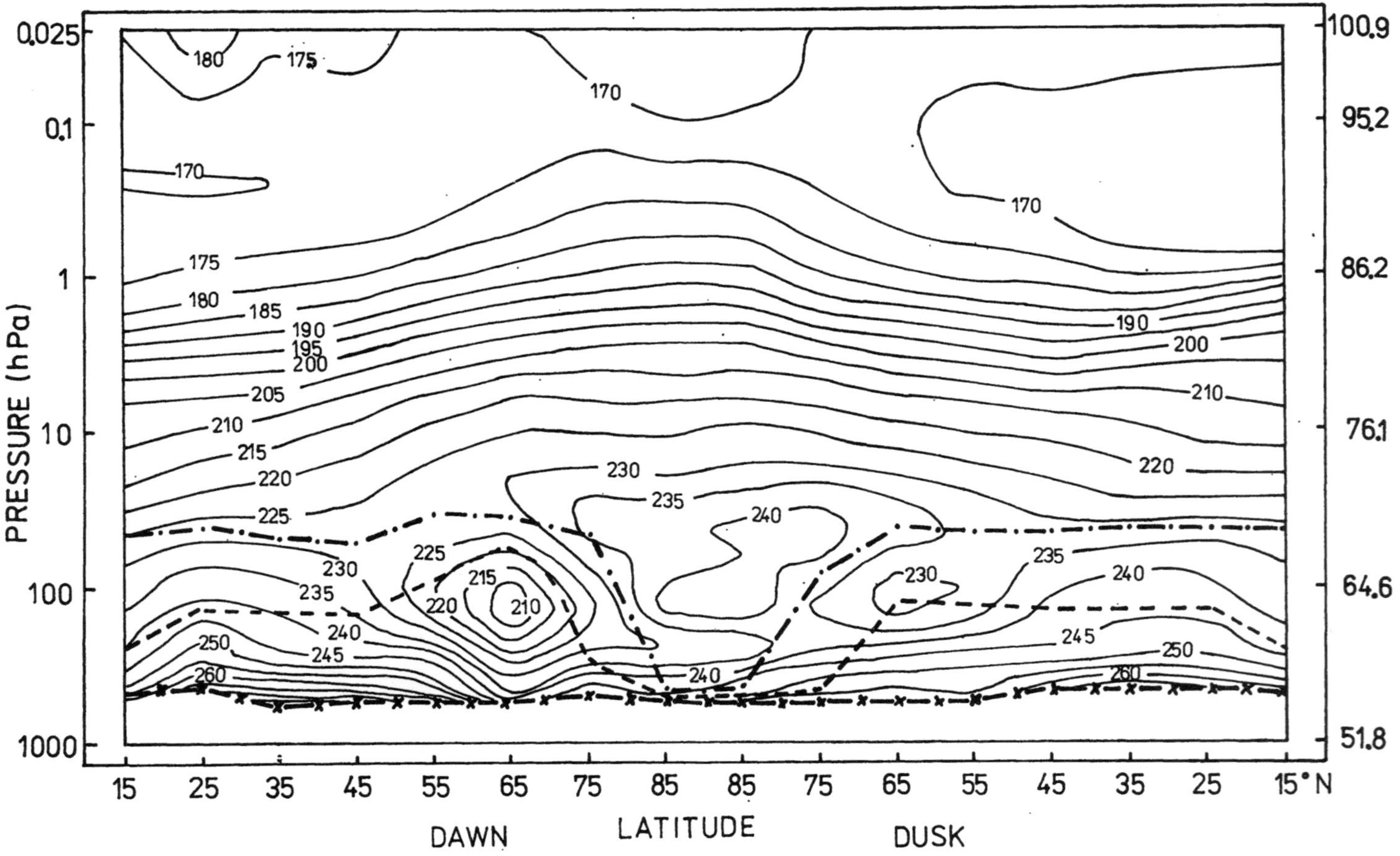

Fig. 12. The retrieved temperature field for the twilight region ($\theta = 80° - 100°$) from the VIFS data. Cloud unit optical depth is indicated for the wave numbers 1153 cm^{-1}, 872 cm^{-1}, and 365 cm^{-1}. The altitude H is valid for 45° latitude and for other regions as an approximation

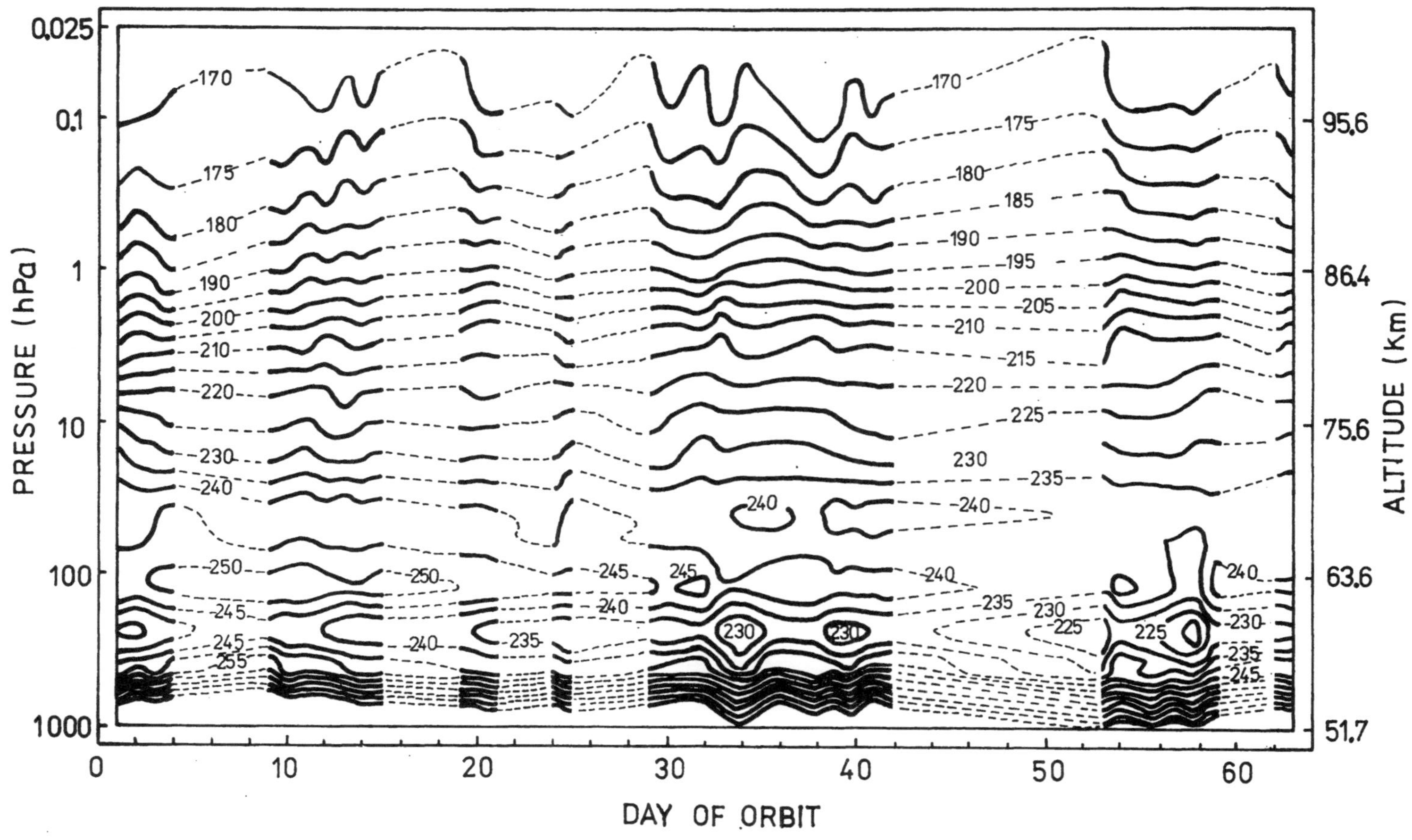

Fig. 13. The retrieved temperature field from VIFS data in the latitude range $\varphi \geqq 85°$ (twilight region $\theta = 80° - 100°$) during the Venera-15 mission from 12 October 1983 till 12 December 1983 as a function of day number. The altitude H is valid for 45° latitude and for other regions as an approximation

perature profiles for two different days are plotted to illustrate long-term variability in the twilight region ($\theta = 80° - 100°$). Fig. 11 shows the retrieved zonal mean temperature field of the VIFS data as isotherms in a latitude pressure diagram using a 10° latitude bin with mean values for the corresponding $\pm 5°$ interval. The retrieved cloud structure is indicated by the altitude of the effective cloud top, defined as unit cloud optical depth above that altitude, for three selected wave-numbers. A comparison of the effective cloud top at the wavenumber 872 cm^{-1} between the OIR and the VIFS mean values reveals a 4 to 5 km higher altitude in the VIFS-atmospheric structure at low and mid-latitudes and a lower cloud top level of about 7 km in the polar region around 85° latitude. A first look on Fig. 11 clearly shows that the cold collar is mainly a temperature feature with its core at about 65° latitude around 100 hPa ($\triangleq$62 km). Fig. 12 provides a general view of the retrieved Venusian mean thermal structure in the twilight region ($\theta = 80° - 100°$) during the VIFS-mission. The strong temperature inversion is fully developed in the sunrise region but nearly absent in the evening (sunset).

Some aspects of the atmospheric variability over the north pole during the Venera-15 mission are discussed in detail by Spänkuch and Dubois [19]. In addition to this discussion the isothermes in the time (day of mission) — pressure diagramm (Fig. 13) give an impression of the long time scale variability of the temperature structure in the middle atmosphere near the north pole — the region of the giant Venusian complex vortex.

Conclusion

The verification of the VIRA [1] middle atmospheric temperature model by the VIFS-temperature sounding reveals satisfactory agreement at the 5 latitudes (30° N, 45° N, 60° N, 75° N, 85° N) in the altitude between 80 and 100 km but systematically lower temperatures (up to 10 K) in the VIFS-mean profiles for the polar region (60°N, 75° N, 85° N) at altitudes below 80 km up to the lower sounding height around 58 km.

References

SEIFF, A., J. T. SCHOFIELD, A. J. KLIORE, F. W. TAYLOR, S. S. LIMAYE, H. E. REVERCOMB, L. A. SROMOVSKY, V. V. KERZHANOVICH, V. I. MOROZ, M. YA. MAROV: Models of the structure of the atmosphere of Venus from the surface to 100 kilometers altitude. Adv. Space. Res. **5** (1985) 11, 3—58.

[2] TAYLOR, F. W., J. T. SCHOFIELD, P. J., VALDES: Temperature structure and dynamics of the middle atmosphere of Venus. Adv. Space Res. **5** (1985) 9, 5—23.

[3] MOROZ, V. I., D. SPÄNKUCH et al.: Venus spacecraft infrared radiance spectra and some aspects of their interpretation. Appl. Opt. **25** (1986) 10, 1710—1719.

[4] CHAHINE, M. T.: Inverse problems in radiative transfer: Determination of atmospheric parameters. J. Atmos. Sci. **27** (1970), 960—967.

[5] SMITH, W. L.: Iterative solution of the radiative transfer equation for the temperature and absorbing gas profile of an atmosphere. Appl. Opt. **9** (1970), 1993—1999.

[6] SPÄNKUCH, D., ZASOVA, L. V., SCHÄFER, K. et al.: Infrared experiment aboard the automatic interplanetary stations Venera 15 and Venera 16. 2. Preliminary results of temperature profile retrieval. Veröffentlichungen des Forschungsbereichs Geo- und Kosmoswissenschaften, Heft 18, 1990.

[7] DÖHLER, W., K. SCHÄFER, D. SPÄNKUCH: Line-by-line gaseous transmittances for Venus standard atmospheres in the 250 cm^{-1} to 1400 cm^{-1} region (1983). Paper presented at the workshop on the Venus Atmosphere and Ionosphere (including VIRA). Hamburg, FRG, Aug. 24–26, 1983.

[8] OERTEL, D., MOROZ, V. I., NOPIRAKOWSKI, J. et al.: Infrared experiment aboard the automatic interplanetary stations Venera 15 and Venera 16. 1. Methodics and first results. Veröffentlichungen des Forschungsbereichs Geo- und Kosmoswissenschaften, Heft 18, 1990.

[9] OERTEL, D. et al.: Infrared Spectrometry of Venus from Venera 15 aund Venera 16. Adv. Space Res. **5** (1985) 9, 25–36.

[10] ZASOVA, L. V., SPÄNKUCH, D., MOROZ, V. I. et al.: Infrared experiment aboard the automatic interplanetary stations Venera 15 and Venera 16. 3. Some conclusions on the structure of clouds, based on the ana'ysis of spectra. Veröffentlichungen des Forschungsbereichs Geo- und Kosmoswissenschaften, Heft 18, 1990.

[11] KNOLLENBERG, R. G., D. M. HUNTEN: Clouds of Venus: A preliminary assessment of microstructure. Science **205** (1979a), 70–73.

[12] KNOLLENBERG, R. G., D. M. HUNTEN: Clouds of Venus: Particel size distribution measurements. Science **203** (1979b), 792–795.

[13] KNOLLENBERG, R. G., D. M. HUNTEN: The microphysics of the clouds of Venus: results of the Pioneer Venus particle size spectrometer experiment. J. Geophys. Res. **85** (1980) A 13, 8039–8058.

[14] POLLACK, J. B., et al.: Distribution and Source of the UV Absorption in Venus' Atmosphere. J. Geophys. Res. **85** (1980) A 13, 8141–8150.

[15] SPÄNKUCH, D., W. DÖHLER, K. SCHÄFER, L. V. ZASOVA and E. A. USTINOV: Spectral Radiances for Venus Standard Atmospheres in the 250 cm^{-1} to 1400 cm^{-1} Region. Report for VIRA Workshop, Hamburg 1983.

[16] TAYLOR, F. W., SCHOFIELD, J. T., S. P. BRADLEY: Pioneer Venus atmospheric observations. — Phil. Trans. R. Soc. London. **A 303** (1981), 215–223.

[17] SCHOFIELD, J. T., F. W. TAYLOR: Measurements of the mean, solar-fixed temperature and cloud structure of the middle atmosphere of Venus. Quart. J. R. Met. Soc. **109** (1983) 57–80.

[18] TAYLOR, F. W.: The Venusian polar dipole. Veröffentlichungen des Forschungsbereichs Geo- und Kosmoswissenschaften, Heft 18, 1990.

[19] SPÄNKUCH, D., DUBOIS, R.: Atmospheric variability over the north pole during the Venera 15 mission. Veröffentlichungen des Forschungsbereichs Geo- und Kosmoswissenschaften, Heft 18, 1990.

[20] JAKOWLEW, O. J. et al.: The Venusian atmosphere in the north polar region from radio-occulation-data by means of the spacecraft Venera-15 and Venera-16 (Russian). Kosm. Issled. **25** (1987) 2, 275–284.

[21] ROTHMANN, L. S. and R. A. MCCLATCHEY: Updating of the AFCRL atmospheric absorption line parameters compilation. Appl. Opt. **15** (1976) 11, 2616.

[22] ZASOVA, L. V., SPÄNKUCH, D., MOROZ, V. I. et al.: Venusian clouds from Venera 15 data. Veröffentlichungen des Forschungsbereichs Geo- und Kosmoswissenschaften, Heft 18, 1990.

8. Observed Cloud Level Circulation on Venus

Temporal Variations and Solar Longitude Dependence

Sanjay Shridhar Limaye

Abstract

Continuing observations of the Venus atmospheric circulation via tracking of cloud features atmospheric entry probes and balloons are revealing a relatively wider range of the basic zonal atmospheric circulation than may be anticipated from the nearly constant nature of the external parameters that control it. Calculations of the mean zonal flow as a balance between the pressure force and the centripetal force due to the zonal motion are also providing a framework within which to interpret the cloud motion observations.

The global hemispheric vortex organization of the clouds and motions has been seen throughout the Mariner 10 (1974) and Pioneer (1979—1986) observations. Yet cloud motions derived from Mariner and Pioneer images show changes of 10—20 m $\times$ s^{-1} in the meridional profile of the average zonal flow on a time scale of about a year. The magnitude of the cloud level mean poleward flow is also slightly different a different observing periods. Such changes could also be due to long term changes in the relative level of the cloud layer along with changes in the atmospheric circulation.

Solar thermal tides and large scale planetary waves are also seen in averaged zero (east-west) and meridional (north-south) cloud motions. The dayside cloud level flow is found to be divergent in low and mid-latitudes with peak divergence downwind of the sub-solar point. The cloud motions are convergent in latitudes where the ultraviolet brightness is markedly higher (and the contrasts) lower), near about 45° latitude. Consistent with the apparent temporal dependence of the mean zonal circulation the tidal signature in the averaged cloud motions is also found to be varying with time, yet the horizontal divergence field appears stable over time. The role of these tides and other planetary scale waves in maintaining the cloud level circulation and any relationship between the large scale circulation and global cloud structure need to be quantitatively determined yet.

1. Introduction

The rapid rotation of the atmosphere on Venus relative to that of the solid planet has now been known for some time. Initially determined from cloud motions measured from earth-based images of Venus taken in ultraviolet light, this circulation has been measured many times by atmospheric entry probes (beginning at about 62 km and down to the surface), cloud motion measurements from not only television and spin scan images, but also from multicolor polarization observations (Limaye, 1984) as well ground-based near infrared observations (Allen and Crawford, 1984; Allen, 1987). Further, the zonal circulation has been inferred from thermal structure (Limaye, 1983; 1985; Newman et al., 1984; Walterscheid et al., 1985, Taylor et al., 1985) and has been the subject of simulation by several numerical models (Covey et al., 1986; Del Genio et al., 1987; Rossow, 1983) as well as diagnostic studies (Leovy, 1987; Hou and Goody, 1984). Yet, the basic processes that support this circulation are not yet well understood. Schubert (1983) gives a good summary of the Venus circulation studies and a summary of circulation observation can be found in Kerzhanovich and Limaye (1986).

Along with the tracking of the atmospheric entry probes, measurements of cloud motions from ultraviolet images of Venus reveal the three dimensional structure of the atmospheric circulation. The results of the cloud motion measurement efforts have revealed the global (two-dimensional) organization of the circulation in the form of hemispheric vortices centered over each pole with the flow being largely zonal with a small poleward component at the cloud level, while the entry probe and the VEGA balloon drift wind profiles provide an inkling of the vertical structure. The Cloud Photopolarimeter (OCPP, Russel et al., 1977) on the Pioneer Venus Orbiter spacecraft has provided extended imaging coverage since December 1978 in several 2 to 12 week-long intervals. Cloud motions measured from these images reveal that although the circulation retains its basic structure, its magnitude is apparently time dependent. In this paper the results of these measurements are summarized and their salient features noted.

2. Cloud motion measurements

Figure 1 shows a typical sequence of four latitude-longitude maps in the ultraviolet generated from images acquired from the Pioneer spacecraft. The images were acquired with the spin-scan imaging technique by using the orbital motion of the orbiter to provide cross-scan coverage rather than stepping the telescope on successive scans. Due to the high eccentricity of the orbit the foot-print of the instantaneous field of view of the telescope gradually changes with every scan and therefore the images are shown in the

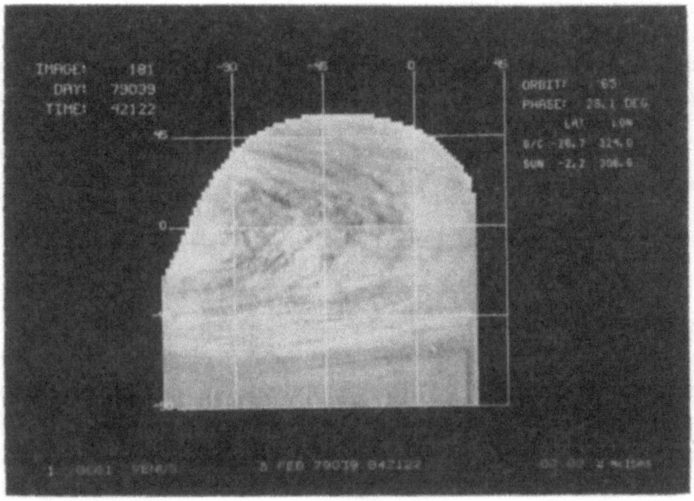

Fig. 1a

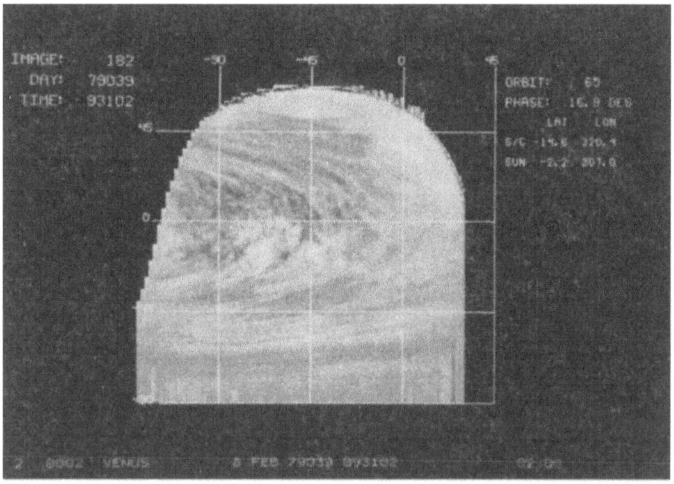

Fig. 1b

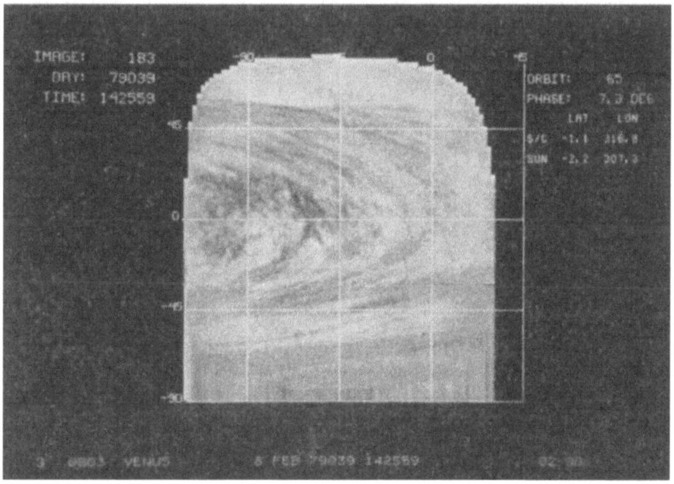

Fig. 1c

Fig. 1 (a—c). Series of 3 views of Venus acquired on one orbit of Pioneer Venus from the Cloud Photopolarimeter shown in a latitude longitude map. The gradual shading due to illumination and viewing geometry has been removed by Minnaert limb darkening law and the data have been digitally filtered to enhance high spatial frequencies

form of latitude longitude maps. Use of such a format also makes tracking features in the images simpler due to constant geometry. The typical time interval between the images is between 4 to 5 hours, so that a maximum of only four images are acquired per orbit as imaging is not performed near periapsis when the spacecraft is too close to the planet.

Sequences of such images are displayed on an interactive display device (McIDAS, Suomi et al., 1983) as time lapse movie loops. Targets are selected by an analyst visually and re-located in succeeding image(s) by stepping through the display frames and recording their locations with a computer generated cursor. From these locations the exact time difference between the two cloud sightings is computed from a knowledge of the spacecraft orbit, the telescope pointing information and the planetary co-ordinates of the cloud in the consecutive images. This time difference and the vector displacement yields the east-west and north-south components of the cloud feature motion which are then assumed to provide an indication of the atmospheric flow.

Limitations

Ideally it is desirable to have measurements of motion over the entire planet and spanning a long enough time interval to study the circulation in a global and time average sense so that it can be properly diagnosed from available meteorological analysis. Clearly such measurements are not possible from a fly-by mission such as Mariner 10 (Murray et al., 1974) due to the relatively short amount of time spent in proximity of the planet. However even with an orbiting spacecraft the orbit orientation can impose limits. Although the Pioneer Venus spacecraft has been orbiting around Venus since 1978, image acquisition has been possible only when the apoapsis portion of the orbit falls in the sunlit hemisphere of the planet as the orbit is fixed in inertial space and is carried around the sun with Venus. This restricts the duration of the "season" during which images can be acquired to no more than about three months every Venus year. In addition, other constraints such as competing deep space mission have limited the period over which images can be acquired.

Thus, our understanding of the atmospheric circulation and in particular variation of its strength with time over "long" term, where long is about a year, is limited due to averages obtained over limited periods ranging from a few days to about two months. Analysis of shorter term variations is possible, but there again some aliasing of the measurements is present due to the variation of the phase angle with time, such that in the beginning of the season the images depict the morning hemisphere whereas the afternoon hemisphere is imaged during the latter half of a typical imaging season. And indeed, such aliasing is what makes a detailed analysis of the solar longitude dependence of the motions possible, so that the short term variations on the order of a few weeks are not readily separable from time-of-day variations (tides), and even shorter term variations (less than a week) are not detectable from the cloud motion measurements due to inadequate sampling.

Finally, the question "to what height or pressure level do these cloud motions belong to?" still cannot be satisfactorily answered due to the diffuse nature of the Venus clouds containing a mix of a deep layer of 1 μ radius sulfuric acid particles and 0.2 to 0.5 μ radius particles, presumably also sulfuric acid particles. Polarization observations from Pioneer Venus do indicate variable amounts of a haze in polar regions but the haze optical thickness in equatorial latitudes is generally very small. The dark features in the ultraviolet images traditionally used for cloud tracking are due to uv absorbing material buried deep within the cloud/haze deck of as yet unknown composition.

3. Longitudinally averaged circulation

Variability in the strength of the longitudinally averaged zonal and meridional circulation at the cloud level at different times is indicated by the measurements made from images acquired from not only the Pioneer Venus but also Mariner 10 images. A comparison of the circulation observed during 1974, 1979 and 1982 is presented by Limaye et al. (1987). Recently additional results of cloud measurements performed from Pioneer images acquired during the 1980 opportunity have been compiled and they continue to show the variability of the atmospheric circulation. The time and longitudinally averaged zonal and meridional flow at the cloud level can be summarized as follows.

3.1. Average Dayside Zonal Flow

Figure 2 shows the average zonal flow as measured from available images during 1974 (circles), 1979 (squares), 1980 (asterisks) and 1982 (triangles). The times and number of vectors obtained during these periods is given in Table I. The ⟨u⟩ and ⟨v⟩ magnitudes in 5° lititude bins are given in Table II. While the cloud level flow is seen to retain its rapid retrograde nature, its equatorial magnitude as well as the strength and width of the mid-latitude jets are seen to vary with time. Similar results have been obtained

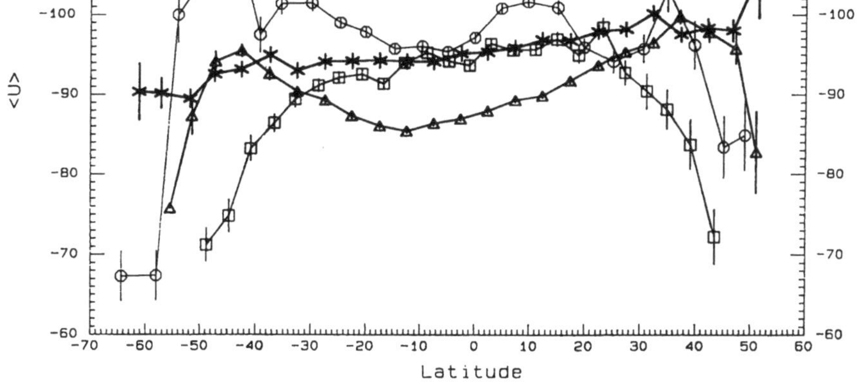

Fig. 2. Comparison of average zonal component during 1974 (circles), 1979 (squares), 1980 (asterrisks) and 1982 (triangles)

Table I.

Summary of Cloud Motion Measurements on Venus

Period		No. of Spacecraft	Images	Vectors	Reference
Feb 6—Feb 9,	1974	Mariner 10	40	2000	Limaye & Suomi (1981), Sidi (1976), Travis (1979)
Feb 4—Apr 12	1979	Pioneer	120	8000	Rossow et al. (1980), Limaye et al. (1982)
Oct 21—Nov 1,	1979	Pioneer	30	1000	Limaye et al. (1982)
Feb 8—Apr 20	1980	Pioneer	200	12000	Limaye (1988), Rossow (1985)*
Feb 17—Apr 26,	1982	Pioneer	250	13000	Limaye et al. (1987), Limaye (1987), Rossow (1985)*
Feb 15—Mar 30,	1983	Pioneer			Currently being analyzed
	1985	Pioneer			Currently being analyzed
	1986	Pioneer			Currently being analyzed

* Zonal component only

Table II.

Comparison of average zonal and meridional day-side flow ($m \times s^{-1}$) over time

Latitude	Average Zonal Component				Average Meridional Component			
	1974	1979	1980	1982	1974	1979	1980	1982
55°—50° N	−75	?	−98	−80	2	11	15	6
50—45	−84	?	−98	−92	0	11	15	16
45—40	−81	−72	−99	−98	2	11	14	14
40—35	−96	−84	−97	−101	3	9	12	14
35—30	−101	−88	−100	−96	5	9	11	14
30—25	−94	−93	−98	−96	3	8	10	13
25—20	−99	−98	−99	−94	3	10	9	11
20—15	−93	−95	−98	−92	1	7	8	8
15—10	−97	−96	−98	−90	1	4	6	6
10—5	−99	−96	−97	−90	0	3	5	5
5—EQ	−99	−96	−97	−89	1	2	3	3
EQ—5° S	−98	−94	−96	−87	−1	2	2	1
5—10	−93	−95	−95	−87	1	−1	1	−1
10—15	−95	−94	−96	−86	1	−1	−1	−3
15—20	−96	−91	−96	−87	0	−3	−2	−5
20—25	−94	−92	−95	−88	−2	−4	−3	−7
25—30	−98	−91	−95	−90	−2	−5	−4	−8
30—35	−94	−89	−94	−91	−3	−6	−5	−8
35—40	−95	−85	−97	−94	−3	−6	−6	−11
40—45	−108	−83	−95	−95	−4	−6	−7	−11
45—50	−108	−72	−94	−93	−4	−8	−9	−12
50—55	−98	?	−90	−89	−1	−6	−8	−10
55—60	−75	?	−94	−76	−1	−2	−4	−2

from digital tracking of Pioneer images (Rossow, 1985). At the same variation is also seen in the large scale wave activity at the cloud level as evidenced by spectra of image brightness (Del Genio and Rossow, 1982; Del Genio et al., 1986), so that a relationship between the atmospheric dynamics and large scale cloud structure is clearly indicated. This relationship could either be presence or absence of equatorial Kelvin waves or a slow modulation of the apparent cloud level with time with the basic circulation remaining essentially unchanged, or, a combination of both. This is discussed further in Section 5.

The time required by a parcel of Venus atmosphere to make a complete circuit around theplanet at a given latitude and moving with the ambient flow is much shorter than on most planets as can be seen in Figure 3. The period (days) corresponding to the average day-side zonal flow at each latitude is shown for 1974 (circles), 1979 (squares), 1980 (asterisks) and 1982 (triangles). At higher latitudes infrared observations have shown that the period is as still about 2.3 days (Schofield and Diner, 1983; Matsygorin, 1989).

3.2. Average Dayside Meridional Flow

The average meridional flow corresponding to the zonal flow profiles shown in Figur 2 is shown in Figure 4. The average cloud level meridional flow is directed towards the pole in respective hemispheres at all time periods except very near the equator, but the rate of increase with latitude is different at different times so that a rough relationship is apparent in that lower mid-latitude average meridional flow may be associated with weaker mid-latitude zonal jets and stronger meridional flow with stronger mid-latitude

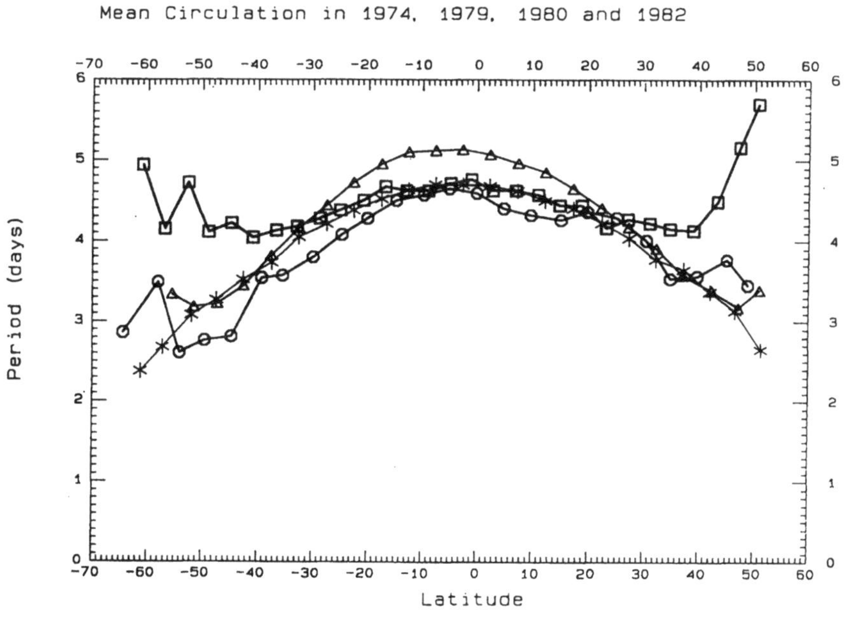

Fig. 3. Cloud level atmospheric rotation (days) corresponding to the zonal flow shown in Figure 2: using same symbols as in Figure 2

zonal flow. In other words, the mean meridional circulation may be important in determining the shape of the zonal component profile with latitude by virtue of weaker or stronger angular momentum transport.

The fact that at the equator the average meridional flow is non-zero is not surprising despite the relatively low obliquity of Venus orbit about the sun or the near coincidence of its equatorial plane with its orbital plane as it is known that the processes that control the atmospheric circulation on Venus are non-linear due to huge thermal and dynamical inertia and that these can lead to asymmetric circulations despite symmetric forcing (Rossow, 1980). Further, it can be seen that the profiles of ⟨v⟩ are not exactly symmetric about the equator or about the latitude of pure zonal flow, and this is similar to the asymmetric profiles of ⟨u⟩ about the equator.

3.3. Planetary Scale Waves

Limaye et al. (1988) have shown that the root mean square deviations of the zonal and meridional components during the different periods measured are different. While some of this is due to measurement and navigation errors, a significant fraction is due to real variability of atmospheric motions in time and longitude in that latitude bin. This variability is due to presence of both large and small scale waves in the atmosphere. While the small scale waves (<2000 km wavelength) cannot be resolved by the cloud motions, the larger (planetary) scale waves can be investigated from the measurements, at least in a time averaged sense, provided the propagation of the waves relative to the flow is slow.

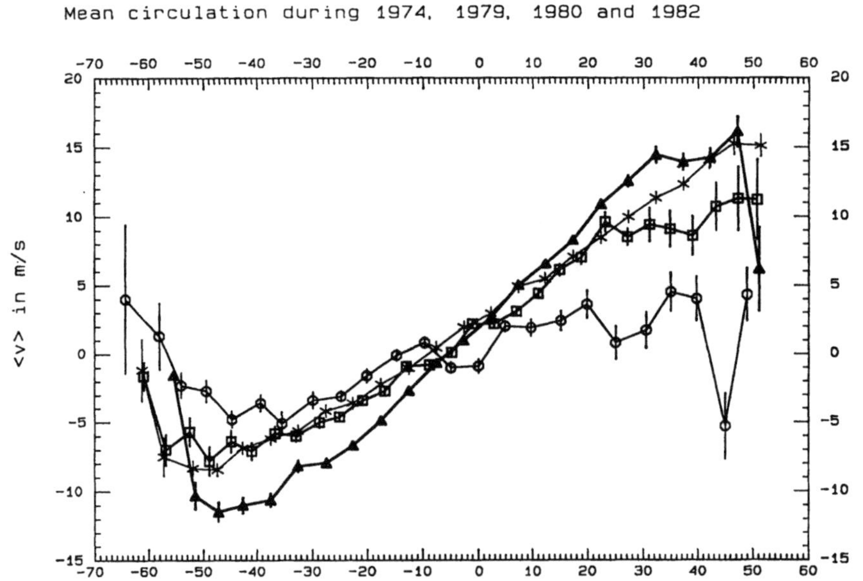

Fig. 4. Comparison of average meridional component of cloud motion during 1974 (circles), 1979 (squares), 1980 (asterisks) and 1982 (triangles)

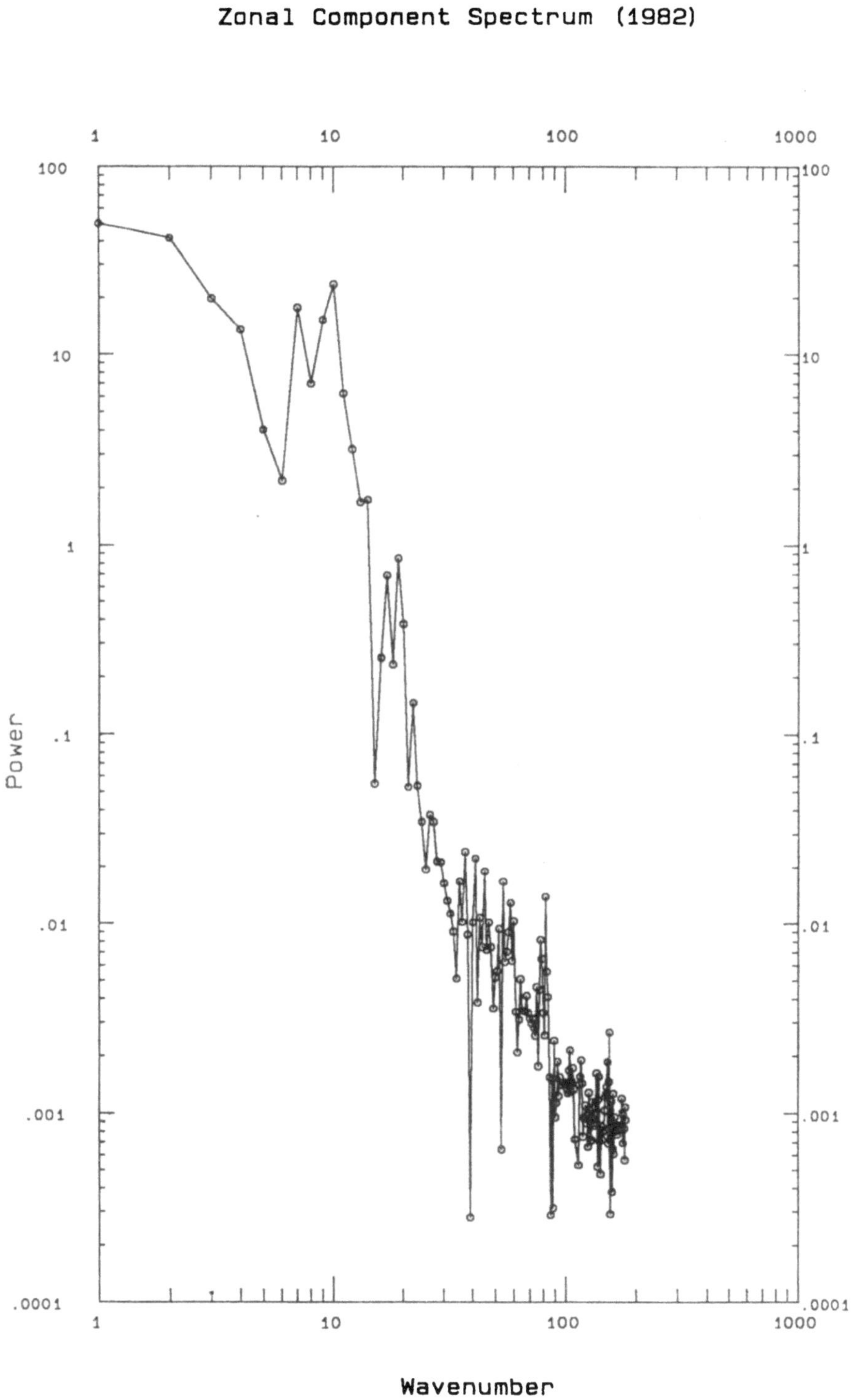
Zonal Component Spectrum (1982)
1
10
100
1000
100
10
1
.1
.01
.001
.0001
Power
Wavenumber

Fig. 5a

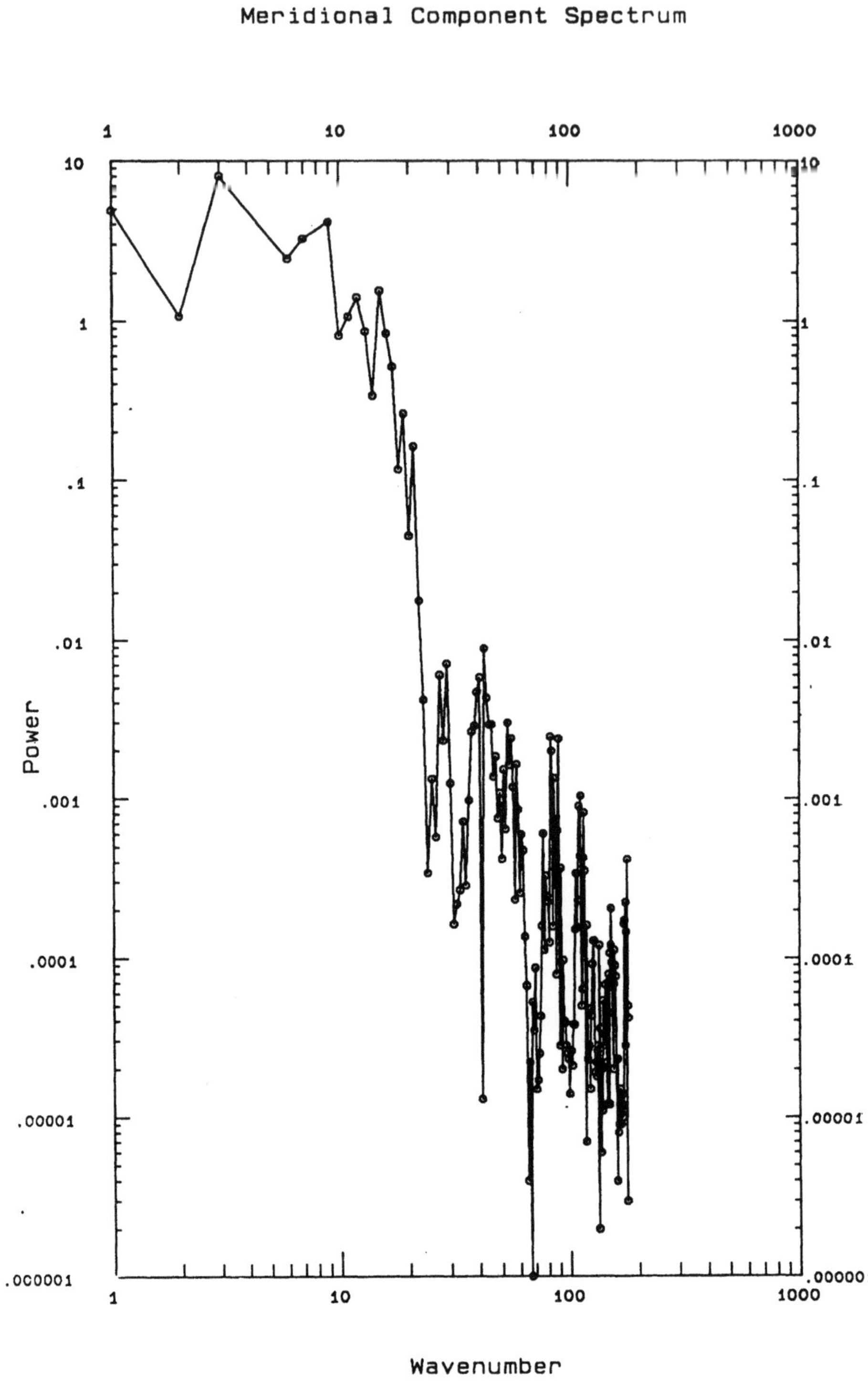

Meridional Component Spectrum
1
10
100
1000
10
1
.1
.01
.001
.0001
.00001
.000001
Power
Wavenumber

Fig. 5b

The limitations posed by the fact that we can measure the cloud motions over only about a third of a latitude circle can be minimized if we assume that over the period taken by the bulk of the Venus atmosphere (at the cloud level) the change in the circulation is small. In such a case the individual cloud motions can be assigned longitudes relative to a fictitious reference point that rotates at the average rotation rate of the ambient flow. If it is further assumed that the flow is relatively stable over many atmospheric rotations, then we can investigate if any stationary waves (i.e. in the rotating frame of reference) exist on Venus. Based on the available observations it is apparent that the temporal changes over a two to three month period in the longitudinally averaged zonal and meridional component profiles with latitude are relatively small, so this seems a worthwhile exercise.

Figure 5 shows the spectrum of waves in the time averaged zonal (a) and meridional component (b) of cloud motions within $\pm 10^\circ$ latitude of the equator. The individual cloud motions were binned in longitude in a reference frame rotating at the observed average zonal flow (1982) at the equator with a bin size of 1° longitude. The longitudinal dependence of the zonal and meridional components was then smoothed by first using 15-point moving average and then an 11-point running average, so that wave numbers higher than about 15 are smeared. No tapering of the data was performed in computing the spectra shown as the effect of the tapering was to primarily to flatten the spectrum at low wavenumbers.

It was found that the results of both the two analysts who measured the cloud motions for the 1982 period showed very similar spectra for both components of motion. Low wave numbers appear to be dominant in both the components in equatorial latitudes with noticeable peaks at wave-numbers 1—2, 8—10 in both the zonal and meridional component. The spectrum for the zonal component is similar to the cloud brightness spectrum obtained also from the OCPP images by Del Genio and Rossow (1982). The slope appears to change near wavenumber 30 for the zonal component spectrum, and near wavenumber 20 for the meridional component spectrum. The meridional component also shows a peak near wavenumber 27 and 50. Such high wavenumbers may correspond to presence of gravity waves (as found in the VEGA balloon results) that either propagate up after being induced by topography in the low level flow (Young et al., 1987), or they may just as likely be caused in the very stable region of the atmosphere, where the ultraviolet features are supposed to occur. Gierasch (1987) has shown that the wavefronts of such gravity waves could be responsible for some of the small structure seen in the ultraviolet images.

Del Genio et al. (1986) and Rossow et al. (1987) have interpreted the cloud brightness

---

Fig. 5. Spectrum of the zonal component of cloud motions during 1982 when averaged in longitude bins in a frame of reference that rotates with the average equatorial zonal flow (a), and for the meridional component (b). The vectors are binned in 1° wide relative longitude boxes within 10° of the equator, and smoothed twice by a 15 point and 11 point running average respectively. Both the zonal and meridional components show a peak at wave numbers 1 and 10, with significant power also present at wave numbers 2—4, 7—9, and 15—19 in both the components. The two lowest probably correspond to solar tides or Kelvin waves, while 7—10 correspond to planetary waves, possibly Rossby-gravity type. The slope appears to change around wavenumber 30

spectra and cloud motion spectra to indicate that in 1980 the planetary waves during 1980 were Kelvin waves. This was based on the fact that no corresponding fluctuations in the meridional component were seen, unlike during 1982. According to this scenario the Kevin wave, present during 1980 disappeared during 1982, revealing only the solar tidlal modes. As both Kelvin waves and solar tides are ultimately gravity waves, the distinction of this crucial and critically depends on the structure in the meridional component. The different equatorial wave-modes interpreted from the observations may be involved in deceleration and acceleration of the equatorial zonal flow that is apparent from the cloud motions. Alternately the large scale level of the tracked features could just as plausibly be affected by dynamical changes leading to a different level of atmospheric circulation being sensed via feature tracking in ultraviolet images. For example, even when the cloud motions do not show a well defined mid-latitude jet, cyclostrophically balanced zonal flow computed from observed thermal structure does show such jets in each hemisphere. How can these two different observations be reconciled? Unfortunately we lack sufficient observations, especially on the night-side to enhance confidence in the interpretation of the waves in the observed circulation and to understand the temporal variation of the zonal flow on a yearly scale.

4. Solar longitude dependence of cloud motions

The solar-locked structure of the cloud motions has been discussed in detail for the 1982 Pioneer Venus observations by Limaye (1988). In brief, the dayside hemisphere is found to be divergent at low latitudes with the peak divergence exceeding 5×10^{-6} sec^{-1} just downwind of the sub-solar point. The flow becomes non-divergent along a boundary between roughly $\pm 30°$ (morning terminator) and $\pm 40°$ (evening terminator) latitudes, poleward of which the atmosphere is weakly convergent to at least the maximum latitudes measured (about 60°). The zonal flow is weakest slightly upwind of the sub-solar point and most polewards at the sub-solar meridian.

This general structure in the cloud motions in solar-locked co-ordinates is the signature of the solar-thermal tides which have been previously detected in the temperature observations (Schofield and Taylor, 1983), and is the engine that derives the atmospheric circulation, i.e. this is where the absorbed solar energy is being converted into zonal mean kinetic energy.

4.1. *Diurnal and Semidiurnal tides*

It is possible to decompose the observed solar-locked average zonal and meridional components into diurnal and semi-diurnal components as follows. The zonal and meridional components at a given latitude and solar-longitude can be represented as:

$$u(\varphi, \theta) = u_0 + u_1 \times \sin(\alpha_u + \theta) + u_2 \times \sin(\beta_u + 2 \times \theta) \tag{1}$$

$$v(\varphi, \theta) = v_0 + v_1 \times \sin(\alpha_v + \theta) + v_2 \times \sin(\beta_v + 2 \times \theta) \tag{2}$$

where φ and θ represent latitude and longitude respectively, subscripts 0, 1 and 2 denote zonal average, diurnal and semidiurnal amplitudes and α and β represent phases of diurnal and semidiurnal components for the zonal and meridional components when denoted by the subscripts u and v respectively. In both the zonal and meridional component of motion thus there are a five unknowns each which can be determined from the available observations by linear regression.

Note that here the zonal averages u_0 and v_0 are true zonal averages in that they incorporate knowledge about the night-side hemisphere and hence are treated as unknows. In view of the fact that we have observations only over about 120° of the day-side hemisphere, the confidence in the regressed quantities is sensitive to errors in the observed fields of $\langle u \rangle$ and $\langle v \rangle$. Figure 6 shows an example of such decomposition at about 12° S latitude, where it is seen that the diurnal amplitude is almost twice as large as the semidiurnal amplitude for both the zonal and meridional components.

Another interesting feature is that at this latitude the fastest zonal flow appears on the night side, the difference between the slowest an fastest zonal components being almost 20 m $\times$ s^{-1}. The only measurements of atmospheric flow on the night side were made below the clouds by Vega balloons and the Pioneer Nightside Probe. Calculations of the tidal structure have been performed by Pechman and Ingersoll (1983) but mostly for the thermal structure as measured by the Pioneer Venus Orbiter Infrared Radio-

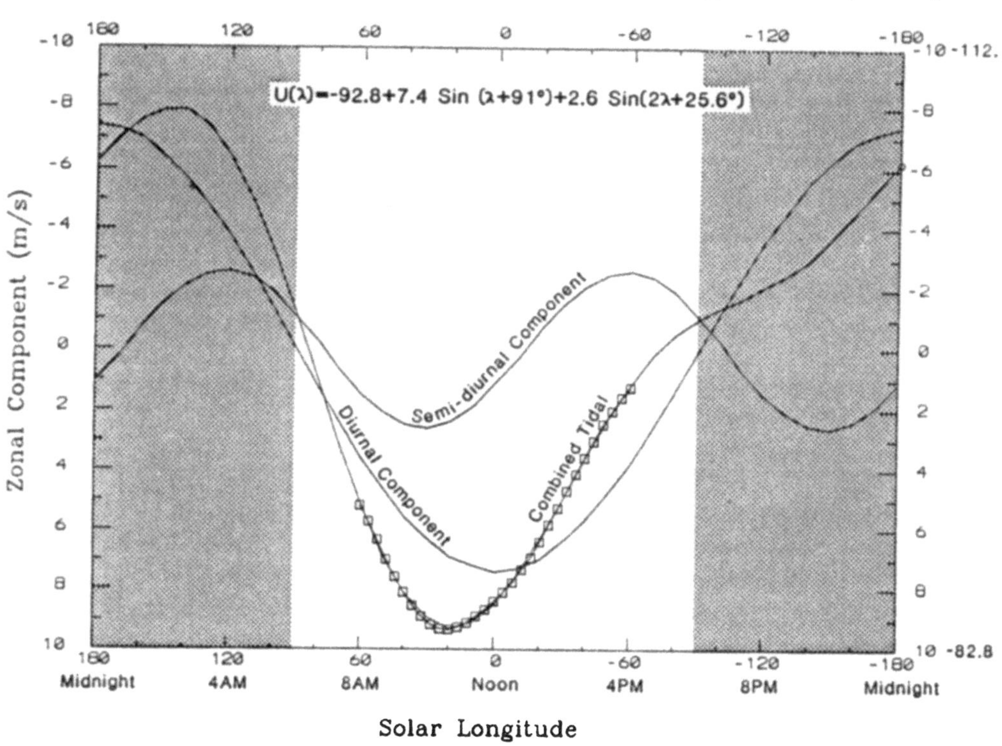

Fig. 6. Diurnal and semidiurnal components at 12° S latitude during 1982 for the zonal component computed by least squares analysis of the observed motions (squares). The fitted profile is extended into the nightside (right hand scale) using the two amplitudes (left hand scale), the two phases and the zonal average. The fit is very sensitive to errors in the average flow in each longitude box

meter. Similar tidal structure has been detected in the Fourier Spectrometer observations from Venera 15 and 16 (Spänkuch et al., 1989; Schäfer et al., 1989). Nevertheless additional calculations regarding the latitudinal structure as well as vertical structure of the tides in the zonal and meridional components are needed. Also useful would be reliable observations of the atmospheric flow on the night side at all levels possible.

4.2. Temporal Dependence or Height Dependence?

The solar longitude dependence presented here has been determined by averaging the vectors measured over the entire 1982 season from images with phase angles less than about 60°. Similar structure is seen when the average is taken over much shorter periods (about 10 days) although the results are not as precise due to lower number of vectors in each time average. The temporal averaging is out of a necessity to increase the sample size whereas the phase angle limit is due to the lower contrasts at higher phase angles. Measurements at higher phase angles, i.e. at higher solar longitudes are possible if the level of detail visible in the images is enhanced by high pass filtering. While this extends the length of the season so that temporal trends could be discerned from short term averages, it also introduces a new effect, namely that due to the much longer slant paths traversed by the incident solar radiation, the level of the visible features that are tracked is likely to be different. Thus it is difficult to separate the time dependence from apparent level dependence at this stage.

Figure 7a shows an example of this dilemma. The zonal component averaged over four days in a 15° latitude and solar-longitude box centered at the equator and 2 PM (circles), Noon (triangles) and 10 am (squares) is shown as a time series. The phase angle at the center of the disk varies from about 75° to 88° from the beginning to the end through zero phase (morning to afternoon) (Figure 7b). Near about day 108 the average zonal component within 15° of the sub-solar point increases rapidly from about -88 m $\times s^{-1}$ to -105 m $\times s^{-1}$ by day 130. Similar behavior is seen within a local time window of 1:00 pm to 3:00 pm LST, but on the morning side (between 9:00 am and 11:00 am LST) the rapid increase begins much earlier, near day 90. This asymmetry between morning and evening displayed by the time series may be reflective of true cloud level variation by measurements made at high phase angles.

A similar question about possible altitude variation with latitude of the tracked features has been raised previously (Limaye, 1985). It was found that if a match is required between the calculated cyclostrophically balanced flow and the measured cloud motions, the features lie at somewhat lower pressures in mid-latitudes compared to low latitudes. The main problem in ascertaining this with greater confidence is the poor knowledge of the zonally averaged thermal structure of the Venus atmosphere.

5. Global Organization

The similarity in the structure of a terrestrial hurricane and the global circulation on Venus despite the scale difference have been previously noted (Limaye, 1977; Suomi and Limaye, 1978), namely their vortex nature. A vortex is characterized in its simplest form as having substantial axial symmetry with radially inward flow at one level and outward flow at another level and with a tangential component of the flow that increases from a low magnitude to a peak value some distance away from the center and then decrease to zero at the center. This global vortex organization of the motions and clouds is preserved during all Pioneer observations, even when pronounced mid-latitude jets are not seen at the cloud level. Figure 8 compares the northern hemisphere of Venus during 1974 and 1982. These polar composites have been constructed from images acquired over 4—5 days and by using the measured zonal circulation to mosaic them

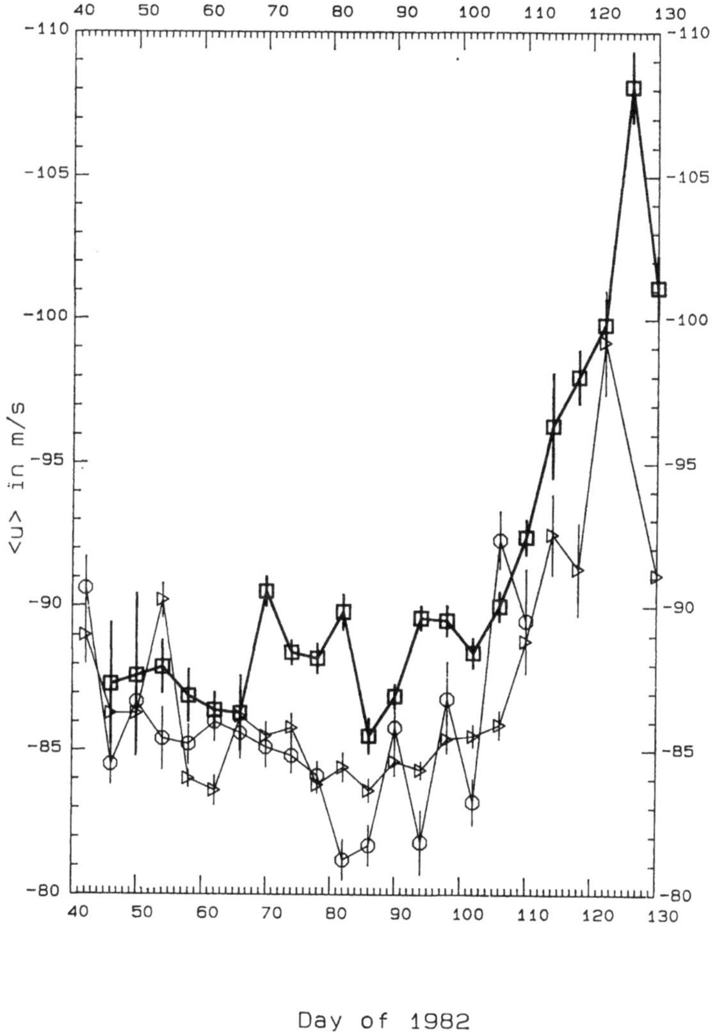

Fig. 7a. Variation of the zonal component with time in a 15° latitude and 30° solar longitude box centered at the equator and noon (triangles), 10:00 am LST (squares), and at 2:00 pm LST (circles). Vectors are averaged in a fourday window to achieve a sufficient number of vectors in the averaging window

appropriately. No major differences are apparent in the large scale structure. The equatorial atmosphere is thought to be rising (on the dayside at least), and slowly spiralling in towards the pole. The flow becomes more southerly (northerly) at mid-latitudes in the northern (southern) hemisphere. In mid-latitudes the convergence of the meridians causes the poleward moving flow to become convergent (rather than by a latitudinal variation of the meridional flow itself) thereby forcing the flow to form a descending branch of the cloud level Hadley cell (Schubert, 1983). The temperature inversions observed near 50—100 mb and between 60—70° latitude from Pioneer and Venera radio occulation and infrared radiometer observations could be manifestations of the sinking motion in the high latitudes in the poleward branch of the Hadley cell.

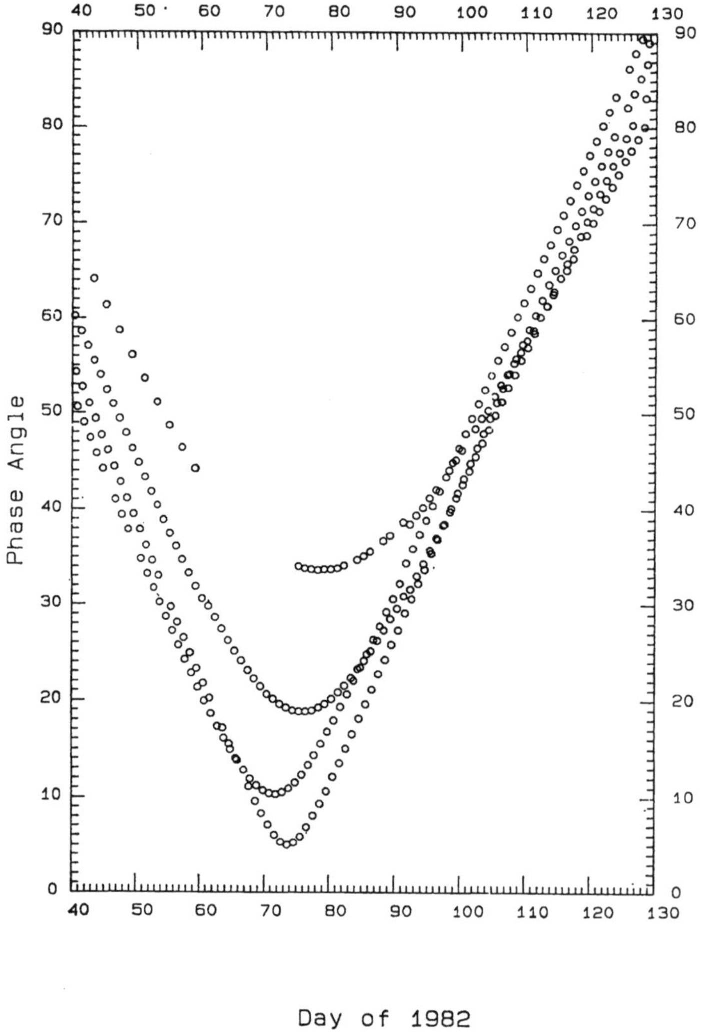

Fig. 7b. Variation of the phase angle at the center of the visible disk of Venus (as seen from the Pioneer Orbiter) of each images used in cloud tracking is also shown in each case

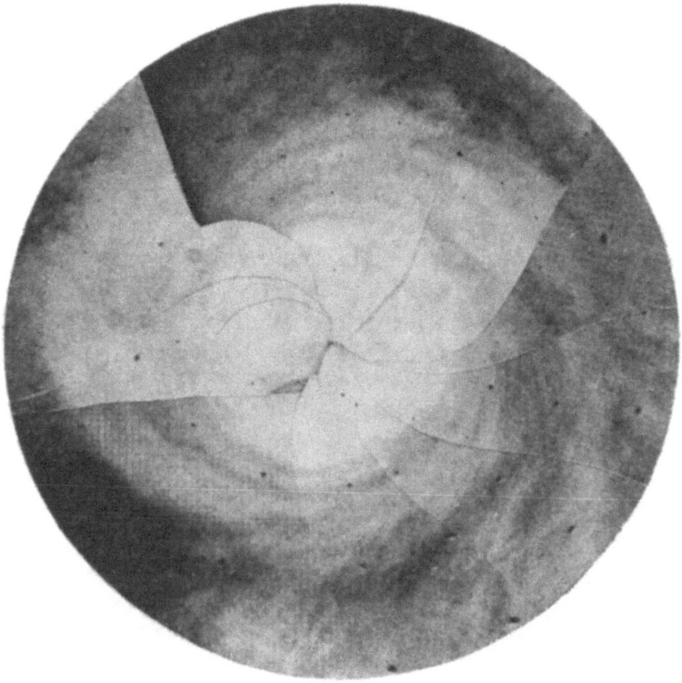

Fig. 8a

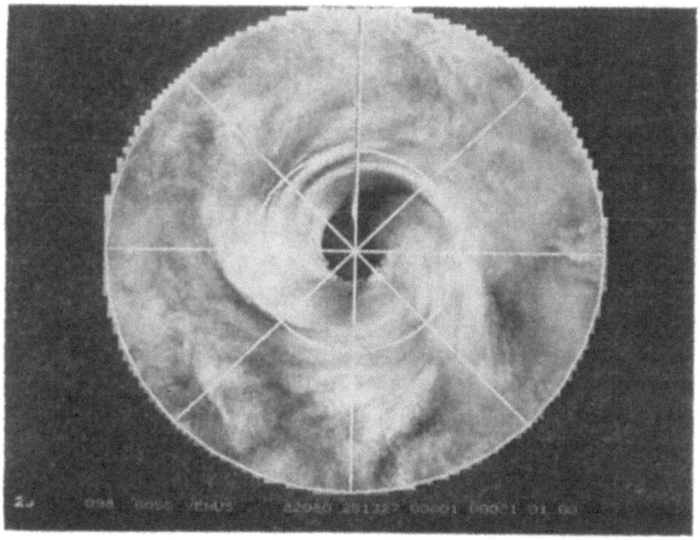

Fig. 8b

Fig. 8. Comparison of the global organization of the Venus clouds during 1974 (a, Mariner 10) and 1982 (b, Pioneer Venus). The mosaics are constructed in polar stereographic projection and are composed of images acquired over 4—5 days after the scattering geometry effects have been removed. The measured zonal flow profile is used to displace the consecutive images in relative longitude

6. Summary

We have accumulated observational evidence of the relatively stable zonal circulation at the cloud level on Venus. On the other hand changes on a yearly time scale appear substantial in both the zonal and meridional components of motion. It is apparent that no equatorward transport of angular momentum has been observed from the cloud motions yet, a key component of explanations for re-supply of angular momentum to the equatorial latitudes tó maintain the rapid retrograde flow. Presence of solar thermal tides in both the components can be identified in the cloud motions averaged over a few weeks. Relative amplitudes of the diurnal and semi-diurnal components appear unequal at most latitudes with the diurnal component having an amplitude almost twice as large as the semi-diurnal component in the zonal speed. It is also apparent that the night-side zonal flow magnitude can be significantly different from the day-side value.

Long waves are also apparent in the zonal and meridional components. These waves are relatively stationary with respect to the average flow over the observation period. The waves at equatorial latitudes have been interpreted as Kelvin waves at some periods and solar tides at others, but both are gravity waves, and the interpretation implicitly assumes structure in the motions to correspond to a level surface. Numerical studies of some cyclostrophically balanced zonal flow profiles indicate that barotropic instability can generate waves at wavenumbers 1—7 with wavenumber 2 having the greatest instability rate (Dethloff and Zelck, 1987). Wave-mean flow interaction could be responsible for changes in the average zonal flow on yearly scale, but other mechanisms are also possible (Leovy, 1987; Limaye et al., 1988).

7. Future Observations

The Pioneer Orbiter is expected to function through 1992 when depletion of the attitude control gas will erode the capability to control the periapsis altitude and earth-pointing of the high gain antenna. At that time the increased drag of the atmosphere on the spacecraft during the periapsis passage will continually lower the periapsis altitude and finally the increase frictional heating is expected to consume the spacecraft. Until that time limited observations are continuing. There appears a possibility that some observations of Venus can be made during 1989 from the Galileo spacecraft on its way via a Venus (along with two earth encounters) gravity assist to Jupiter. Some limited imaging observations also seem possible from the Hubble Space Telescope expected to be in earth orbit during 1988. Unfortunately none of these seem capable of observing the night-side of Venus, especially to determine the nightside atmospheric circulation, which probably adequately contains many clues to the maintenance of the global circulation on Venus.

Acknowledgements

This work was largely funded by NASA Grants NAGW-937 and NAG-2—265 and would not have been possible without the efforts of Mike Kuetemeyer, Chris Grassotti, John Lee and Michelle Bridson. Dave Santek provided valuable programming assistance. I also thank COSPAR for enabling participation in the symposium.

References

ALLEN, D. A., and J. W. CRAWFORD: Cloud structure of the dark side of Venus, Nature **305** (1984) 222–224.

ALLEN, D. A.: The dark side of Venus. Icarus **69** (1987) 221–229.

COVEY, C., E. J. PITCHER, and J. P. BROWN: General circulation model simulations of superrotation in slowly rotating atmospheres. Icarus **66** (1986) 380–396.

DEL GENIO, A. D., and ROSSOW, W. B.: Temporal Variability of Ultraviolet Cloud Features in the Venus Stratosphere. Icarus **51** (1982) 391–415.

DEL GENIO, A. D., and R. SUOZZO: A comparative study of rapidly and slowly rotating dynamical regimes in a terrestrial general circulation model. J. Atmos. Sci. **44** (1987) 973–986.

DETHLOFF, K., and M. ZELCK: Barotropic instability in the middle atmosphere of Venus on the basis of measured zonal winds. Z. Meteorl. **87** (1987) 220–224.

GIERASCH, P. J.: Waves in the atmosphere of Venus. Nature **328** (1987) 510–512.

HOU, A. Y.: Axisymmetric circulations forced by heat and momentum sources: a simple model applicable to Venus. J. Atmos. Sci. **41** (1984) 3437–3455.

HOU, A. Y., and R. M. GOODY: Diagnostic requirements for the superrotation on Venus. J. Atmos. Sci. **42** (1985) 413–432.

KERZHANOVICH, V. V., and S. S. LIMAYE: Circulation of the atmosphere from surface to 100 km. In Venus International Reference Atmosphere, Adv. Space Res. **5** (1985) 11, 59–83.

LEOVY, C. B.: Zonal winds near Venus' cloud tops. Icarus **69** (1987) 193–201.

LIMAYE, S. S.: Venus: Cloud level circulation during 1982 as determined from Pioneer Cloud Photopolarimter images. Part II. Solar longitude dependent circulation. Icarus **73** (1988) 212–226.

LIMAYE, S. S.: Venus atmospheric circulation: observations and implications of the thermal structure. Adv. Space. Res. **5** (1985) 51–62.

LIMAYE, S. S.: Morphology and movements of polarization features as seen in the Pioneer Venus orbiter cloud photopolarimeter (OCPP) data. Icarus **57** (1984) 362–385.

LIMAYE, S. S.: Venus atmospheric circulation from near cloud top level to the surface. Paper presented at the IUGG/COSPAR Venus International Reference Atmosphere Workshop (VIRA), Hamburg, W. Germany, August 25–26, 1983.

LIMAYE, S. S.: Venus stratospheric circulation: a diagnostic study. Ph. D. thesis, Dept. of Meteorology, University of Wisconsin-Madison, 104 pp, 1977.

LIMAYE, S. S., C. GRASSOTTI, and M. J. KUETEMEYER: Venus: Cloud level circulation during 1982 as determined from Pioneer Cloud Photopolarimter images. Part I. Time and zonally averaged circulation. Icarus **73** (1988) 193–211.

LIMAYE, S. S., C. J. GRUND, and S. P. BURRE: Zonal mean circulation at the cloud level on Venus: Spring and Fall 1979 OCPP observations. Icarus **51** (1982) 416–439.

LIMAYE, S. S., and V. E. SUOMI: Cloud motions on Venus: Global structure and organization. J. Atmos. Sci. **38** (1977) 1220–1235.

MATSYGORIN, I. A.: Sun coupled waves in the field of the thermal radiation of Venus from Venera 15 infrared spectrometry. Veröffentlichungen des Forschungsbereichs Geo- und Kosmoswissenschaften, Heft 18, 1990.

MURRAY, B. C., M. BELTON, G. DANIELSON, M. DAVIES, D. GAULT, B. HAPKE, B. O'LEARY, R. STROM, V. SUOMI, and N. TRASK: Venus: Atmospheric motion and structure from Mariner 10 pictures. Science **183** (1974) 1307–1315.

NEWMAN, M. E., G. SCHUBERT, A. J. KLIORE, and I. PATEL: Zonal winds in the middle atmosphere of Venus from Pioneer Venus radio occulation data. J. Atmos. Sci., **41** (1984) 1901–1913.

PECHMAN, J. B.: Thermal tides in the atmosphere of Venus. Ph. D. thesis, California Institute of Technology, Pasadena, California, 286 pp, 1983.

PECHMAN, J. B., and A. P. INGERSOLL: Thermal tides in the atmosphere of Venus: comparison of model results with theory. J. Atmos. Sci. **41** (1984) 3290–3313.

Rossow, W. B.: Atmospheric circulation of Venus. In Advances in Geophysics, S. Manabe ed., Volume 28 A, Academic Press, New York, 347–379, 1985.

Rossow, W. B.: A general circulation model of a Venus-like atmosphere. J. Atmos. Sci. **40** (1983) 273–302.

Rossow, W. B., and G. P. Williams: Large scale motion in the Venus stratosphere. J. Atmos. Sci. **36** (1979) 377–389.

Rossow, W. B., A. D. Del Genio, S. S. Limaye, L. D. Travis, and P. H. Stone: Cloud morphology and motions from Pioneer Venus images. J. Geophys. Res. **85** (1980) 8107–8128.

Rossow, W. B., A. D. Del Genio, and A. Abrams: Planetary scale waves on Venus. Abstract, Bull. Amer. Astron. Soc. **19** (1987) 846.

Russel, E., L. Watts, S. Pellicori, and D. Coffeen: Orbiter cloud photopolarimeter for the Pioneer Venus mission. Proc. Soc. Photo-Opt. Instrum. Eng. **112** (1987) 28–44.

Schäfer, K., V. M. Linkin, K. Dethloff et al.: Temperature and thermal wind field of the middle atmosphere of Venus from Venera 15 data. Veröffentlichungen des Forschungsbereichs Geo- und Kosmoswissenschaften, Heft 18, 1990.

Schofield, J. T., and F. W. Taylor: Measurements of the mean, solar-fixed temperature and cloud structure of the middle atmosphere of Venus. Quart. J. R. Met. Soc. **109** (1983) 57–80.

Schubert, G., C. Covey, A. Del Genio, L. S. Elson, G. Keating, A. Seiff, R. E. Young, J. Apt, C. C. Counselman III, A. J. Kliore, S. S. Limaye, H. E. Revercomb, L. A. Sromovsky, V. E. Suomi, F. Taylor, R. Woo, and U. von Zahn: Structure and circulation of the Venus atmosphere. J. Geophys. Res. **85** (1980) 8007–8025.

Schubert, G.: General circulation and the dynamical state of the Venus atmosphere. In Venus, D. M. Hunten, L. Colin, T. M. Donahue, and V. O. Moroz, eds., University of Arizona Press, Tucson, AZ, USA, 1143 pp, 1983.

Sidi, C.: Stereoscopic observations of winds on Venus. Icarus **28** (1976) 359–366.

Spänkuch, D., L. V. Zasova, K. Schäfer et al.: Infrared experiment aboard the automatic interplanetary stations Venera 15 and Venera 16. 2. Preliminary results of temperature profile retrieval. Veröffentlichungen des Forschungsbereichs Geo- und Kosmoswissenschaften, Heft 18, 1990.

Suomi, V. E., and S. S. Limaye: Further evidence of vortex circulation. Science **201** (1978) 1009 to 1011.

Suomi, V. E., R. Fox, S. S. Limaye, and W. L. Smith: McIDAS III: A modern interactive image analysis system. J. Clim. and Appl. Meteor. **22** (1983) 766–778.

Taylor, F. W., J. T. Schofield, and P. J. Valdes: Temperature structure and dynamics of middle atmosphere of Venus. Adv. Space Res. **5** (1985) 5–23.

Travis, L. D.: Nature of the atmospheric dynamics on Venus from power spectrum analysis of Mariner 10 images. J. Atmos. Sci. **35** (1978) 1584–1595.

9. Temperature and thermal wind field of the middle atmosphere of Venus from Venera 15 data

K. Schäfer, V. M. Linkin, K. Dethloff, D. V. Pacaev, R. Dubois, V. V. Kerzhanovich, D. Spänkuch, L. V. Zasova, D. Oertel

Abstract

The temperature field retrieved from Venera 15 infrared spectra of Venus in the time interval from Oct. 12 to Dec. 14, 1983 for the 60 to 95 km altitude region covers both hemispheres and the inferred thermal wind field also. Detailed temperature structures as hot spots and cold collar are strong variable in latitude and time. Because temperature inversions exist clearly in the night-morning sector (11 K maximum) but only weakly in the evening-night sector (5 K maximum) the cold collar is a solar fixed feature around the morning terminator of the northern hemisphere. Global day-night temperature differences are in the order of some K and 5 K maximum. However the temperature differences between morning and evening are twice of that order. This is a hint to wavenumber 2 longitutinal structure of global temperature field. Distinct wind jets with peak velocities of about 90 (evening) to 125 m/s (morning) are located near 70 km altitude and 50° latitude in the thermal wind field. A linear barotropic model has been used to investigate these zonal wind profiles on instability. Assuming small friction, longitudinal waves with wavenumbers between 1 and 6 occur due to barotropic instability of the zonal wind. Wavenumber 2 has the greatest instability rates.

1. Introduction

Two years ago the first version of Venus Reference Atmosphere (VIRA) had been published. A definitive data set for the mean thermal structure of the Venusian middle atmosphere had been given for the first time [1]. Results on a modelling study to interpret the measured thermalfield in terms of the global dynamics are there included. The data basis are observations of the Pioneer Venus Orbiter Infrared Radiometer instrument for 72 days in late 1978 and early 1979. Temperature profiles from 60 to 105 km altitude have been retrieved. They cover the whole northern hemisphere and reach also 10° southern latitude. It has been found that the observed solar tides are dominated by the semi-diurnal component in agreement with theory.

Moreover temperatures inferred from radio occulation data aquired by the Pioneer Venus Orbiter between December 1978 and November 1983 in the pressure range from 1 350 to 2.1 hPa and latitude range from 10 to 85° of both hemispheres are included in [1]. Also this temperature field has been used to derive cyclostrophic zonal winds [2, 3].

2. Thermal structure

The global thermal field of both hemispheres is shown in Fig. 1. It was retrieved from the spectra of Venera 15 for the 60 to 105 km altitude region and 0.01 to 200 hPa pressure region, respectively. In this and all other plots the relation between pressure

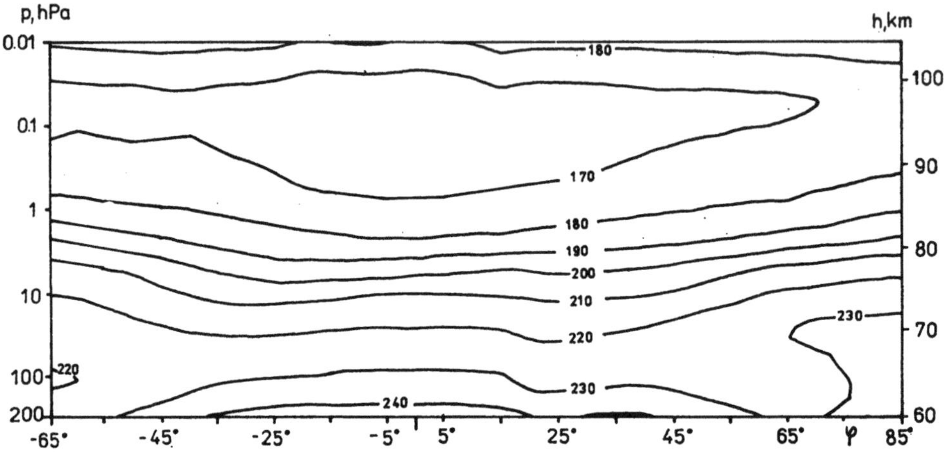

Fig. 1. Isotherms from all retrieved temperature profiles for northern and southern hemisphere neglecting time and longitude dependence in a pressure/altitude—latitude plot

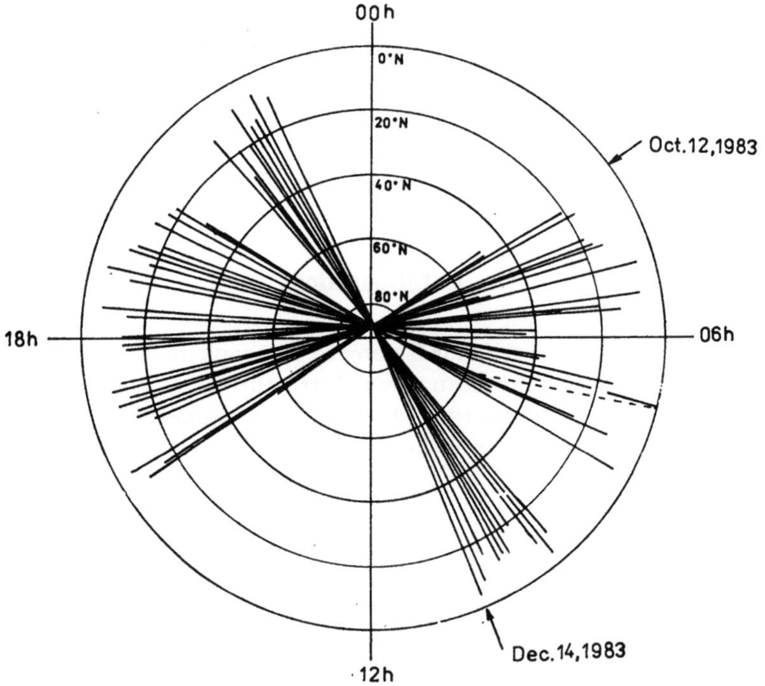

Fig. 2. Positions of the measuring routes during the orbits of Venera 15 in coordinates of latitude and local solar time. The centre of the coordinates system is the northpole

and altitude, which is normally latitude dependent, was taken from a mean Venusian atmosphere model [4]. All retrieved temperature profiles (1500 — i.e. as the number of spectra recorded during the time interval from October 12 to December 14, 1983) from [5] have been smoothed with neglect of time and longitude dependence and by approximation with a polynom, and then tabulated every 5° latitude. Differences between temperature profiles at equal latitudes of northern and southern hemisphere are only small so that we are not able to conclude about differences in the thermal regime between both hemispheres. This is in accordance with other measurements [2]. The mean error of temperature retrieval is 2 K maximum what has been tested by numerical investigations. The influence of aerosol at the edges of the wings of the 667 cm^{-1} CO_2 absorption band reaches an order of 10 K temperature error. Till now the temperature field was retrieved with the CO_2 transmission function only so that the temperature at pressure levels above 50 hPa is systematically too low.

In Fig. 2 the positions of the measuring routes during the orbits of Venera 15 in coordinates of latitude and local solar time are shown. By turn of spacecraft Venera 15 one measuring route over the southern hemisphere was possible which is also shown in this figure as dashed line. The distances between retrieved temperature profiles on these

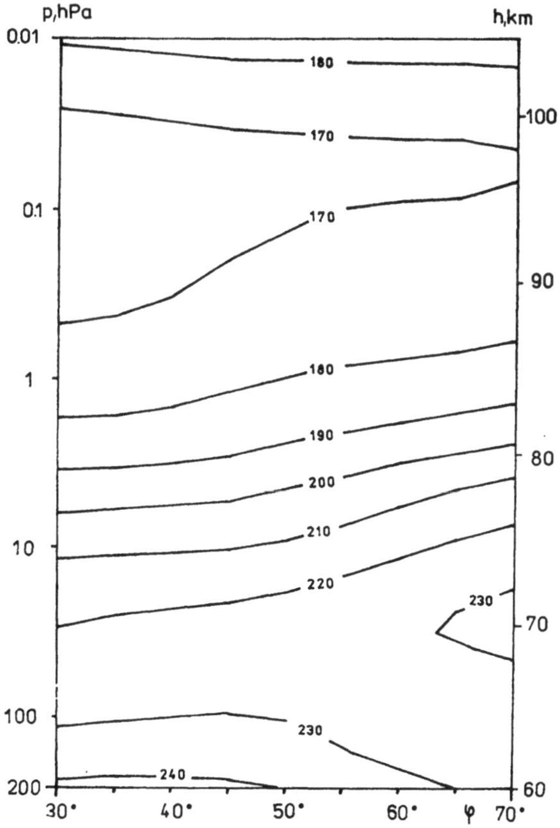

Fig. 3. Isotherms from all retrieved temperature profiles in the 04—06 h sector neglecting time and longitude dependence in a pressure/altitude—latitude plot

measuring routes are about 2 to 3° latitude or about 200 km. This figure shows the good possibility to investigate the latitude dependent temperature field in 4 sectors of local solar time: late night (04—06 h), morning (06—10 h), evening (16—18 h), and early night (18—22 h).

The Figs. 3—6 show the global thermal field in the 4 local time sectors. Since the accuracy of solar zenith angle of the measuring points is in the order of some degrees temperature profiles in the solar zenith angle interval from 80 to 100° have been disregarded in these figures. But the influence of this inaccuracy on the thermal field is in the order of temperature error. An interesting feature of global thermal field is the temperature inversion near cloud top in the cold collar region between 60 and 80° latitude [1, 2, 3, 5]. The maximum temperature difference of the inversion is in the morning — here are 19 K, the minimum is in the evening with 7 K, and in early and late night are 9 and 8 K. That means the cold collar is a solar fixed feature and exists in the northern hemisphere around the morning terminator. 5 years earlier the cold collar was a characteristic feature in the day- and night-time and the temperature inversion was only weak near the terminator [1]. The temperature differences between day and night in the global thermal field are in the order of some K for all latitudes and altitudes and 5 K

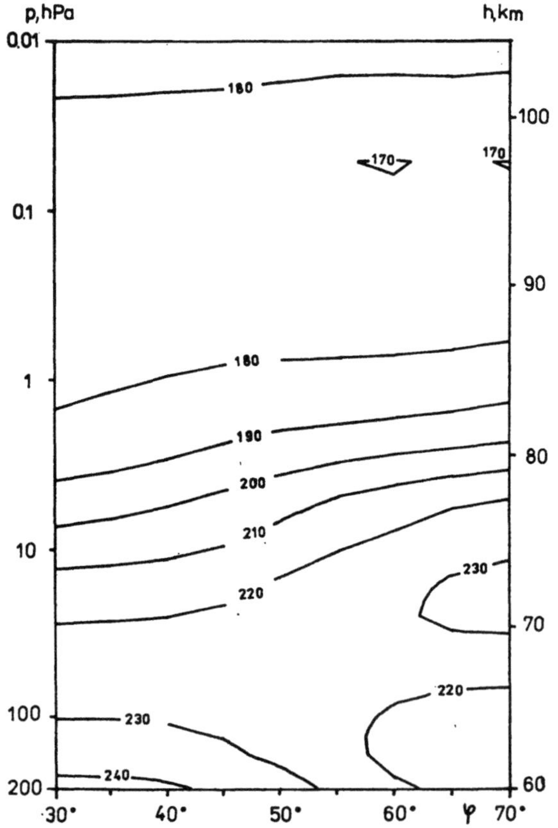

Fig. 4. Isotherms from all retrieved temperature profiles in the 06—10 h sector neglecting time and longitude dependence in a pressure/altitude—latitude plot

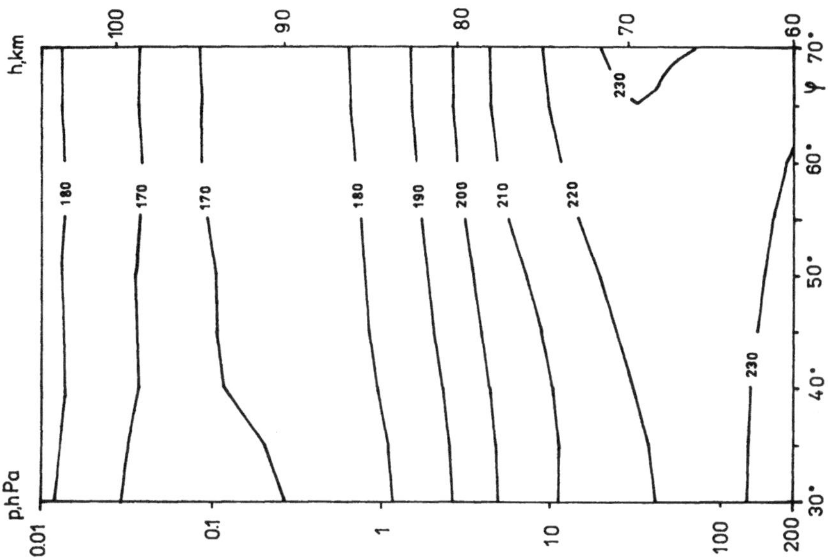

Fig. 6. Isotherms from all retrieved temperature profiles in the 18—22 h sector neglecting time and longitude dependence in a pressure/altitude—latitude plot

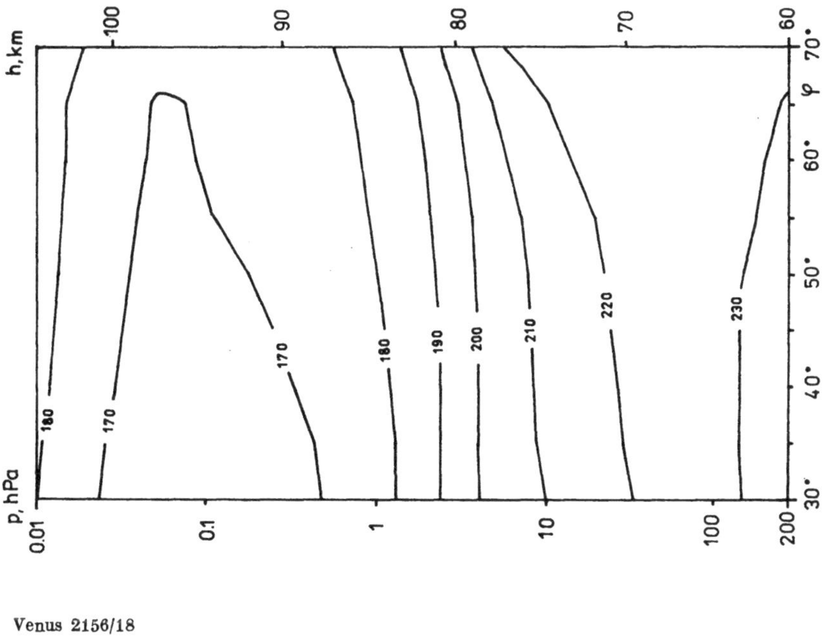

Fig. 5. Isotherms from all retrieved temperature profiles in the 16—18 h sector neglecting time and longitude dependence in a pressure/altitude—latitude plot

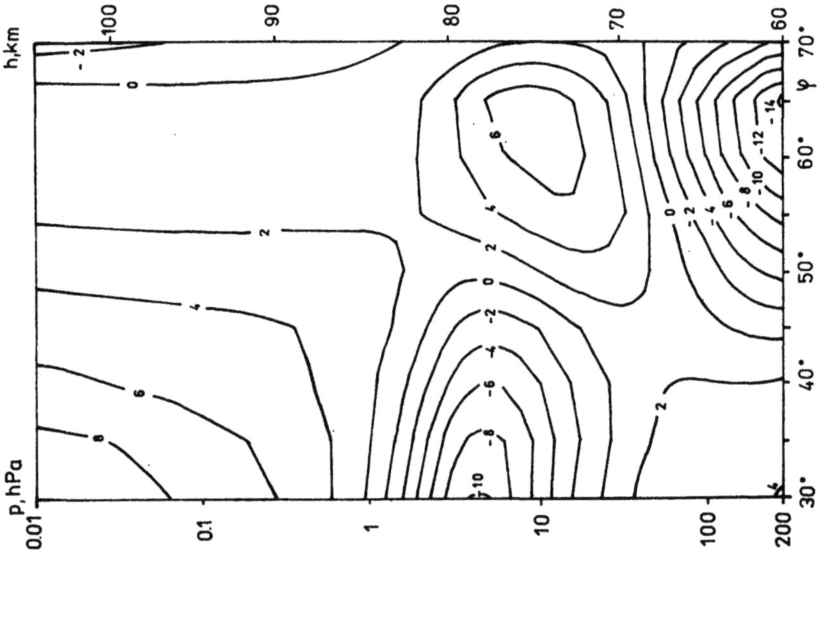

Fig. 7. Temperature differences between thermal fields in the 06—10 h sector (Fig. 4) and 16—18 h sector (Fig. 5) in a pressure/altitude—latitude plot

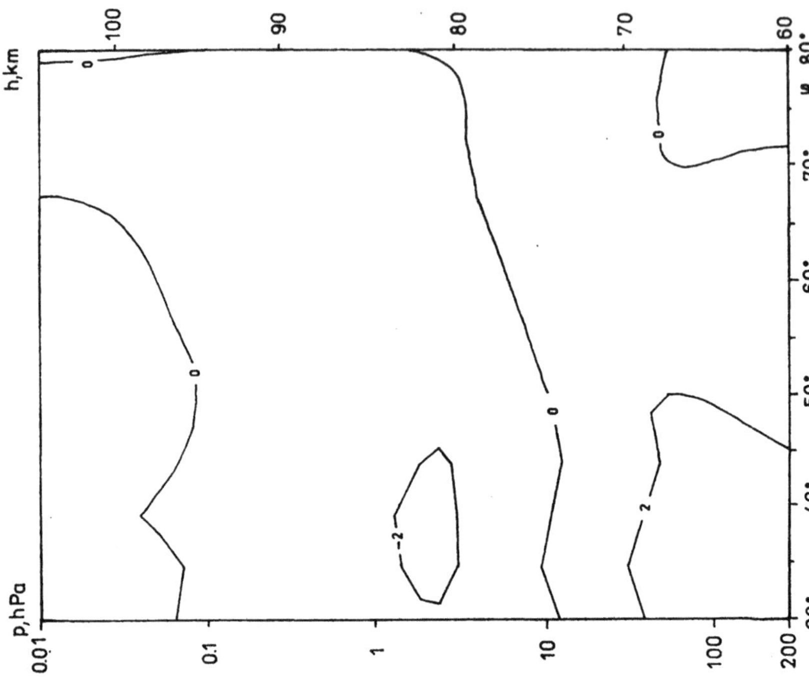

Fig. 8. Temperature differences between thermal fields in global scale (Fig. 1) and in the 04—06 h sector (Fig. 3) in a pressure/altitude—latitude plot

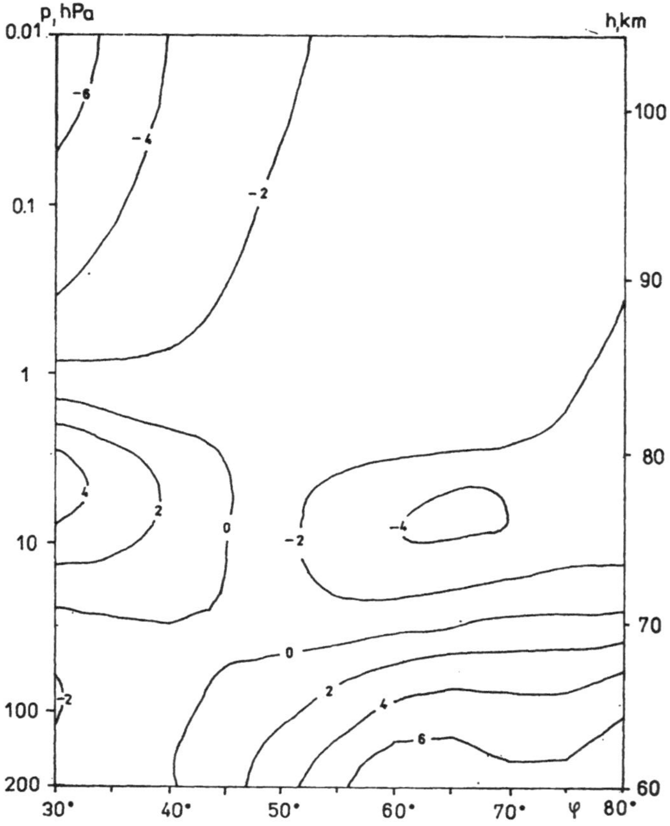

Fig. 9. Temperature differences between thermal fields in global scale (Fig. 1) and in the 06—10 h sector (Fig. 4) in a pressure/altitude—latitude plot

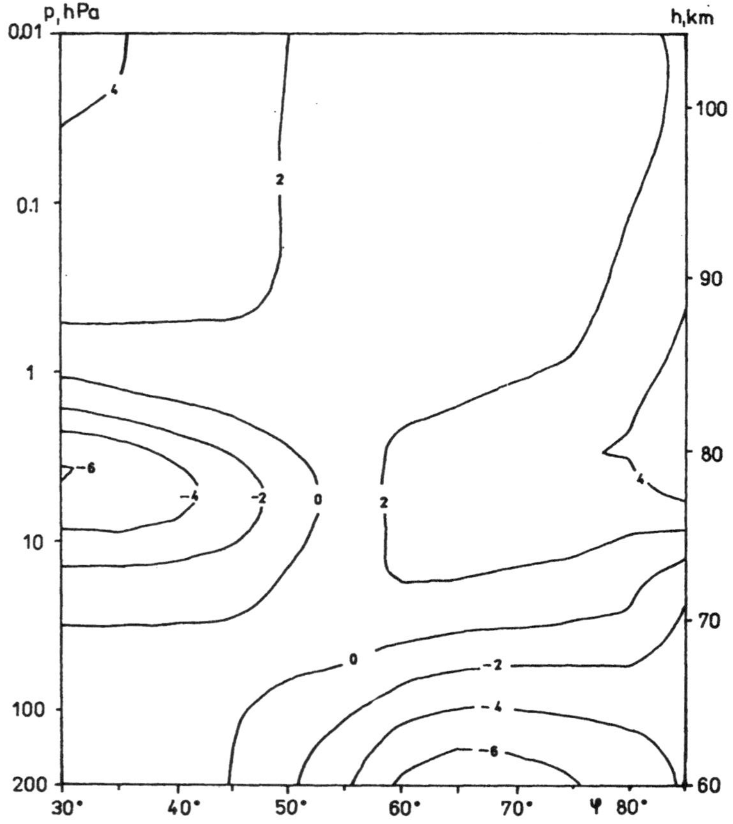

Fig. 10. Temperature differences between thermal fields in global scale (Fig. 1) and in the 16—18 h sector (Fig. 5) in a pressure/altitude—latitude plot

maximum. The temperature differences between morning and late night and also between evening and early night are greater and 8 K maximum. But 10 K maximum the morning — evening temperature differences (see Fig. 7) reach in contradiction to temperature differences between early and late night which are 5 K maximum.

A shortening of the morning and early night sector to 06—08 h and 18—20 h time intervall gave the conclusion that the different size of local solar time sectors has no influence on these results. The properties of thermal field at different local solar times can be completed by investigations of temperature differences between the thermal field in global scale and that in different local solar time sectors as shown in Figs. 8—11. These temperature differences are nearly opposed in comparison between morning and evening (Fig. 9 and 10) and also between both night sections (Figs. 8 and 11) independently from altitude and latitude. This is a hint to a wavenumber 2 longitudinal structure of the thermal field. An additional wavenumber-1-structure can be concluded from a 7 K maximum of these temperature differences in the day-time and a 3 K maximum in the night-time.

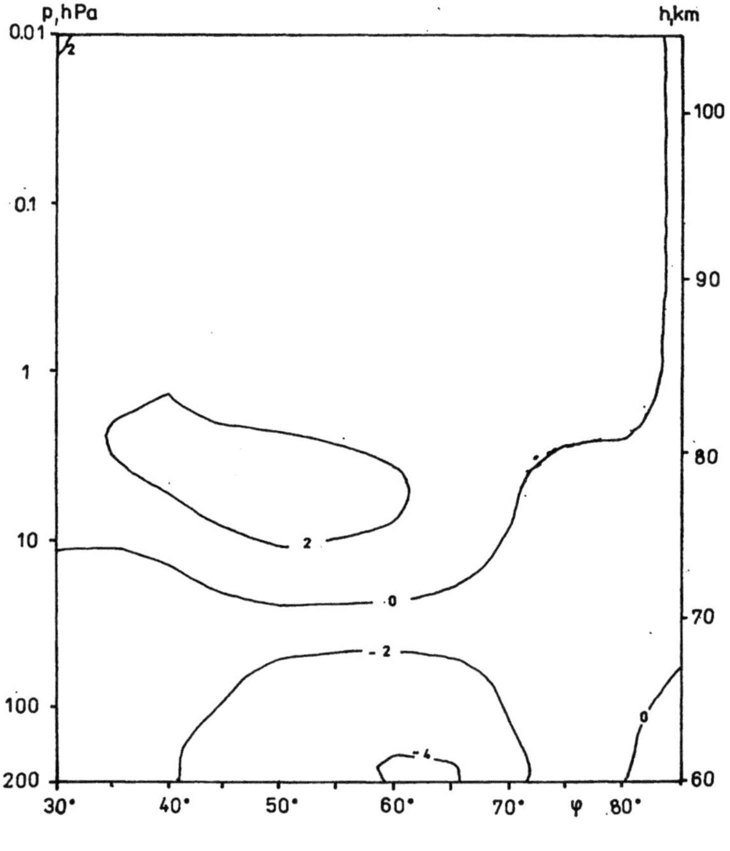

Fig. 11. Temperature differences between thermal fields in global scale (Fig. 1) and in the 18—22 h sector (Fig. 6) in a pressure/altitude—latitude plot

3. Thermal wind field

Thermal winds in the middle atmosphere of Venus are characterized by velocities of about 100 m/s at cloud top altitudes, a decrease to very low values above about 90 km altitude, and a strong mid-latitude jet of 120 m/s which circles the planet approximately during two days [1, 2, 3]. This zonal wind field can be computed by the use of zonal mean temperatures in the meridional momentum equation which describes the balance between the equatorward latitudinal component of the centrifugal force on a zonally rotating atmospheric parcel and the poleward meridional pressure gradient force per unit mass:

$$-\frac{u^2}{a \times \cos\theta} \times \sin\theta = -\frac{1}{a \times \varrho} \times \frac{\partial p}{\partial \theta},$$

where θ is latitude, u zonal wind velocity, ϱ atmospheric density, p pressure, and a planetary radius. From this cyclostrophic belance follows the thermal wind equation:

$$u^2(\theta, z) = u_0{}^2(\theta, z_0) - \frac{R}{\tan\theta} \times \int\limits_{z_0}^{z} \frac{\mathrm{d}T(\theta, z')}{\mathrm{d}\theta} \times \mathrm{d}z'$$

with $z = \ln \frac{p_0}{p}$, $p_0 = 94900\,\mathrm{hPa}$ (surface pressure),

$$z_0 = \ln \frac{p_0}{200\,\mathrm{hPa}}, \qquad R = 190.34\,\frac{\mathrm{J}}{\mathrm{kg} \times \mathrm{K}} \text{ (gas constant)},$$

and T is temperature. Real winds cannot be obtained near the equator and near the pole in this approximation because of $\tan\theta$. This formula has been integrated vertically between the levels of retrieved temperature profiles in linear fashion at 5° latitude intervals using a basic trapezoidal differencing formula so that the wind values are given in the middle of these latitude intervals. The wind velocity at the lower boundary (200 hPa pressure level or 60 km altitude) was taken from [2]: constant 80 m/s wind in lower and middle latitudes and a cos-like decrease till zero between 65 and 90° N. The results depend strongly on the lower boundary wind speed and are independent on the lower boundary altitude.

In Fig. 12 the global thermal wind field of both hemispheres derived from the thermal field of Fig. 1 is shown. The wind velocity strong decreases with altitude. Distinct wind jets with peak velocities of about 110 to 120 m/s are located near 70 km altitude between 40 and 55° latitude in both hemispheres. The second jet in the southern hemisphere cannot be discussed because the data basis is only one measuring route over this hemisphere.

The global pattern of the wind field in the day- and night-time is similar and jet velocities differ only by 15 m/s. Also the global pattern of wind fields in the local solar time sectors are similar but the jet velocities differ by 35 m/s between morning (Fig. 13) and evening (Fig. 14).

So we can mention two correlations:

— The great temperature differences of 10 K maximum between morning and evening for all altitudes and latitudes are correlated with a jet velocity difference of 35 m/s.

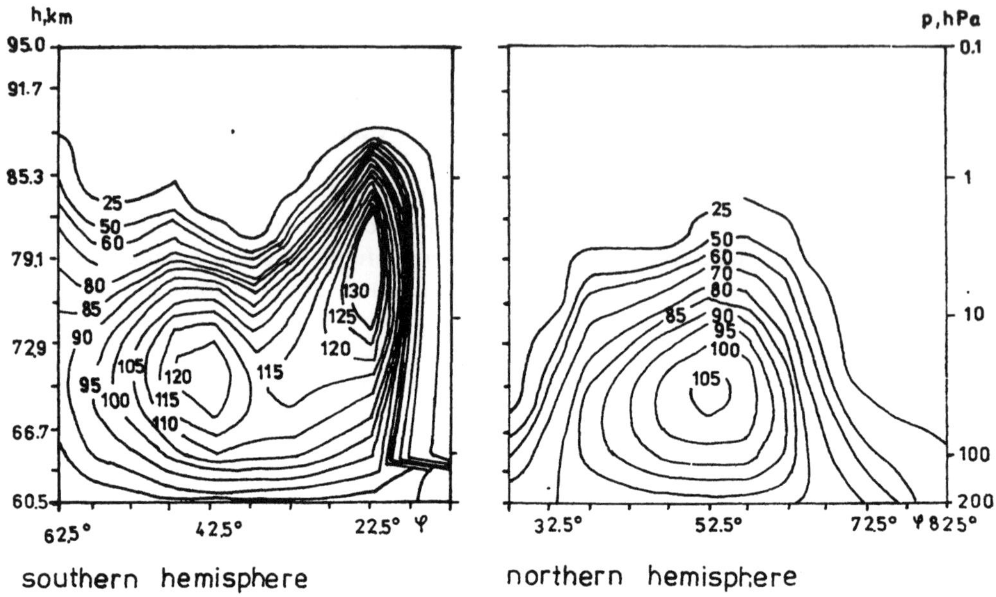

Fig. 12. Zonal wind velocity isolines (m/s) in an altitude/pressure — latitude plot derived under cyclostrophic balance assumption from the thermal field of Fig. 1

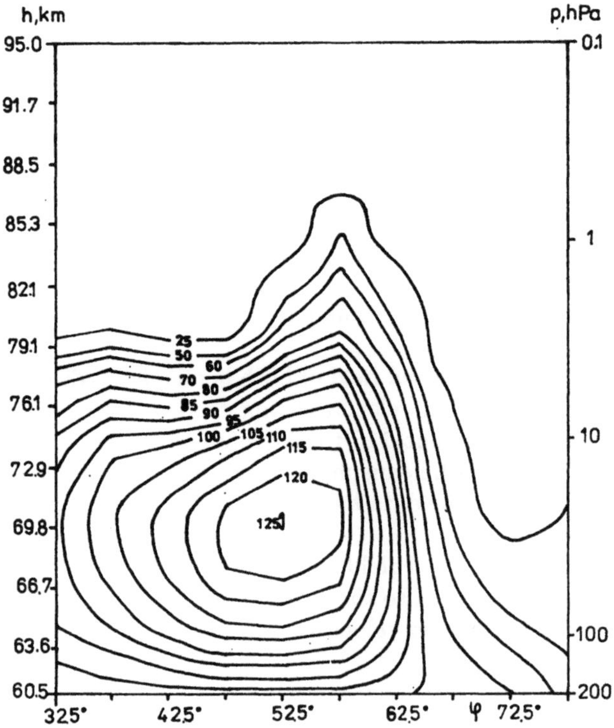

Fig. 13. Zonal wind velocity isolines (m/s) as in Fig. 12 for the morning sector derived from the thermal field of Fig. 4

– The strength of wind jet is correlated with the strength of temperature inversion of the cold collar and both are maximum in the morning. This is in correspondence with the suggestions in [2, 3] although we must mention a nearly constant position of jet and inversion maximum – in the night-time the jet is located 5° more northward than in the day-time.

It is interesting to mention that the minimum zonal wind velocity in 1982 was in equatorial latitudes at 10:40 h local solar time [6].

4. Barotropic model-results

The measured zonal wind profiles have been examined for stability with a barotropic model. This follows the approach of [7], uses the linearized barotropic vorticity equation but includes a linear friction term and was described in [8]. Fig. 15 shows the latitudinal profile of zonal wind in Venusian northern atmosphere from Fig. 12 at 70 km altitude. In equatorial regions due to a lack of data we made different assumptions on the variation of the zonal wind in correspondence with [2] and [3], denoted as wind profiles P2, P3, P4.

Fig. 16 presents the real and imaginary parts of the complex Eigen values μ_r and μ_i for the zonal wind profile P2 in dependence on the wavenumber. At each wavenumber

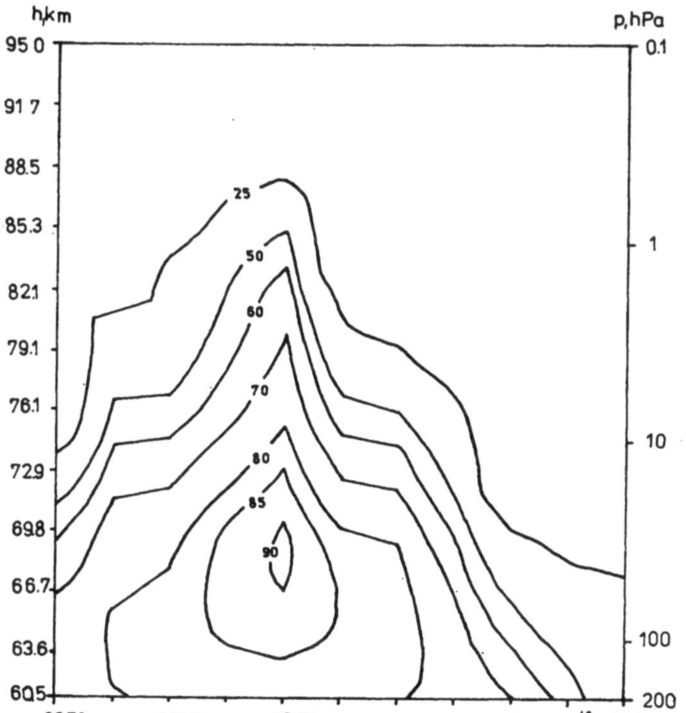

Fig. 14. Zonal wind velocity isolines (m/s) as in Fig. 12 for the evening sector derived from the thermal field of Fig. 5

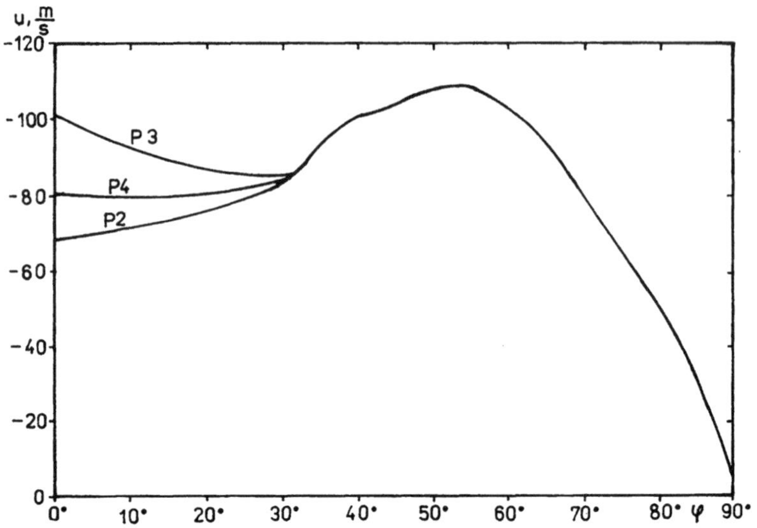

Fig. 15. Latitudinal profile of zonal wind in northern hemisphere from Fig. 12 at 70 km altitude

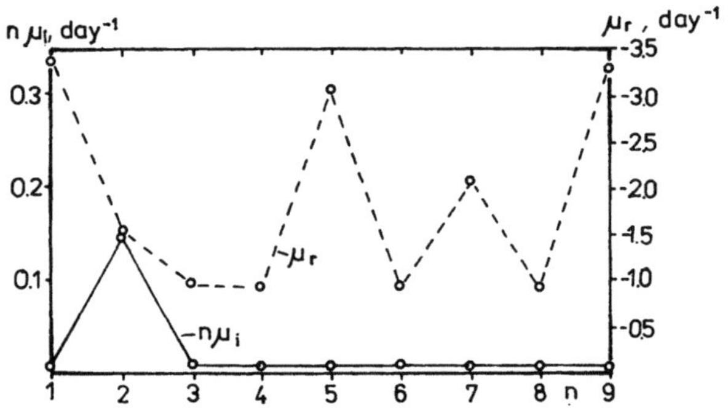

wave number n	t_i (days)	t_r (days)
1	-10^8	-1.85
2	43	-4.45
3	-10^8	-7.65
4	-10^8	-7.65
5	-10^8	-2.03

Fig. 16. Growth rate $n \times \mu_i$ and angular phase speed μ_r in dependence on wavenumber n for wind profile P2 and a table for the adequate growth time $t_i = \frac{2 \times \pi}{n \times \mu_i}$ and phase period $t_r = \frac{2 \times \pi}{\mu_r}$

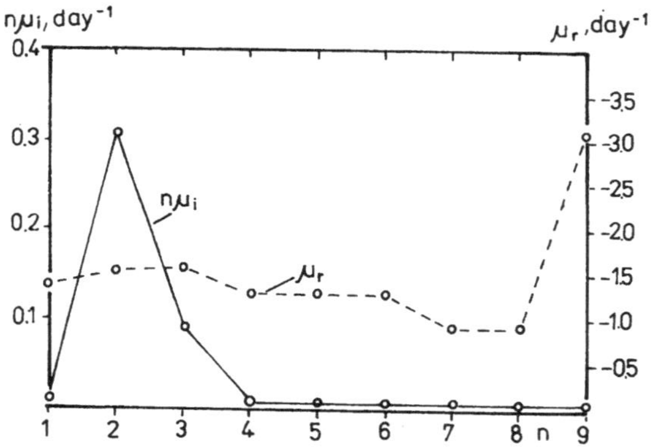

wave number n	t_i (days)	t_r (days)
1	2299.0	-4.7
2	20.5	-4.7
3	71.7	-3.9
4	1089.8	-4.7
5	1165.8	-4.7

Fig. 17. The same as in Fig. 16 for wind profile P3

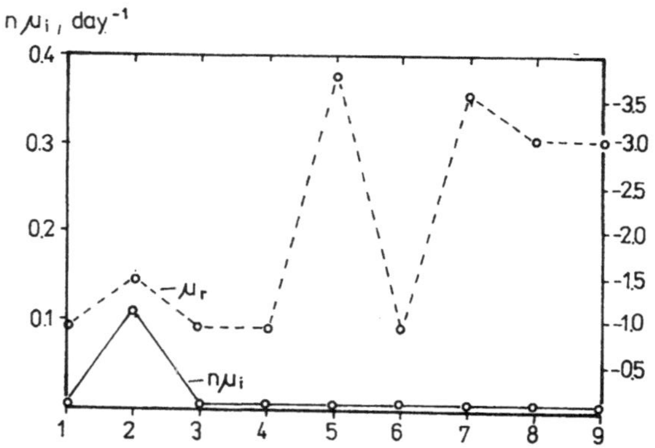

wave number n	t_i (days)	t_r (days)
1	$>10^8$	-7.6
2	57.1	-4.3
3	$>10^8$	-7.6
4	$>10^8$	-7.6
5	$>10^8$	-1.7

Fig. 18. The same as in Fig. 16 for wind profile P4

the most instable barotropic disturbance was selected. The instability rates show a maximum for wavenumber 2 and very small instability rates for all other waves. The computed growth time and phase period shows that the mode 2 is the dominating with a phase period of −4.49 days.

Fig. 17 presents the growth rate $n \times \mu_i$ and the phase speed μ_r for wind profile P3 and Fig. 18 for wind profile P4. In both cases the highest instability rates occur for wavenumber. 2 In the zonal wind profile P3 also a strong wavenumber 3 influence can be seen.

The phase periods are in the range of −3.9 to −4.7 days. In the data analyses of [9] especially wavenumber 2 and 3 were observed. The rotation periods of the Venus atmosphere of [10] of −5.3 + 0.4 days would be in good agreement with the barotropic instabilities as discussed in our investigations.

Zonal wind changes in equatorial regions have no qualitative influences on the dominating barotropic instable wave modes. Wavenumber 2 has in each case the highest instability rates with a phase period around −4.3 to −4.7 days. This picture was discovered also in data analyses and we conclude that barotropic instability may explain some observed characteristica of the polar dipole. In future investigations a computation of the Eigen functions and comparison with analysed amplitudes of planetary waves on Venus may lead to improved insights.

References

[1] F. W. TAYLOR, J. T. SCHOFIELD, and P. J. VALDES: Temperature structure and dynamics of the middle atmosphere of Venus. Adv. Space Res. **5** (1985) 9, 5–23.

[2] M. E. NEWMAN, G. SCHUBERT, A. J. KLIORE, and I. R. PATEL: Zonal winds in the middle atmosphere of Venus from Pioneer Venus radio occultation data. J. Atmos. Sci. **41** (1984) 12, 1901–1913.

[3] R. L. WALTERSCHEID, G. SCHUBERT, M. NEWMAN, and A. J. KLIORE: Zonal winds and the angular momentum balance of Venus' atmosphere within and above the clouds. J. Atmos. Sci. **42** (1985) 19, 1982–1990.

[4] V. I. MOROZ: Working model of Venus' atmosphere: Preprint IKI AN SSSR (1979).

[5] R. DUBOIS, L. V. ZASOVA, D. SPÄNKUCH, V. I. MOROZ, K. SCHÄFER, E. A. USTINOV, D. OERTEL, V. M. LINKIN, W. DÖHLER, and J. GÜLDNER: Thermal structure of the middle atmosphere of Venus from Venera 15 data. Veröff. d. Fo. b. Geo- u. Kosm. W., Heft 18, 1990.

[6] S. S. LIMAYE: Atmospheric dynamics on Venus and Mars. Paper C. 3.2.8. presented at the XXVI. Planery Meeting of COSPAR, Toulouse, France, Juni 30–July 11, 1986.

[7] L. S. ELSON: Barotropic instability in the upper atmosphere of Venus. Geophys. Res. Lett. **5** (1978), 603–605.

[8] K. DETHLOFF and M. ZELCK: Barotropic instability in the middle atmosphere of Venus on the basis of measured zonal winds. Z. Meteorol. **37** (1987), 220–224.

[9] F. W. TAYLOR, R. BEER, M. T. CHAHINE, D. J. DINER, L. S. ELSON, R. D. HASKINS, D. J. MCCLEESE, J. V. MARTONCHIK, P. E. REICHLEY, S. P. BRADLEY, J. DELDERFIELD, J. T. SCHOFIELD, C. B. FARMER, L. FROIDEVAUX, J. LEUNG, M. T. COFFEY and J. C. GILLE: Structure and meteorology of the middle atmosphere of Venus. Infrared Remote Sensing from the Pioneer Orbiter. J. Geophys. Res. **85** (1980) A 13, 7693–8006.

[10] J. APT, J. LEUNG: Thermal periodicities in the Venus' atmosphere. Icarus **49** (1982), 427 to 437.

10. Sun coupled waves in the field of the thermal radiation of Venus from Venera 15 infrared spectrometry

I. A. Macygorin

Abstract

Modes 1 and 2 of sun coupled waves in the field of Venusian thermal radiation are defined on the base of analysis of infrared spectra obtained on board of Venera-15 by means of Fourier-spectrometer. Their amplitudes and phases are complicated functions of spectral wavenumbers and latitude.

Infrared spectra of Venus with spatial resolution in the order of 100 km were observed first time by Venera-15 infrared Fourier spectrometer in October—December 1983 [1]. These observations (including about 2000 spectra between 250 and 1600 cm^{-1} covered latitudes mainly between 20° and 85° N (Fig. 1).

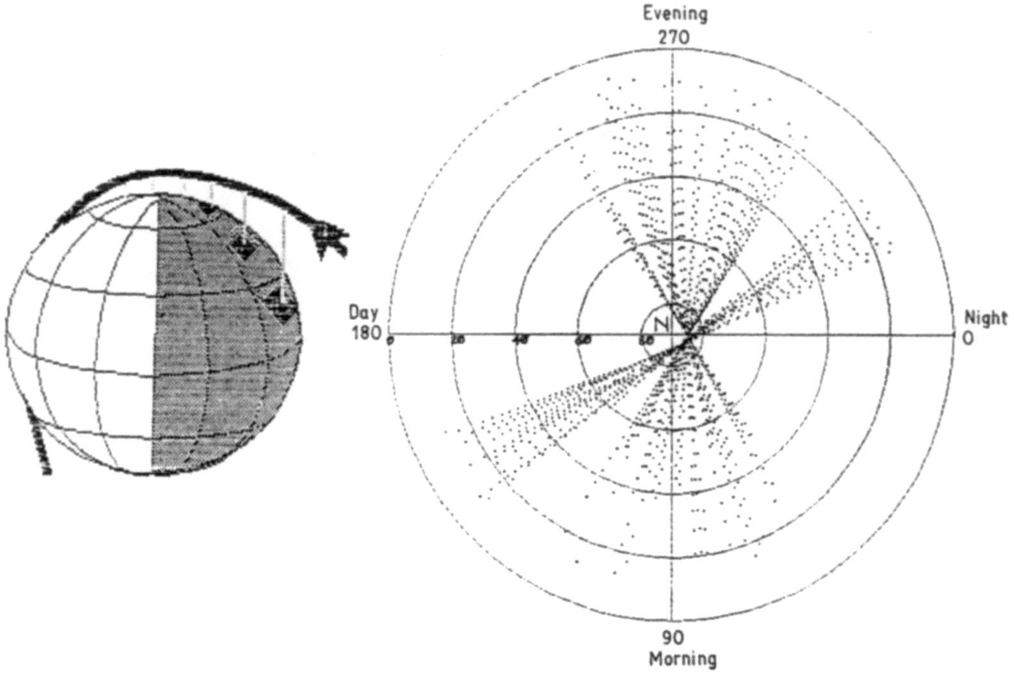

Fig. 1. Positions of observed points of planets. North pole in the centre. Direction of rotation of the planet and its atmosphere is clockwise

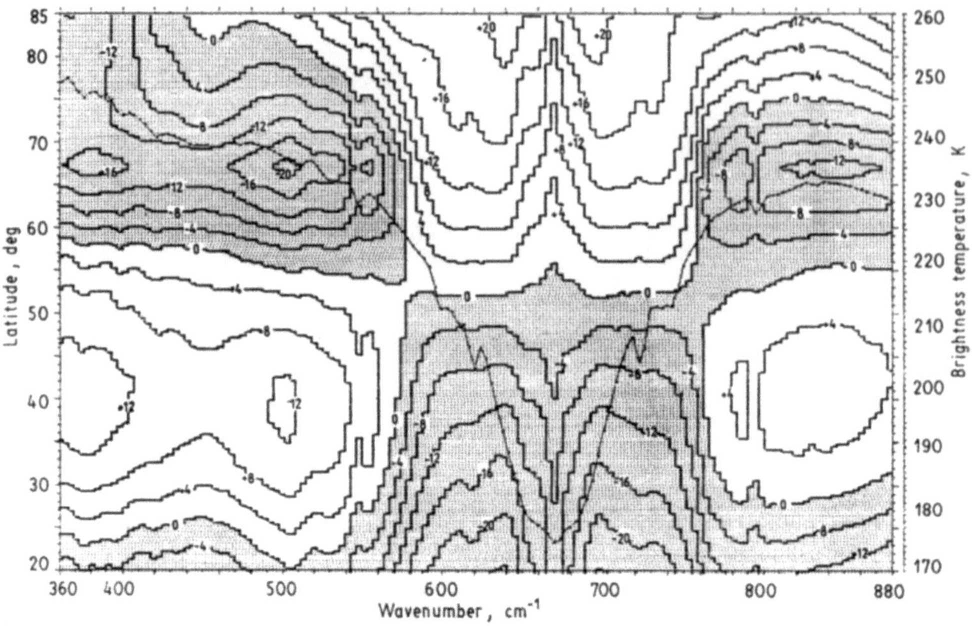

Fig. 2. Map of deviations ΔT of bin averaged spectra from global mean (given by full line, scale on right). X-axis is spectral wavenumber. Y-axis is latitude. Isotherms are given with 4 K step. Negative deviations presented by dottes

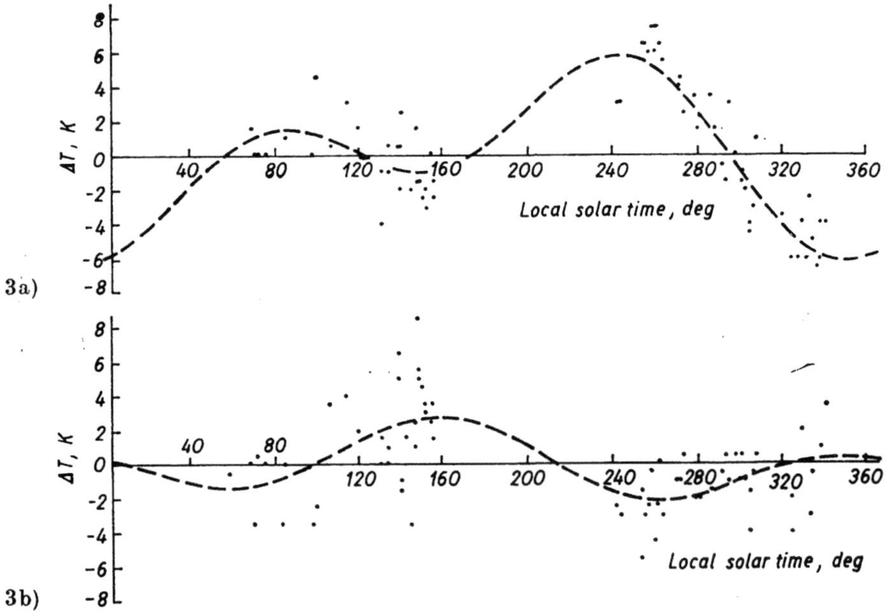

Fig. 3. An example of deviations of brightness temperature ΔT_i and approximation by the model. X-axis is local solar time (equivalent solar longitude, degree), Y-axis — deviation of ΔT_i from the mean value. Data for the bin 35°—40° and spectral wavenumbers 705 and 375 cm^{-1} are presented. Full line is r.m.s. approximation by (1). Criterium $c = 26$ for 3a, and $c = 7$ for 3b

The level of the optical depth $\tau = 1$ is changing very strong in this spectral interval convering heights approximately from 60 to 90 km. Consequently analysis of longitude distribution of the infrared brightness of different spectral wavenumbers could reflect sun coupled phenomena in the different heights.

We attempt here to present results of such ananalysis supposing that longitude distri-

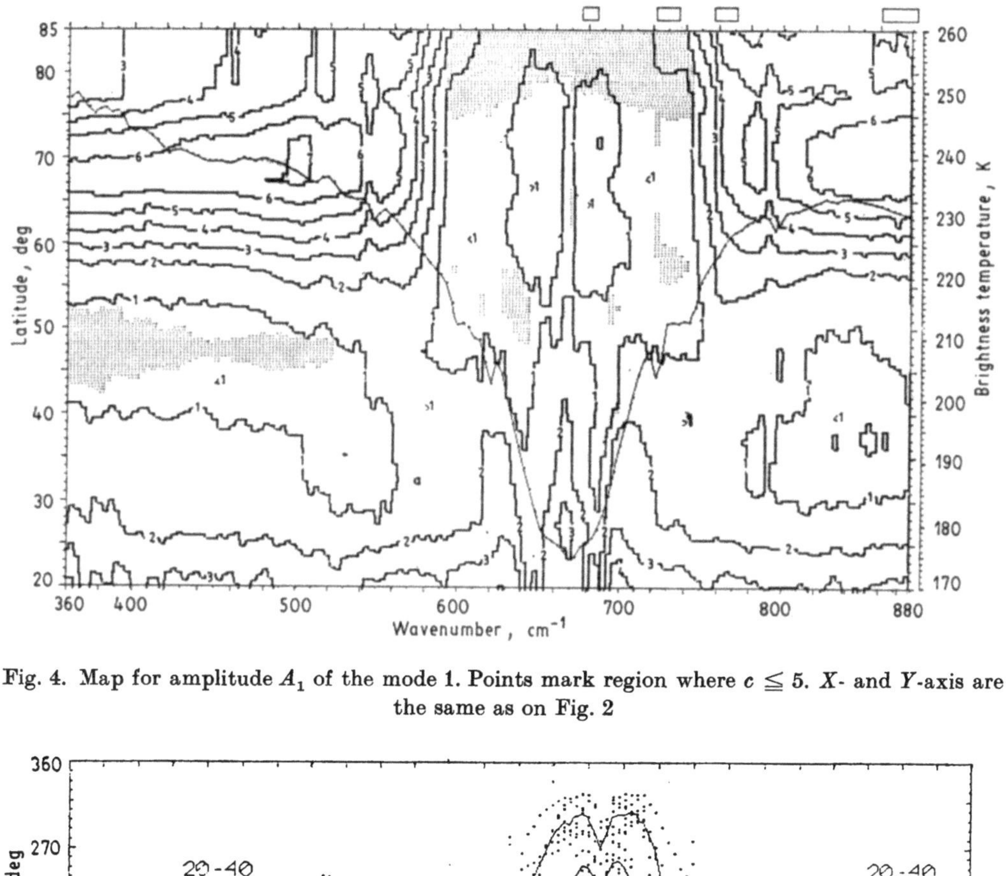

Fig. 4. Map for amplitude A_1 of the mode 1. Points mark region where $c \leq 5$. X- and Y-axis are the same as on Fig. 2

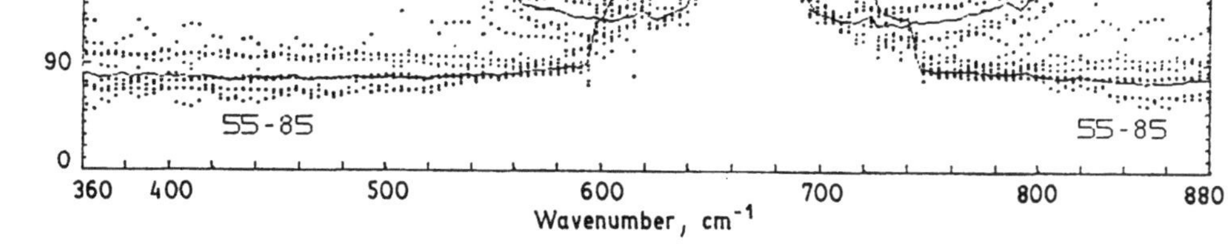

Fig. 5. Boundaries of clusters for phases ψ_1 of mode 1. X-axis is spectral wavenumber, Y-axis is phase measured from midnight clockwise. Latitude zone of clusters are given near corresponding bands

bution can be approximated by longitude wavenumbers ('mode') 1 and 2. It is naturally to suppose that mode 1 is explained by solar heating and mode 2 by atmospherical tides.

We will discuss below brightness temperature field in the spectral interval between 360 and 880 cm^{-1}. There are 3 steps in our calculation: 1) averaging of spectra in latitude

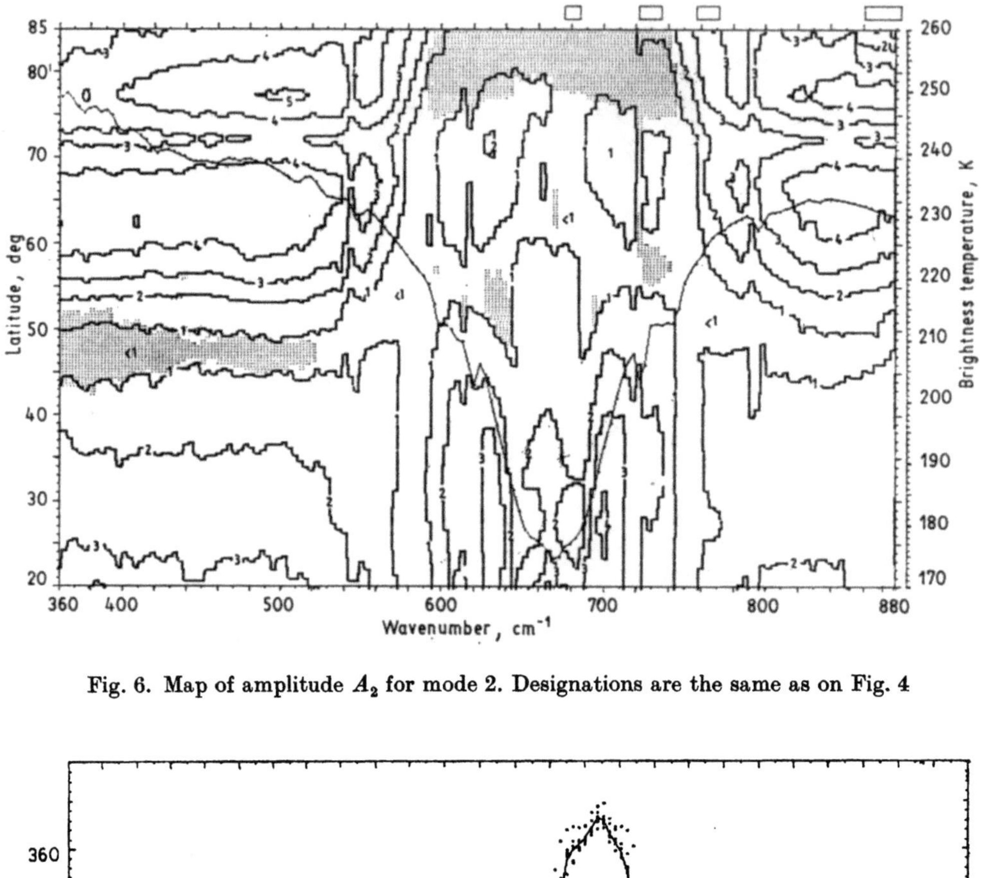

Fig. 6. Map of amplitude A_2 for mode 2. Designations are the same as on Fig. 4

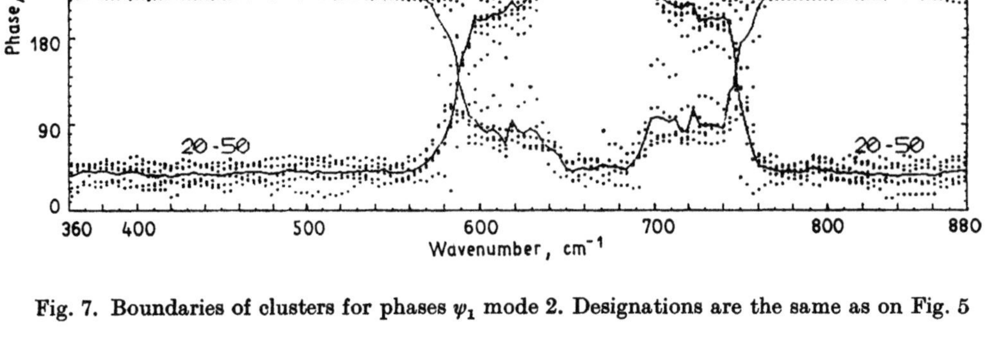

Fig. 7. Boundaries of clusters for phases ψ_1 mode 2. Designations are the same as on Fig. 5

bins 20°—25°, 25°—30° etc., 2) calculation of deviations ΔT of measured spectra from the bins averaged (Fig. 2), 3) introduce an approximation:

$$\Delta T_{\mathrm{m}} = A_1 \times \cos(\tau + \psi_1) + A_2 \times \cos(2 \times \tau + \psi_2), \quad (1)$$

where A_1, A_2 are amplitudes of modes 1 and 2, K,
ψ_1, ψ_2 are their phases, deg,
τ is solar longitude defined as solar local time, deg.

Parameters A_1, A_2, ψ_1, ψ_2 were defined by the r.m.s. method with standard error:

$$\sigma = \sqrt{\frac{\sum_{i=1}^{n} (\Delta T_{\mathrm{i}} - \Delta T_{\mathrm{m}})^2}{n \times (n-1)}},$$

where ΔT_{i} is deviation of brightness temperature from mean for given bin, K
n is quantity of spectra in this bin ($40 \leqq n \leqq 220$).

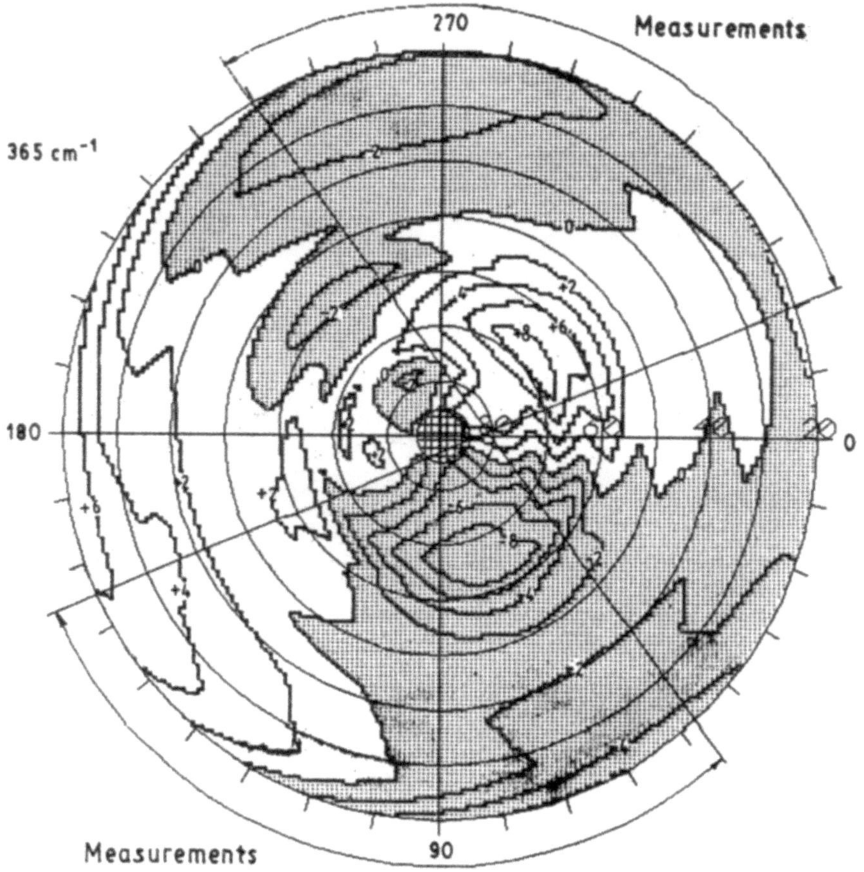

Fig. 8. Map of ΔT for deepest level of clouds (361—368 $\mathrm{cm^{-1}}$) X-axis is the solar longitude, Y-axis is latitude. Direct lines give boundaries of the really observed part of the planet. Isolines are given with the step 2 K. Points mark negative values of ΔT

Criterium of reliability of the model (1) can be introduced as $c = \sqrt{(A_1^2 + A_2^2)}/\sigma$. We suppose that $c \leq 5$ is enough for reliabile description of observations by this model. Examples of approximations are given on Fig. 3.

Isolines of A_1 as a function of the latitude and spectral wavenumber are presented on Fig. 4 and regions where $c \leq 5$ are marked by points. Phases of the mode 2 have strong tendention to the clustering. This clusters are shown on Fig. 5. Within 15 μm band of CO_2 properties of clusters are very different from those outside of band. The same type of presentation is given for mode 2 on the Fig. 6 and 7.

On Fig. 8—11 maps of ΔT model presentation are given in polar coordinates for different spectral intervals:

361—368 cm^{-1} (deepest level of cloud radiation)

861—882 cm^{-1} (upper level of cloud radiation)

625 and 701 cm^{-1} (maximum of the weighting function on ≈80 km)

666 cm^{-1} (maximum of the weighting function on ≈ 95 km).

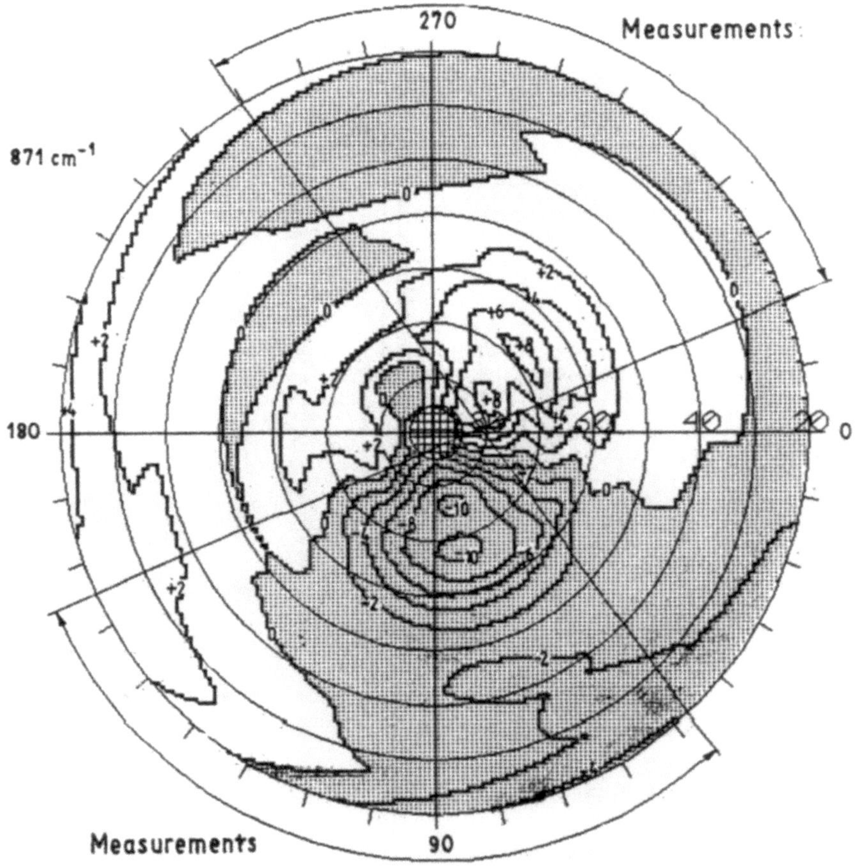

Fig. 9. The some as Fig. 8 but for upper level of clouds (861—882 cm^{-1})

Four characteristic latitudinal zones can be distinguished:

- low latitudes, 20° to 40°;
- middle latitudes, 40° to 50°;
- 'cold collar', 50° to 70°;
- polar region, higher than 70°.

Low latitudes (20° to 40°)

In this latitude range the brightness temperature at the cloud layer height is higher everywhere in the day time than at night. The long-wave spectrum portion is characterized by a maximum of the brightness temperature which forwarded subsolar point by about 25°. The value of this maximum is decreased with distance from the equator. At the level of the upper boundary of clouds the day-time maximum is blurred and more coincided with the subsolar point. In the CO_2-band wings above the cloud layer two wave systems are well-expressed which are in antiphase and located at various heights of about 80 to 85 km and 90 to 95 km respectively (approximately). These systems are characterized by a noticeable mode 2 structure.

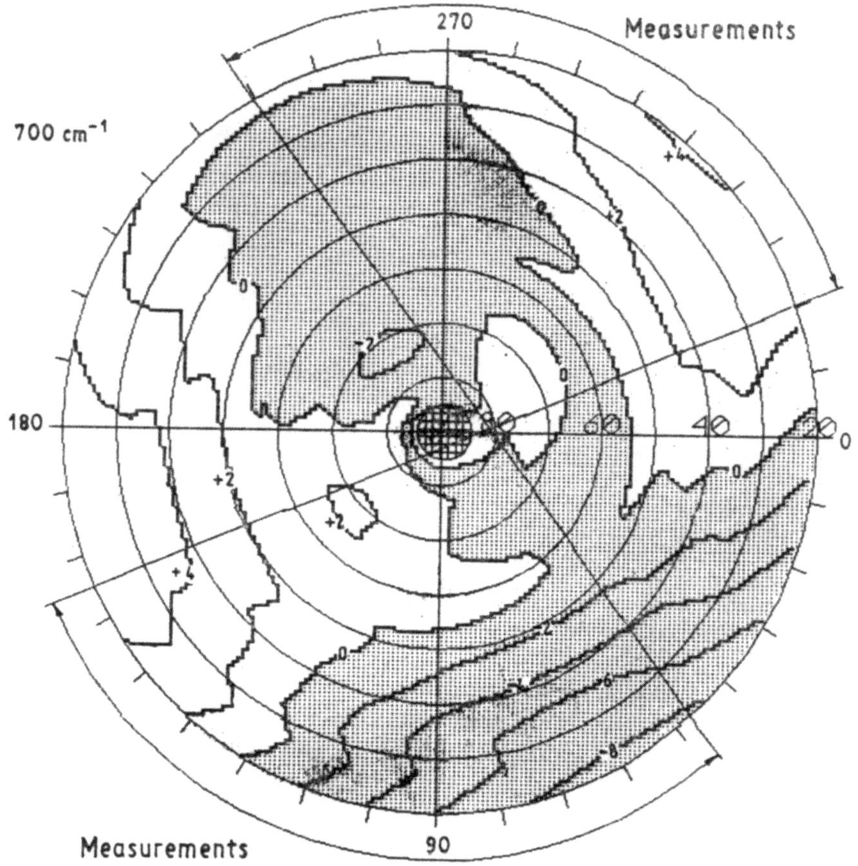

Fig. 10. Map of ΔT for wavenumbers 625 and 701 cm^{-1} (≈ 80 km)

Middle latitudes (40° to 50°)

In this latitudinal range there are no observed temperature contrasts, discernable between the day and night time values. The wave systems in the CO_2-band wings have the same structure as at low latitudes.

'Cold collar' (50° to 70°)

The 'cold collar' is best-defined at latitudes 55° to 65°. The amplitudes of the minimum and the maximum of the brightness temperature here exceed 10 K. In the CO_2-band wings the system of sun-related waves is rearranged. The structure at ≈ 70 to 80 km is in antiphase with the 'cold collar'. That wave system which was observed at ≈ 85 to 90 km at lower latitudes disappear gradually.

Polar region (> 70°)

The feature of this region is the prevalence of a mode 2 structure at the cloud level and low value of *c*. The latter can be explained by the effect of dipole which rotates in this region arround the pole at the cloud level. The wave system which has formed in the 'could collar' zone remains at ≈ 80 and ≈ 90 km.

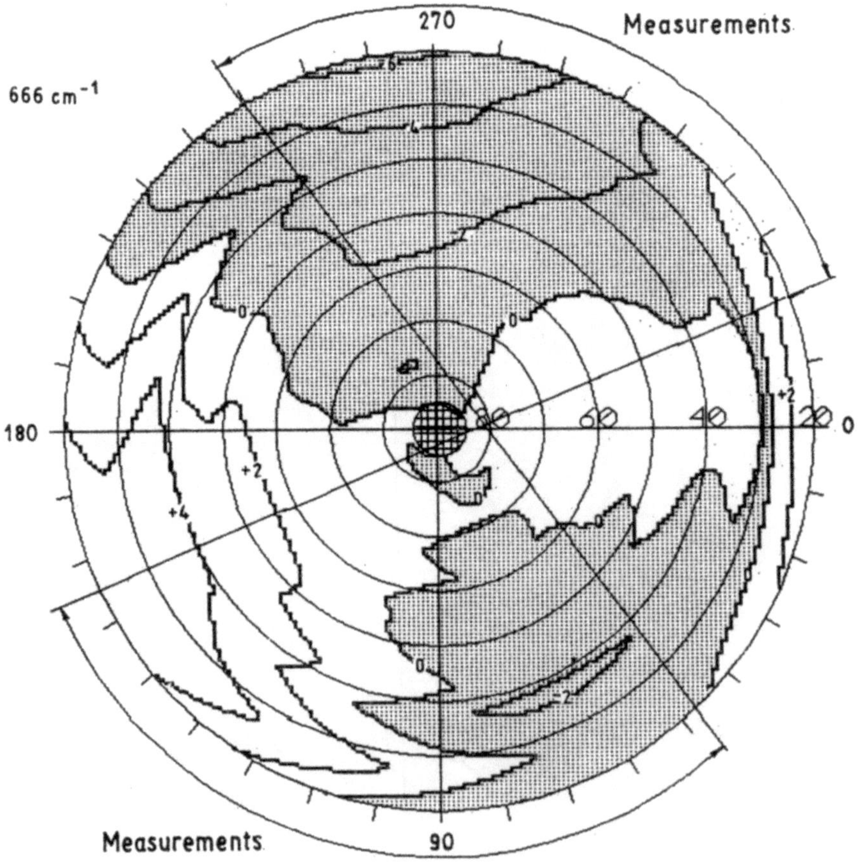

Fig. 11. Map of ΔT for wavenumber 666 cm^{-1} (≈ 95 km)

Comparison of map upper clouds (Fig. 9) with Pioneer Venus OIR data [2] show that in main features both observed discriptions are nearly identical.

We hope that empirical model description presented here can be some additional source for future studies of dynamical properties of the Venusian atmosphere.

References

[1] OERTEL, D., MOROZ, V. I., NOPIRAKOWSKI, J. et al.: Infrared experiment aboard the automatic interplanetary stations Venera 15 and Venera 16. 1. Methodics and first results. Veröffentlichungen des Forschungsbereichs Geo- und Kosmoswissenschaften, Heft 18, 1990.

[2] TAYLOR, F. W., BEER, R., CHAHINE, M. T. et al.: Structure and meteorology of the middle atmosphere of Venus: Infrared remote sensing from the Pioneer Orbiter. J. Geophys. Res. 85 (1980) A 13, 7963—8006.

11. Atmospheric Variability Over the North Pole During the Venera 15 Mission

D. Spänkuch and R. Dubois

Abstract

Infrared radiation flux variability near the North Pole is — according to the observations of the Pioneer Venus Orbital Infrared Radiometer and of the Venera-15 Infrared Fourier Spectrometer — extremely high in the dipole and cold collar region. This study describes significant changes in outgoing radiation at the very North Pole. During the nearly two months of observations made by Venera-15 a significant decrease of about 30% in the infrared radiation flux was observed for latitudes $\varphi \geqq 85°$. This decrease is, as the infrared spectra show, caused by simultaneous changes of temperature and cloud properties in the 60 to 70 km altitude range. Differences in equivalent brightness temperatures are about 20 K outside the ν_2- CO_2 fundamental band (around 668 cm^{-1}) for the October and December spectra (Fig. 3). On the other hand, above the clouds there is an increase in temperature of up to about 5 K from October to December 1983. Hence, there are spectral intervals with unchanging outgoing radiation. These intervals are at 595 cm^{-1} (16.8 μm) and 725 cm^{-1} (13.8 μm).

Outgoing longwave flux and cloud top temperature are closely correlated. The total radiance L for the spectral interval 275 to 1600 cm^{-1} at quasi-zenith direction ($\Theta \lesssim 10°$) in $W \times m^{-2} \times sr^{-1}$ and the absolute spectral brightness temperature T_B at 1152 cm^{-1} are connected by $L = 0.625 \times T_B$ (Fig. 5). The decrease in outgoing long-wave flux is traceable also in the latitude belt 60°—75° despite a difference in outgoing flux during evening and early night, on the one hand, and late night and morning, on the other, of about 15 per cent. In contrast, the outgoing long-wave flux in mid-latitudes (30—50°) is nearly constant during the mission and seems to be only slightly dependent on the Venusian day time at an increase of about 5 per cent during the transition from night to day. This increase is the net result of an increase in brightness temperature of about 2.5 K outside the ν_2- CO_2 band as well as in its centre and a decrease by about the same amount in the intervals 595—638 cm^{-1} and 691—752 cm^{-1}.

1. Introduction

The first description of atmospheric variability in the Venusian polar regions was concerned with the polar haze. This variability was derived from polarimetric measurements (Dollfus et al. 1979, Kawabata et al. 1985) and is, therefore, related to the altitude range of 70—90 km. For the altitude range below 70 km the measurements in the 'cloud channel' (872 cm^{-1}) of the Pioneer Venus Orbital Infrared Radiometer and in the corresponding spectral regions of the Venera-15 Infrared Fourier spectrometer show high nfrared radiation flux variability at about 75° to 85° N for the dipole and at about 60°

to 75° N for the cold collar region (Taylor et al. 1980, Linkin et al. 1985, Moroz et al. 1986).

However, scarcely any information on infrared radiation flux variability has been reported for the immediate North Pole region surely due to the extremely striking feature of the warm dipole and its rotation. This study describes significant changes in outgoing long-wave radiation directly over the North Pole, or more precisely speaking, for the region $\varphi \geqq 85°$ N, during the nearly two months of observations from the Venera-15 spacecraft. All these measurements from October 12, 1983 to December 14, 1983 were made during the Venusian night, but day-to-night contrasts should be of no importance there.

2. Decrease in outgoing infrared radiation flux for $\varphi \geqq 85$ °N

Fig. 1, containing — as a measure of outgoing infrared radiation flux (OIRF) — for three latitude belts the total radiance of the spectral range covered by the Fourier spectrometer (275—1600 cm^{-1}) versus day of orbit, shows its decrease by nearly 30% for latitudes $\varphi \geqq 85°$ N during the two-month mission. The maximum at the pole was observed on day 9 (20 October, 1983) with nearly 175 $W/m^2 \times sr$, the mimimum on day 58 (8 December, 1983) with about 131 $W/m^2 \times sr$. There are day-to-day variations (related to the Earth day) but with a considerably smaller amplitude. Decreasing outgoing infrared radiation flux is also observed in the latitude belt 60° to 75° and that for the Venus morning as well as for the Venus evening by about 20 $W \times m^{-2} \times sr^{-1}$, which is also of the order of about 30%. The thermal tide at those latitudes, demonstrated by the difference in OIRF between evening and morning on the average by roughly 20 $W \times m^{-2} \times sr^{-1}$ is superimposed by the decrease in OIRF. At those latitudes the day-to-day variations are larger than at the pole and depend to a large extend on the orbit's crossing or not crossing the cold polar collar. Interdiurnal variations in brightness temperature were also reported for the collar and dipole from the Pioneer Venus measurements with no more than 15 K (Taylor et al. 1980). Nevertheless, the OIRF decrease is, by and large, parallel in both latitudes, though day-to-night contrasts are significant at the lower latitudes. The nearly constant OIRF in mid-latitudes in the belt 30° to 50° N is one of the arguments for the observed decrease in OIRF at high latitudes being a real outcome rather than merely an effect of ceasing instrument sensitivity. We note at mid-latitudes larger OIRF in Venus evening than in late night and a sudden increase about one hour after sunrise at 0700 local time in the morning by about 5 per cent. Although the data set is too small to draw any definite conclusion, there is some indication to a steady decrease in OIRF from 0330 local Venus time to sunrise at 0600 local Venus time by about 7 $W \times m^{-2} \times sr^{-1}$ or 7 per cent.

The decline in OIRF at the pole can also not be explained by the position of the orbital track to the dipole axis as shown in Fig. 2. There the brightness temperature at wave number 864 cm^{-1} (11.57 μm) is given for each individual measurement as a function of the 'normalized' longitude which takes into account a dipole rotation with a period of 3.2 days derived from our measurements (Macygorin 1987). Spectral radiances or equivalent brightness temperatures at an individual suitable wave number are a direct and unequivocal measure for the OIRF in polar latitudes as brightness temperatures are there nearly constant outside the ν_2-CO_2-fundamental absorption at 667 cm^{-1}

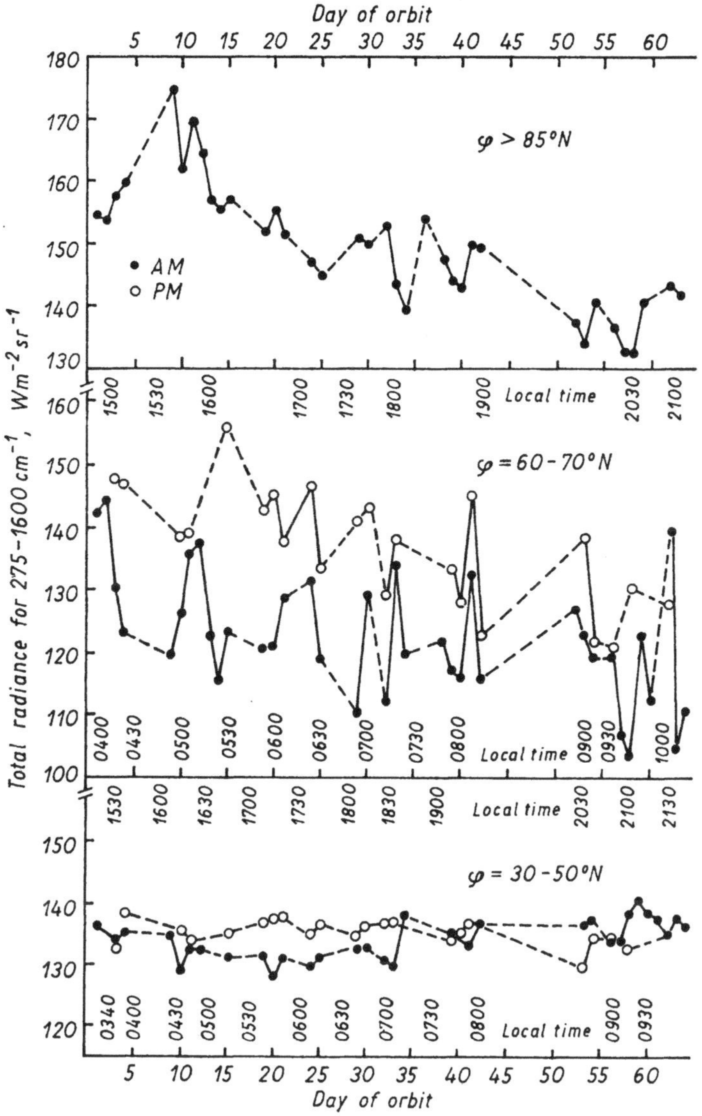

Fig. 1. Total radiance of the spectral region 275—1600 cm^{-1} vs day of orbit for three latitude belts. The radiance values are not normalized to unitary measurement geometry (zenith angle). Hence, the values of different latitude belts are not comparable speaking in absolute terms. The temporal behaviour is correctly represented as there are no changes in measurement geometry with time. For the latitude belts 30°—50° (below) and 60°—75° (centre) separation is made between Venus morning and evening •——• (o——o) measurements. Local Venus time is given for morning (lower abscissa) and evening (upper abscissa). Broken lines indicate time intervals without measurements

due to a considerably thick isothermal layer in the corresponding height interval (Spänkuch et al. 1984, Oertel et al. 1985, Zasova et al. 1985, Moroz et al. 1986, see also Fig. 3). The maximum number of measurements per day for $\varphi > 85°$ is 5 spanning a longitude interval of about 100°. Though the position of the measurement relative to the dipole axis exerts some effect on the OIRF, no more than 5 K in the spectral brightness temperature T_{864} as seen in the points labelled with the day number, the differences in T_{864} of 20 to 22 K between the days 9 to 12, on the one side and, e.g., 53 and 58, on the other, exceed by far this 'position influence'. The averaged brightness temperature spectra for the days with extreme differences in OIRF at $\varphi > 85°$ are shown in Fig. 3 demonstrating a brightness temperature difference of about 20 K for the whole spectral region outside the ν_2-CO_2-fundamental band. Constant brightness temperatures for spectral regions with large differences in cloud optical depth as, e.g., between 1150 cm^{-1} (maximum) and 350 cm^{-1} (minimum) are typical of an isothermal layer where the outgoing radiation comes from (see, e.g., in this respect also the difference as compared with the mid-latitude spectra of Fig. 3 which were added mainly to show that the great variability in OIRF is restricted to polar latitudes). In December 8, even a sligth temperature inversion can be recognised as indicated by a decline in brightness temperature at the edges of the ν_2-CO_2 band, e.g., between 550 and 570 cm^{-1} (see Moroz et al. 1986 for a more detailed discussion of this effect). Cloud top heights are in December much higher than in October (by about 9 km). These changes in cloud top heights are linked with changes in the vertical temperature structure. This can directly be inferred from the

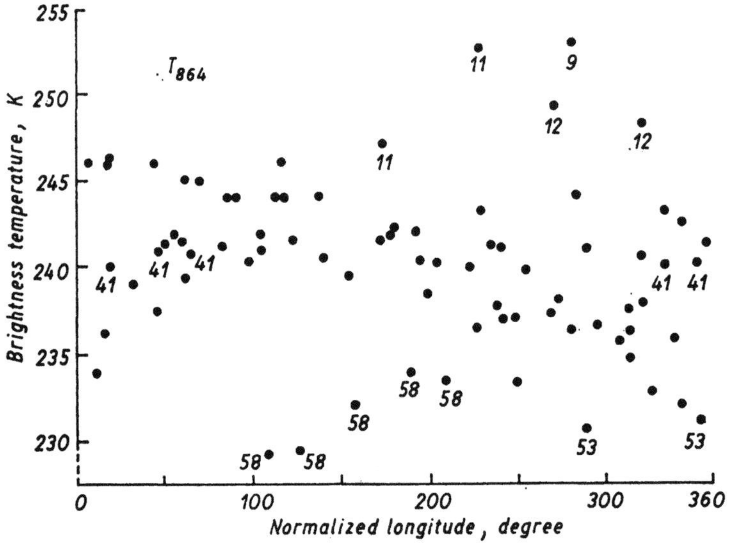

Fig. 2. Spectral brightness temperature $T_{864\,\mathrm{cm}^{-1}}$ for measurements north of 85° N as a function of 'normalized' longitude. The normalized longitude $\lambda' = \frac{360°}{\tau_d}(n-1) + \lambda_♀$ reduces the influence of the position of the measurement relative to the dipole. τ_d-dipole rotation period (~3.2 days during the mission), n — day number, $\lambda_♀$ — Venusian longitude. Some measurements are labelled with day number

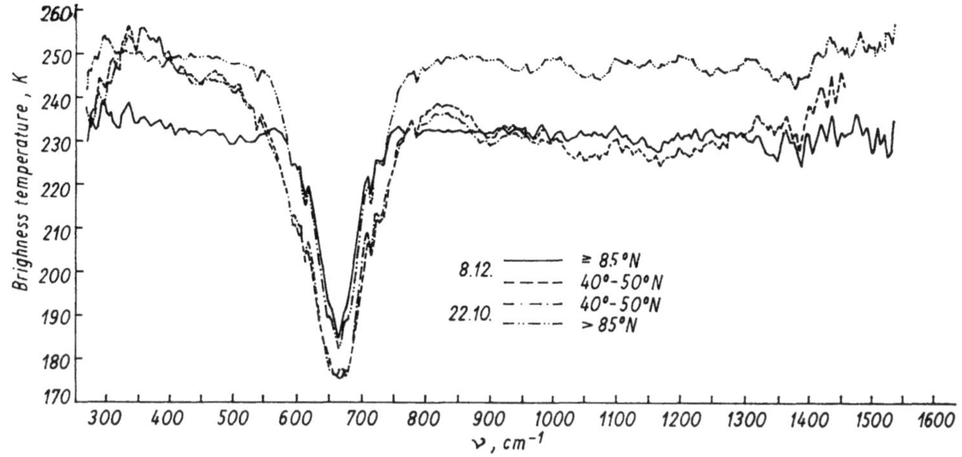

Fig. 3. Averaged brightness temperature spectra for days with extreme differences in outgoing infrared radiation flux for latitudes $\varphi \geqq 85°$ N. Averaged mid-latitude spectra are added to show the much less variability there. No zenith angle reduction (see Fig. 1)

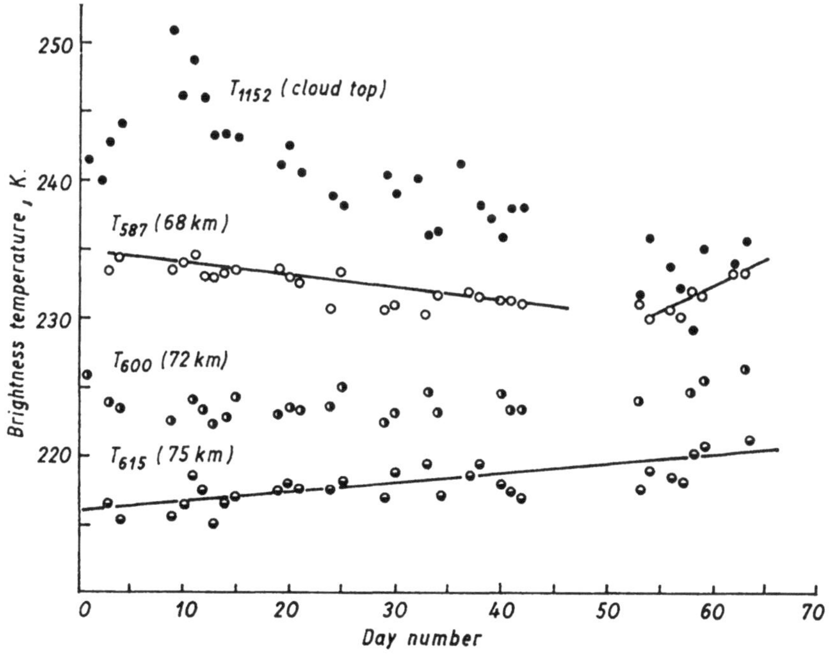

Fig. 4. Spectral brightness temperatures at $\varphi \geqq 85°$ N for selected wave numbers vs day number starting with Oct 12, 1983. T_{1152} represents pretty well the cloud top temperature. T_{587}, T_{600} and T_{615} correspond to about 45 hPa (68 km), 20 hPa (72 km) and 12 hPa (75 km). Tendency is, if drawn, by eye

brightness temperature spectra shown in Fig. 3. The isothermal layer cooled and lifted in altitude together with the cloud top height from October to December seen in the less deep CO_2 band. On the other hand, the temperature increased by about 5 K above the clouds from October to December. A more detailed view of the progress within the time of the mission is obtained from the time dependence of spectral brightness temperatures at properly selected wave numbers as shown in Fig. 4. The ir cloud optical depth has its maximum at 1152 cm^{-1} (see. e.g., Ksanfomaliti 1977, Goroskova and Petrova 1983, Zasova et al. 1985, Kondrat'ev et al. 1987). As there is no gaseous absorption, the measured brightness temperature at this wave number gives a good estimate of the true cloud top temperature and, thus, reflects the change in cloud top height with time. The weighting functions for $\nu = 587$ cm^{-1} and $\nu = 615$ cm^{-1} have their maxima for nadir viewing geometry at about 45 hPa corresponding to 68 km and at about 12 hPa corresponding to 75 km (Spänkuch et al. 1986). A comparison of the time dependence of the OIRF (Fig. 1) with the time dependence of the brightness temperature at 1152 cm^{-1} shows that cloud top temperature and OIRF are closely connected and that the observed decrease in OIRF is explained by a corresponding decrease in cloud top temperature caused by the lifting of the cloud top height. This lifting of the cloud top height is simultaneously connected with a cooling of the isothermic layer. This corresponds very well with estimates from the simultaneous temperature-cloud-retrieval scheme with a cloud top height of 58 km in October and of 67 km in December. A decrease in brightness temperature is also observed for 587 cm^{-1} for most of the period (up to day 55) with an increase afterwards to nearly the initial brightness temperature. The coinciding brightness temperature at 1152 cm^{-1} and at 587 cm^{-1} during the end of the period indicate a cloud top height of about 45 hPa (68 km) at that time and significantly lower cloud top heights in October. Between the total radiance in the

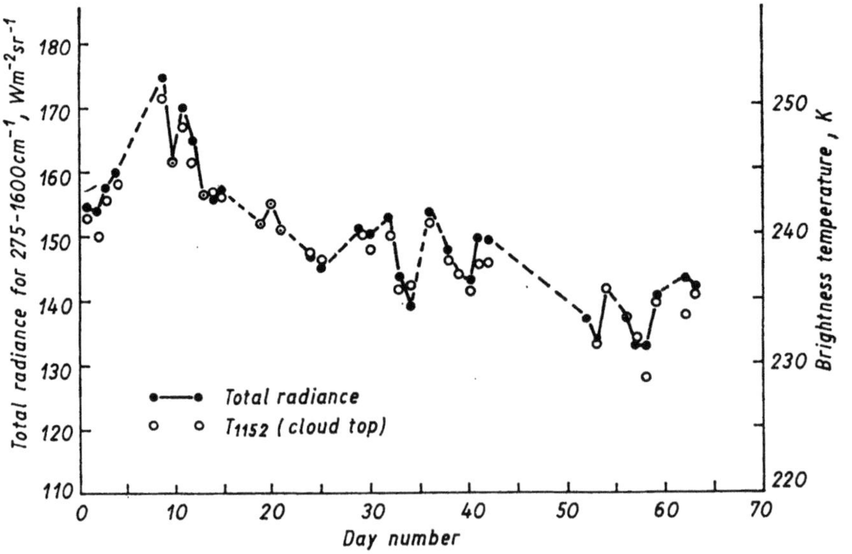

Fig. 5. Total radiance of the spectral interval 275 to 1600 cm^{-1} and spectral brightness temperature T_{1152} representing the cloud top temperature versus day of orbit

spectral interval 275 to 1600 cm^{-1} at zenith angle $\Theta \lesssim 10°$ and the absolute spectral brightness temperature at 1152 cm^{-1} there is the relation 275—1600 cm^{-1}

$$L(\Theta \leqq 10°) = 0.625 \times T_B{}^{1152\,\mathrm{cm}^{-1}},$$

with L in W $\times$ m^{-2} $\times$ sr^{-1} and T_B in Kelvin (Fig. 5). Well above the clouds at about 12 hPa ($\sim$75 km) a steady temperature increase of about 5 K altogether was observed on the average with a maximum interdiurnal variation of 2 K. Both layers, the nearly isothermic cloud top layer with, on the average, decreasing temperatures and the layer above it with increasing temperatures during the mission are separated by a level with nearly no temperature trend. This level is situated at about 20 hPa (72 km) as shown by the temporary behaviour of the spectral brightness temperature at 600 cm^{-1} where the weighting function maximum is at 20 hPa (Spänkuch et al. 1986).

3. Temporal Changes at $\varphi \geqq 85$ °N

The atmospheric variability at the North Pole described in the previous chapter is, not surprisingly, linked with the atmospheric variability at lower latitudes. The intensity and location of the cold polar collar and the warm dipole, the most striking features in

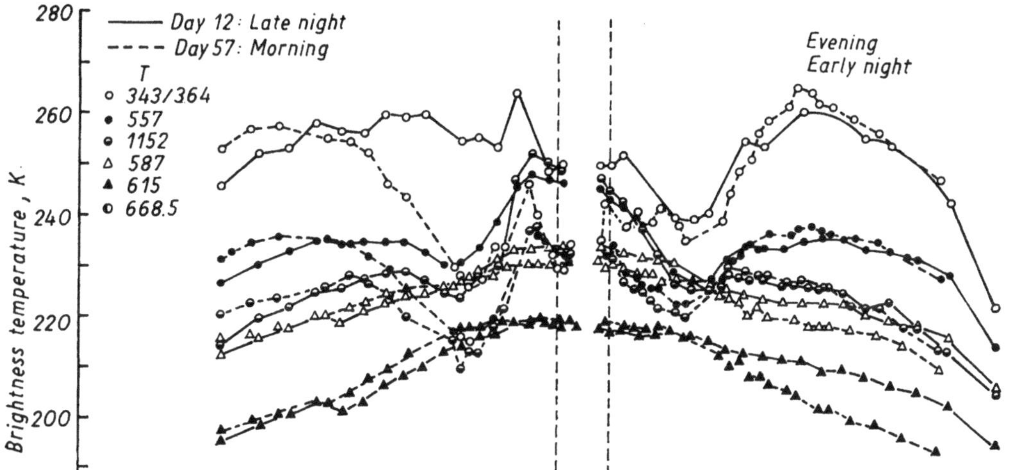

Fig. 6. Spectral brightness temperatures vs latitude for selected wave numbers
$T_{343/364}$ is the arithmetic mean of T_{343} and T_{364}
$T_{66.5}$ corresponds to about 1 hPa (86 km). For the other wave numbers see Fig. 4
The values are not normalized to a unitary measurement geometry. See comments on Fig. 1.
Position of measurements to solar time is labelled at the abscissa
for 23 Oct 1983 (day 12) —— and for 7 Dec 1983 (day 57) ----

infrared images of the Venusian atmosphere (Taylor et al. 1980), must naturally affect the atmospheric circulation in the immediate surrounding of the pole.

Fig. 1 reveals that the decrease in OIRF can be traced up to the 60° to 75° latitude belt although there is an extremely large variability at these latitudes due to the occurrence of the warm dipole and the cold collar.

Here are some further details on the atmospheric state at lower latitudes solely on the basis of the analysis of spectral brightness temperature data. As seen in Fig. 6 — where spectral brightness temperatures at selected wave numbers are plotted versus latitude for two typical days, i.e. day 12 (23 October 1983) and day 57 (7 December 1983)[1]) — significant differences in brightness temperatures between the two days occur already at about $\varphi = 50°$ at wave numbers with no (1152 cm^{-1} and 343/364 cm^{-1}) or rather small CO_2 opacity (557 cm^{-1}). Hence, these wave numbers mainly represent cloud features. Note that throughout the paper the differences in observational geometry (different zenith angles) have not been eliminated. The collar is much more pronounced in December with much colder brightness temperatures for wave numbers outside the ν_2-CO_2 band than in October, which is indicated by a sharp drop in the 'cloud channels', e.g., at 343/364 cm^{-1} by about 30 K. Evidence of a temperature inversion is available at both sides of the pole in December with $T_{557} < T_{615} < T_{587} \sim T_{343/364}$ in the morning and $T_{557} < T_{587}$ in the early night, whereas in October isothermal stratification is found in the evening $T_{557} \sim T_{587}$ but not indication of temperature inversion or isothermal stratification in the late night. The more distinct collar in December is also recognizable

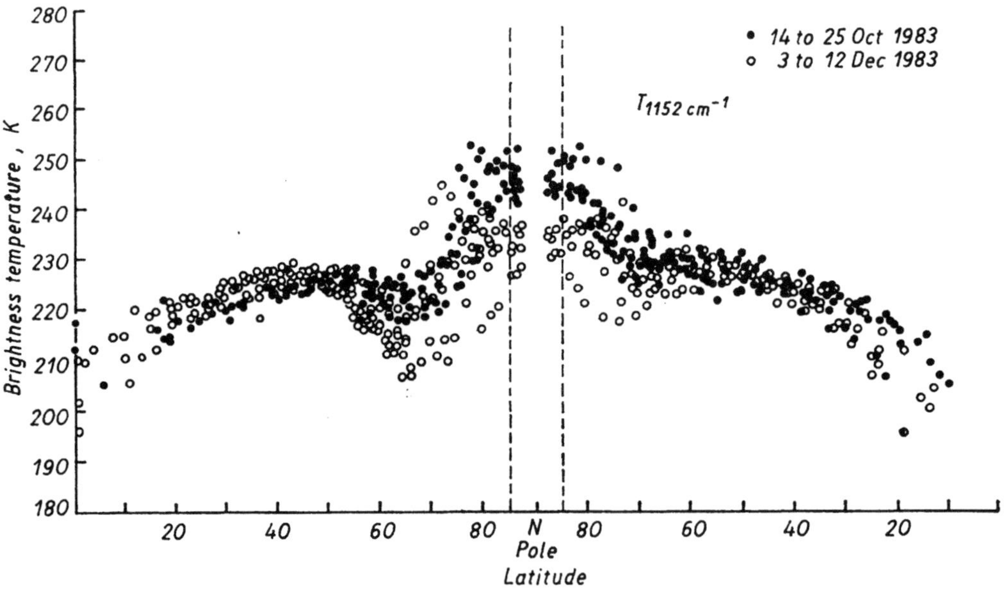

Fig. 7. Spectral brightness temperature T_{1152} representing cloud top temperature vs latitude for two periods
14 to 25 Oct 1983 (day 3 to 13), 13 to 14 Dec 1983 (day 53 to 64)

[1]) Other examples are given in Oertel et al. (1985) and Linkin et al. (1985).

for the cooler period at that time at the pole since the increase in spectral brightness temperature outside the ν_2- CO_2 band from the collar toward the pole is of the same order of magnitude for both periods. The Fig., as another version of Fig. 3, also shows that the isothermal temperature structure is at the pole in October below 45 hPa, in December, however, above that level. Remarkable differences in the atmospheric state between October and December can also be seen from the T_{615} curves roughly representing the 12 hPa (75 km) level. We find asymmetry to the pole in October with brightness temperatures about 10 K cooler in the late night than in the evening at 40° to 50° latitude as compared with symmetry in December between morning and early night resulting in a more distinct polar plateau for $\varphi \gtrsim 60°$ N in December. A survey of both periods from day 3 to 13 (14 to 25 October) and day 53 to 64 (3—14 Dec) is obtained from Fig. 7 showing the brightness temperature at 1152 cm^{-1} and thus the cloud top temperature as a function of latitude. Cloud top temperatures between the two periods tend to be different already at 60° despite the considerable scatter in the latitude belt where the dipole is situated. T_{1152} is at the pole consistently lower in December than in October with differences of 5 to 20 K. This cooling, observed during the mission is, however, not a small-scale feature restricted to latitudes $\varphi > 85°$ but a general trend at polar latitudes $\varphi \gtrsim 60°$ which superposes a diurnal (Venus day) wave. Additionally, Fig. 7 shows that possible minor uncertainties in locating the position of the measurements are negligible and do not affect the findings.

4. Outlook

It should be mentioned that the same effect of a cooling trend at the pole was also observed in January and February 1979 by the Pioneer Venus Orbital Infrared Radiometer and that it is of the same order of magnitude (Taylor 1987). It seems therefore, that the middle atmosphere of the polar regions exhibit changes in temperature and cloud properties lasting at least for more than one month. This mechanism could be an effective mean to set off imbalances in the energy budget of the Venusian atmosphere. Once again we emphasize that the effect described cannot be explained by the variability in polar haze since it is observed in a lower atmospheric layer and would need much more change in haze concentration as has been reported so far. The phenomenon described here is perhaps also be linked to irregularities in the rotation of the dipole, the secular decrease in period and/or the sinusoidal component (Schofield and Diner, 1983).

5. Conclusions

1. During the 64 days of the Venera-15 mission a decrease in outgoing infrared radiation flux (OIRF) was observed at the Venusian North Pole ($\varphi \gtrsim 85°$) to be about 30%. This decrease is traceable up to 60° N. In the 60° to 72° latitude belt the decrease in OIRF superimposes variations with local Venus time.
2. The decrease in OIRF is caused by simultaneous changes of temperature and cloud properties in the 60 to 70 km altitude range.
3. However, above the clouds at about 75 km an increase in temperature by about 5 K altogether is measured.

4. OIRF and cloud top temperature are at the pole closely correlated. Between the spectral radiance from 275 to 1600 cm^{-1} at zenith angle $\Theta \lesssim 10°$ and the spectral brightness temperature T_B at 1152 cm^{-1} exists the relation

$$L^{275-1600\mathrm{cm}^{-1}}(\Theta \leqq 10°) = 0.625 \times T_B{}^{1152\mathrm{cm}^{-1}}$$

with L in W $\times$ m^{-2} $\times$ sr^{-1} and T_B in Kelvin (Fig. 5).

References

[1] DOLLFUS, A., AURIERE, M., SANTER, R.: Wavelength dependence of polarization. 37. Regional observations of Venus. Astron. J. **84** (1979), 1419–1436.

[2] GOROSHKOVA, N. V., PETROVA, E. V.: Ob ocenke koncentracii aerozolejj v verkhnem sloe oblakov Venery. Kosm. Issled. **21** (1983), 106–110.

[3] KAWABATA, K., COFFEEN, D. L., HANSEN, J. E. et al.: Cloud and haze properties from Pioneer Venus polarimetry. J. Geophys. Res. **85** (1980) A 13, 8129–8140.

[4] KONDRAT'EV, K. A., KRUPENIO, N. N., SELIVANOV, A.: Planeta Venera. Gidroizdat, Leningrad, 1987.

[5] KSANFOMALITI, L. V.: Oblaka Venery: Raschjotnye svojjstva aerozol'nojj sernokislotnojj sredy v infrakrasnom teplovom diapazone. Astr. Zh. **54** (1977), 1110–1117.

[6] LINKIN, V. M., SCHÄFER, K., MACYGORIN, I. A. et al.: Infrared experiment aboard the automatic interplanetary stations Venera 15 and Venera 16. 5. Preliminary results of the analysis of fields of brightness temperatures and thermal fluxes. Veröffentlichungen des Forschungsbereichs Geo- und Kosmoswissenschaften, Heft 18, 1990.

[7] MACYGORIN, I. A.: Sun coupled waves in the field of the thermal radiation of Venus from Venera 15 infrared spectrometry. Veröffentlichungen des Forschungsbereichs Geo- und Kosmoswissenschaften, Heft 18, 1990.

[8] MOROZ, V. I., SPÄNKUCH, D., LINKIN, V. M. et al.: Venus spacecraft infrared radiance spectra and some aspects of their interpretation. Appl. Opt. **25** (1986), 1710–1719.

[9] OERTEL, D., SPÄNKUCH, D., JAHN, H. et al.: Infrared spectrometry of Venus from "Venera-15" and "Venera-16". Adv. Space Res. **5** (1985) 9, 25–36.

[10] SCHOFIELD, J. T., and DINER, D. J.: Rotation of Venus's polar dipole. Nature **305** (1983), 116–119.

[11] SPÄNKUCH, D., OERTEL, D., MOROZ, V. I. et al.: Venus spectra obtained from Venera spacekraft 15 and 16. In: FIOCCO, G. (Ed.): IRS'84: Current problems in atmospheric radiation. A. Deepak Publ., Hampton, Virginia, USA, 1984, 373–376.

[12] SPÄNKUCH, D., ZASOVA, L. V., SCHÄFER, K. et al.: Infrared experiment aboard the automatic interplanetary stations Venera 15 and Venera 16. 2. Preliminary results of temperature profile retrieval. Veröffentlichungen des Forschungsbereichs Geo- und Kosmoswissenschaften, Heft 18, 19 90.

[13] TAYLOR, F. W., BEER, R., CHAHINE, M. T. et al.: Structure and meteorology of the middle atmosphere of Venus: Infrared remote sensing from the Pioneer Orbiter. J. Geophys. Res. **85** (1980) A 13, 7963–8006.

[14] TAYLOR, F. W.: Discussion of the Symposium on Middle Atmosphere and clouds of Venus. June 22–26, 1987, Potsdam, German Democratic Republic.

[15] ZASOVA, L. V., SPÄNKUCH, D., MOROZ, V. I. et al.: Infrared experiment aboard the automatic interplanetary stations Venera 15 and Venera 16. 3. Some conclusions on the structure of clouds, based on the analysis of spectra. Veröffentlichungen des Forschungsbereichs Geo- und Kosmoswissenschaften, Heft 18, 1990.

12. IR Cooling Rates of Venus' Middle Atmosphere

R. HAUS and H. GOERING

Abstract

The present paper deals with thermal radiation of the atmosphere of Venus in the region 55 to 100 km. Radiances, radiation fluxes, and cooling rates are calculated from 0 to 2000 cm^{-1} taking into account cloud absorption and scattering for a mean standard temperature profile, and retrieved temperatures from the Fourier spectrometer experiment aboard Venera 15, respectively.

Concerning the gaseous component of the atmosphere, the 15 μm CO_2 band and the rotational band of H_2O (> 20 μm) act as significant energy sinks. Transmittances of these constituents are calculated with a line-by-line technique, transmittances as well as single-scattering albedo of the clouds are determined on the basis of a Mie algorithm. Cloud opacity clearly dominates at wavelengths outside the strong absorption bands of gases.

For the standard temperature profile T_0 there is a cooling around 95 km of about 25 deg/Earth day. At altitudes above 65 km clouds enhance cooling up to 3.5 deg/day in contrast to a pure gaseous atmosphere, whereas below 65 km they diminish cooling and give a significant heating below 63 km.

Above 60 km and in the wavelength region being under consideration scattering is of minor importance. Multiple-scattering can be neglected for energetical purposes.

Above 90 km equatorial latitudes show the strongest cooling rates, below 90 km the polar regions dominate infrared cooling with a strong maximum around 75 km.

1. Introduction

From the beginning of planetary exploration the thermal structure and the dynamical processes including their manifold interactions have been an essential research field of the physics of planetary atmospheres. The entire energy balance of a planet's atmosphere first of all is dominated by radiative processes. The absorption of solar energy is the only significant energy source. All energy conversions after all result in an energy release by radiation as the only quantitative significant energy sink.

In the middle atmosphere of Venus (100—55 km) the absorption of solar radiation by carbon dioxide, water vapor, and sulphur dioxide, which causes a heating of the atmosphere, and the thermal emission by carbon dioxide and water vapor, which mainly leads to a cooling, are the general terms of the radiative energy balance as well as below 80 km absorption and scattering of radiation in the sulphuric acid clouds.

The aim of the present paper is to analyse one component of the radiative energy balance, namely the thermal emission in the infrared spectral region below 2000 cm^{-1}

($\lambda > 5\ \mu$m). On the basis of detailed calculations on radiative transfer radiances, radiation fluxes, and cooling rates are determined taking into account cloud absorption and scattering.

Transmittances of carbon dioxide and water vapor have been calculated with an accurate line-by-line procedure. The method proposed by Fels and Schwarzkopf [1981] for finding transmittances of carbon dioxide for any meteorological conditions from three standard transmission matrices has been incorporated. Transmittances as well as single-scattering albedo of the clouds have been calculated on the basis of a Mie algorithm.

After a short presentation of the theoretical fundamentals to handle the problem (section 2.), section 3. deals with the input data concerning temperature models, mixing ratios, and particle densities, respectively. Results for cloud efficiencies, cross-sections and single-scattering albedo are presented, too. In section 4. the results of emission and cooling calculations will be shown and discussed.

2. Theorie

The infrared cooling rates for a plane-parallel, hydrostatic atmosphere can be obtained from

$$\frac{\mathrm{d}T}{\mathrm{d}t} = \frac{g}{c_p} \times \frac{\mathrm{d}F^{\mathrm{net}}}{\mathrm{d}p}, \tag{1}$$

where T: temperature [Kelvin], t: time [s], p: pressure [hPa = mb], g: acceleration due to gravity [cm/s^2], c_p: specific heat at constant pressure [J/kg $\times$ grd]. F^{net} is the frequency- (or wavenumber-) integrated net radiative flux [erg/s $\times$ cm^{-2} = 10^{-3} W $\times$ cm^{-2}] which is the difference of the upwelling and downwelling fluxes,

$$F^{\mathrm{net}}(p) = F^{+}(p) - F^{-}(p). \tag{2}$$

The fluxes can be obtained from the solution of the radiative transfer equation for a cloudy atmosphere, which for monochromatic radiances has the following differential form (the index ν has been omitted for abbreviation),

$$\mu \frac{\mathrm{d}I(u, \Omega)}{\mathrm{d}u} = I(u, \Omega) - \tilde{\gamma}(u) \times B(u) - \tilde{\omega}(u) \times J^{\mathrm{s}}(u, \Omega), \tag{3}$$

where I: radiance [erg/s $\times$ cm^2 $\times$ cm^{-1} $\times$ sr], B: Planck function [erg/s $\times$ cm^2 $\times$ cm^{-1} $\times$ sr], J^{s}: scattering source function [erg/s $\times$ cm^2 $\times$ cm^{-1} $\times$ sr], u: optical depth of extinction, $\tilde{\omega}$: albedo of the atmosphere, $\tilde{\gamma} = 1 - \tilde{\omega}$, $\Omega \equiv (\mu, \varphi)$, $\mu = \cos\vartheta$, ϑ: zenith angle, φ: azimuth angle.

At altitudes below 115 km LTE (local thermodynamic equilibrium) prevails and the Planck function B is enough for describing the emission character of atmospheric constituents.

The optical depths of gaseous compounds u°, aerosol absorption u^{a}, aerosol scattering u^{s} and the entire atmosphere u^{tot}, respectively, have been calculated as follows,

$$u^{0}(p_{\mathrm{r}}, p) = A \times \int_{p_r}^{p} k^{0}(\tilde{p}) \times \mathrm{d}\tilde{p}, \tag{4a}$$

$$u^{a}(p_r, p) = \int_{p_r}^{p} [1 - \omega_0(\tilde{p})] \times W(\tilde{p}) \times d\tilde{p}, \tag{4b}$$

$$u^{s}(p_r, p) = \int_{p_r}^{p} \omega_0(\tilde{p}) \times W(\tilde{p}) \times d\tilde{p}, \tag{4c}$$

$$u^{tot}(p_r, p) = u^{0}(p_r, p) + u^{a}(p_r, p) + u^{s}(p_r, p), \tag{4d}$$

$$W = 10^2 \times k^{ext} \times R \times T/(g \times p) \quad [mb^{-1}], \tag{5}$$

$$k^{ext} = k^{a} + k^{s} = 10^{-3} \times \pi \times \sum_{i=1}^{L} N_i \times \langle Q \times r^2 \rangle_i^{ext} \quad [km^{-1}], \tag{6}$$

$$\omega_0 = \frac{k^s}{k^{ext}} = \frac{\sum_{i=1}^{L} N_i \times \langle Q \times r^2 \rangle_i^{s}}{\sum_{i=1}^{L} N_i \times \langle Q \times r^2 \rangle_i^{ext}}. \tag{7}$$

In these equations k^0, k^a, and k^s are the gas absorption coefficient, aerosol absorption coefficient, and aerosol scattering coefficient, respectively. A is a conversion constant, R is the particular gas constant, $R^{Venus} = 0.1914\ W \times s \times grd^{-1} \times g^{-1}$, N_i the number density of aerosol particles [cm^{-3}] of mode i.

The quantities $\langle Q \times r^2 \rangle_i^{s}$, $\langle Q \times r^2 \rangle_i^{ext}$ are the scattering and extinction cross-sections [μm^2] of mode i, resulting from the corresponding efficiency factors Q^s and Q^{ext}. A numerical Mie-scattering programme, developed by the authors, had been used to compute these efficiency factors. ω_0 in equation (7) is the single-scattering albedo of the particle ensemble. It is connected with the albedo $\tilde{\omega}$ of the atmosphere by equation (8),

$$\tilde{\omega}(p_r) = \frac{du^s(p_r, p)/dp}{du^{tot}(p_r, p)/dp} = \frac{\omega_0(p_r)}{1 + \frac{1}{W(p_r)} \times \frac{du^0(p_r, p)}{dp}}. \tag{8}$$

p_r in (4) and (8) is the pressure of an arbitrary reference level. The scattering source function in equation (3) has the form

$$J^s(u, \Omega) = \frac{1}{4\pi} \times \int_{4\pi} \varphi(u, \Omega, \Omega') \times I(u, \Omega') \times d\Omega', \tag{9}$$

where $\varphi(u, \Omega, \Omega')$ is the phase function of aerosol scattering. Under the assumption of isotropic internal sources equation (9) can be written as

$$J^s(u, \pm\mu) = \frac{1}{2} \times \int_{0}^{1} \varphi(u, \pm\mu, \mu') \times I(u, \mu') \times d\mu'. \tag{10}$$

Now the source function can be transformed to a finite sum of Legendre polynomials (Sekera and Stowe [1973]),

$$J^s(p, \pm\mu) = \sum_{n=0}^{N} a_n(p) \times P_n(\pm\mu) \times J_n(p) \tag{11}$$

with

$$\varphi(\mu) = \sum_{n=0}^{N} a_n \times P_n(\mu). \tag{12}$$

The a_n are the coefficients of multiple-scattering, the meaning of J_n is explained later on.

There are many different phase functions and a lot of work has been done in literature to claculate them or compare results for different functions. In the present paper, the Henyey-Greenstein function has been used,

$$\varphi(\mu) = \frac{1 - g^2}{(1 + g^2 - 2 \times g \times \mu)^{3/2}}, \tag{13}$$

where g is the so-called asymmetry factor. The corresponding coefficients a_n have the form

$$a_n = (2 \times n + 1) \times g^n. \tag{14}$$

The first term of (12) is just the contribution of single-scattering. The contributions of multiple-scattering are small for as long as g is sufficient small, and this holds true in the spectral region $\nu < 2000\ \text{cm}^{-1}$.

After Sekera and Stowe [1973], the radiance can be split up into two components I^{E} and I^{s},

$$I(u, \Omega) = I^{\text{E}}(u, \Omega) + I^{\text{s}}(u, \Omega). \tag{15}$$

The corresponding solutions of the radiative transfer equation (3) in pressure coordinates have the following form,

$$I^{\text{E}}(p_\text{r}, \pm\mu) = \int_{p_\text{r}}^{p_\text{g}} \tilde{\gamma}(p) \times B(p) \times \exp\left[-\frac{u^{\text{tot}}(p_\text{r}, p)}{\mu}\right] \times \frac{1}{\mu} \times \frac{\text{d}u^{\text{tot}}(p_\text{r}, p)}{\text{d}p} \times \text{d}p$$
$$+ \tilde{\gamma}(p_\text{g}) \times B(p_\text{g}) \times \exp\left[-\frac{u^{\text{tot}}(p_\text{r}, p_\text{g})}{\mu}\right], \tag{16}$$

$$I^{\text{s}}(p_\text{r}, \pm\mu) = \int_{p_\text{r}}^{p_\text{g}} \tilde{\omega}(p) \times J^{\text{s}}(p, \pm\mu) \times \exp\left[-\frac{u^{\text{tot}}(p_\text{r}, p)}{\mu}\right] \times \frac{1}{\mu} \times \frac{\text{d}u^{\text{tot}}(p_\text{r}, p)}{\text{d}p} \times \text{d}p$$
$$+ \tilde{\omega}(p_\text{g}) \times J^{\text{s}}(p_\text{g}, \pm\mu) \times \exp\left[-\frac{u^{\text{tot}}(p_\text{r}, p_\text{g})}{\mu}\right]. \tag{17}$$

p_r is the reference level, p_g is the pressure of lower (ground) and upper (top) boundary, respectively.

The radiative fluxes for finite spectral intervals $\Delta\nu$ can be written as

$$F^{\pm}_{\Delta\nu}(p_\text{r}) = 2 \times \pi \times \int_0^1 I_{\Delta\nu}(p_\text{r}, \pm\mu) \times \mu \times \text{d}\mu, \tag{18}$$

$$F^{\pm\text{E}}_{\Delta\nu}(p_\text{r}) = \pi \times \int_{\tau_{\Delta\nu}^*(p_\text{r}, p_\text{g})}^{1} \tilde{\gamma}_{\Delta\nu}(\tilde{p}) \times B_{\Delta\nu}(\tilde{p}) \times \text{d}\tau^*_{\Delta\nu}(p_\text{r}, p) + \text{boundary}, \tag{19}$$

$$F^{\pm\text{s}}_{\Delta\nu}(p_\text{r}) = 4 \times \pi \times \sum_{n=0}^{N} \int_{p_\text{r}}^{p_\text{g}} a_n(p) \times F_n(p_\text{r}, p) \times J_n(p) \times \omega_0(p) \times W(p) \times \text{d}p \tag{20}$$

with

$$\tau^*_{\Delta\nu} = 2 \times E_{i_3}[u^{\text{tot}}_{\Delta\nu}(p_r, p)] \quad \text{(flux transmittance)}, \tag{21}$$

$$E_{i_3}(x) = \int_0^1 \exp\left[-\frac{x}{\mu}\right] \times \mu \times \mathrm{d}\mu \quad \text{(exponential integral of 3. order)}, \tag{22}$$

$$F_n(p_r, p) = \frac{1}{2} \times \int_0^1 P_n(+\mu) \times \exp\left[-\frac{u^{\text{tot}}_{\Delta\nu}(p_r, p)}{\mu}\right] \times \mathrm{d}\mu. \tag{23}$$

The $J_n(p)$ in equation (20) have to be calculated by an iterative method,

$$J_n(p) = \sum_{i=1}^{L} J_n^{(i)} \left[1 - \frac{a_n(p) \times \omega_0(p)}{z_n + 1}\right]^{-1}, \tag{24}$$

$$J_n^{(0)} = f_n, \tag{25a}$$

$$J_n^{(i)} = L_n\{J_m^{(i-1)}\}, \qquad i \geqq 1, \tag{25b}$$

where the function f_n has the form

$$f_n(p_r) = \int_{p_r}^{p_t} F_n(p_r, p) \times \frac{\mathrm{d}}{\mathrm{d}p} [B(p) \times \bar{\gamma}(p)] \times \mathrm{d}p + \int_{p_r}^{p_b} F_n^*(p_r, p) \times \frac{\mathrm{d}}{\mathrm{d}p} [B(p) \times \bar{\gamma}(p)] \times \mathrm{d}p + \underbrace{B(p_r) \times \gamma_0(p_r)}_{\text{for } n\, =\, 0 \text{ only}} \tag{26}$$

and the operator L_n is defined by

$$L_n\{J_m(p_r)\} = \sum_{m=0}^{N} \left\{\int_{p_r}^{p_b} F_{m,n}(p_r, p) \times \frac{\mathrm{d}}{\mathrm{d}p} [a_m(p) \times \bar{\omega}(p) \times J_m(p)] \times \mathrm{d}p + \int_{p_r}^{p_t} F^*_{m,n}(p_r, p) \times \frac{\mathrm{d}}{\mathrm{d}p} [a_m(p) \times \bar{\omega}(p) \times J_m(p)] \times \mathrm{d}p\right\} \tag{27}$$

with

$$F_{m,n}(p_r, p) = \frac{1}{2} \times \int_0^1 P_m(+\mu) \times P_n(+\mu) \times \exp\left[-\frac{u^{\text{tot}}_{\Delta\nu}(p_r, p)}{\mu}\right] \times \mathrm{d}\mu. \tag{28}$$

The starts in equations (26) and (27) indicate that $P_n(+\mu)$ must be replaced by $P_n(-\mu)$, p_t is the top pressure, p_b the pressure of the lower boundary (ground).

From these equations it is clear, that the calculation of infrared cooling rates of the Venus atmosphere is very complicated. Above all it demands a high expense of computer time, particularly as for as the optical depths and transmission functions of gas absorption have been determined with a line-by-line integration technique. The necessity of highly accurate transmittances results from the fact, that the radiation fluxes contain the pressure derivative of transmittance and the transmittance itself varies only weakly with pressure, especially in the region above the clouds. Details of the numerical calculation of infrared transmittances as well as cooling rates for a clear atmosphere have been published elsewhere (Haus [1985, 1986]).

3. Input data

Figure 1 shows some of the temperature profiles used in the calculations. The thick solid line is the standard profile T_0, the thick broken lines indicate the profiles $T_0 \pm 30$ K necessary for the calculation of carbon dioxide transmittances after Fels and Schwarz-

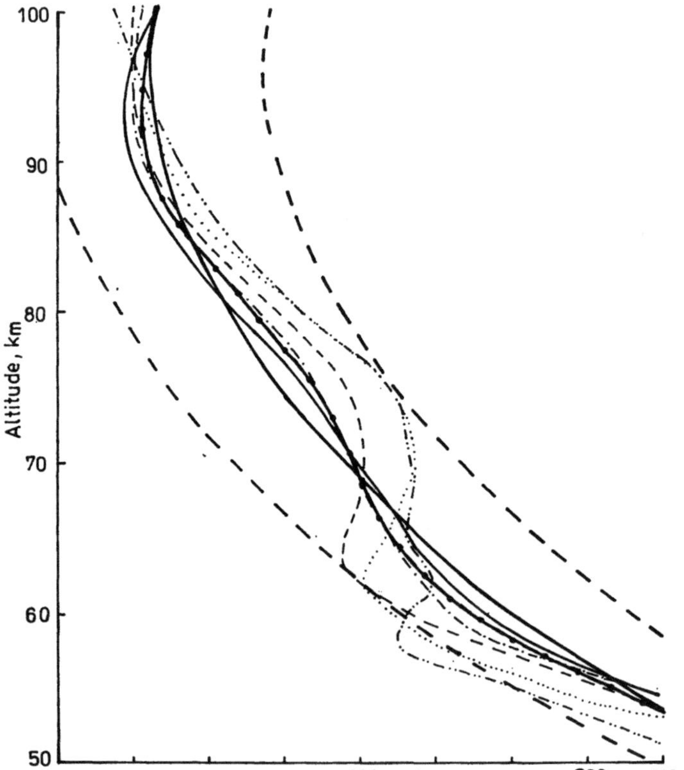

Fig. 1. VIRA temperature models of the Venus middle atmosphere (—) 0°–30°, (-··-) 45°, (---) 60°, (.....) 75°, (-···-) 85° (——) standard profile T_0, (- - -) $T_0 \pm 30$ K, (-·-) profile after Schofield, Taylor (1983)

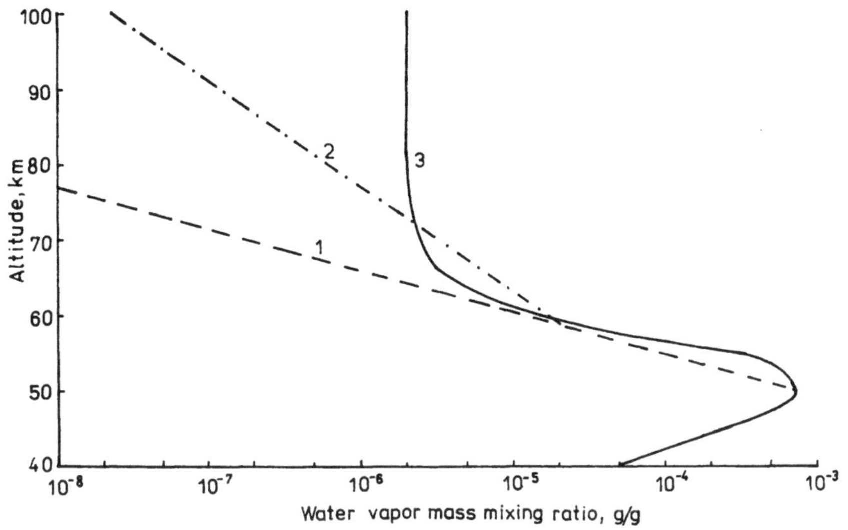

Fig. 2. Water vapor models of the Venus middle atmosphere
1: Knollenberg, Hunten [1980], 3: Schofield, Taylor [1982] (OIR night-side), 2: "mean"

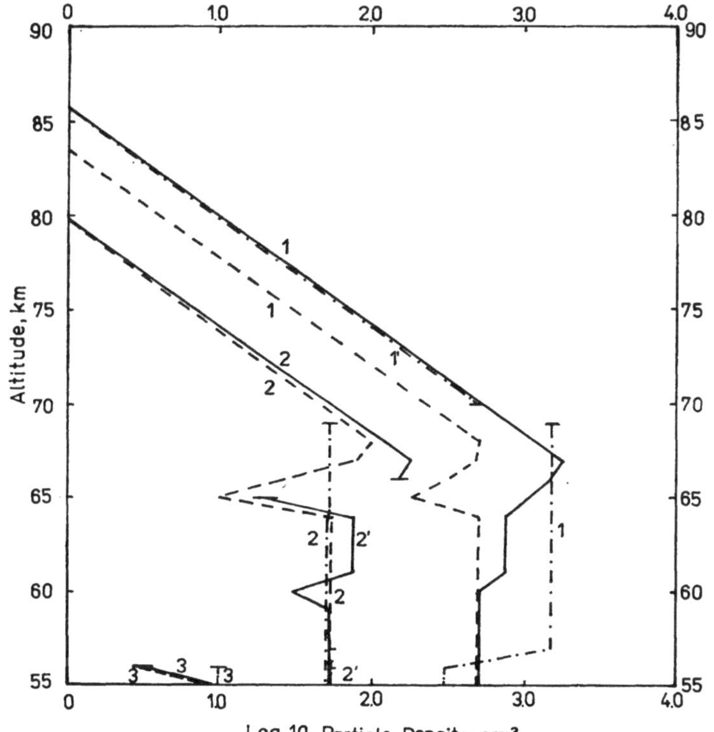

Fig. 3. Model distributions of aerosols in the Venus clouds (---) model 1 after Zasova [1985], (—) model 2 after Zasova [1985], (–·–) model 3 after Knollenberg, Hunten [1980]. The numbers refer to the single modes used

kopf [1981]. The solid line with points is the mean temperature model of Schofield, Taylor [1983]. The other lines correspond to the VIRA profiles for the latitudes indicated in the figure. With the exception of the polar region at altitudes between 60 and 55 km all these profiles are located within the temperature range put up by $T_0 \pm 30$ K. The profiles for latitudes greater than 50° exhibit clear inversions.

The mass mixing ratio of CO_2 used in all calculations and for all latitudes is 0.98.

Figure 2 shows three different water vapor profiles. Model 1 corresponds to the "dry" profile of Knollenberg and Hunten [1980]. Model 3 is the OIR night-side-profile of Schofield et al. [1982]. Model 2 can be regarded as mean profile between the other ones above 70 km. A latitude dependence of water vapor mixing ratio has not been considered till now.

For the Venus clouds three different aerosol models have been used as it is shown in Figure 3. Models 1 and 2 correspond to the particle densities from Zasova [1985] for a starting model (1) and an adapted model (2), respectively. Both models contain particle sizes in four different modes. All the modes have been described by a Log-normal-size distribution function. Particles of mode 3 with a mean diameter of about 3.85 μm are

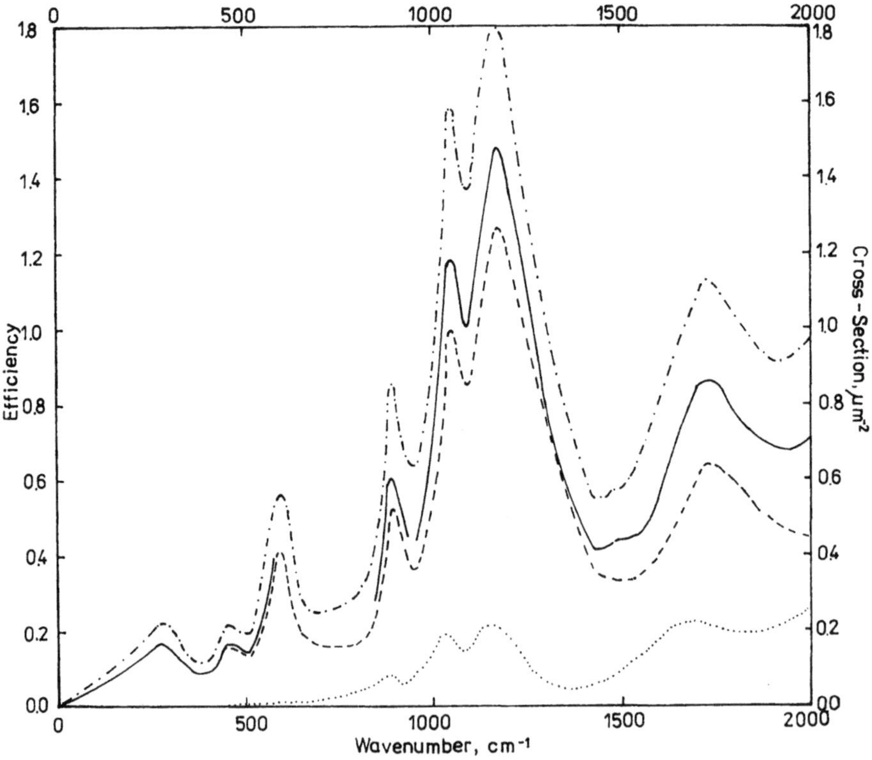

Fig. 4. Spectral dependence of particle efficiencies Q and extinction cross-section $\langle Q \times r^2 \rangle^{ext}$ from Mie theory for cloud model 2 and mode 2 — particles (—) Q^{ext}, (---) Q^{abs}, (....) Q^{sca}, (-·-) $\langle Q \times r^2 \rangle^{ext}$

assumed to exist only below 56 km, therefore, they have only little influence on the results in the altitude region of interest. Model 3 is the density model proposed by Knollenberg and Hunten [1980]. It also contains four different modes with different size distribution functions in each case. Mode 1′ characterizes the upper haze above the main cloud deck.

As it has been mentioned above, a Mie-scattering programme has been developed which calculates the efficiency factors Q^s, Q^a, Q^{ext}, and the corresponding cross-sections for scattering, absorption, and extinction, respectively. The data for the complex refractive index for a 75% H_2SO_4 aerosol have been taken from Palmer and Williams [1975], Jones [1976].

Figure 4 shows the three efficiency factors and the extinction cross-section for mode 2 particles of cloud model 2. Significant maxima are found at 280, 600, 890, 1050, 1170 and 1720 cm^{-1}. In the spectral region considered aerosol absorption is the deminating part in the aerosol extinction.

The corresponding single-sccattering albedo ω_0 of the particle ensemble as well as the resulting albedo $\tilde{\omega}$ of the atmosphere for the cloud model 2 and the standard temperature model T_0 are shown in Figure 5 for two different altitudes (70 km, 59 km). The altidude dependence of the quantities ω_0 and $\tilde{\omega}$ results from the use of multimodal aerosol distributions.

At 70 km, where the clouds cause the strongest deviation of cooling rates from those for a clear atmosphere, as will be shown later on, the albedo ω_0 varies from 0 to 0.4, significant maxima are met at 540, 840, 990, and 1600 cm^{-1}.

From the lines for $\tilde{\omega}$ the influence of gas absorption bands can clearly be seen. Where as the rotational band of water vapor for $\nu < 500$ cm^{-1} causes only weak changes of $\tilde{\omega}$ in comparison to ω_0, the strong 15 μm carbon dioxide band completely covers the influence of clouds in the spectral region from 580 to 570 cm^{-1}. With other words, in the centre of the 15 μm band the clouds can be neglected.

From Figure 5 another quantity can be interpreted, the asymmetry parameter $\bar{g}$ (mean value from all modes), which determines the role of multiple-scattering (cp. equation (14)). From the small values of $\bar{g}$ it is evident, that multiple-scattering can not have a great influence on the resulting cooling rates, especially in the regions of the gas absorption bands.

4. Results

In Figure 6 the spectral dependence of atmospheric transmittances above the 35 mb (or hPa) level which corresponds to about 70 km is presented for clouds, carbon dioxide, water vapor and total atmosphere. Cloud opacity clearly dominates at wavelengths outside the strong 15 μm carbon dioxide band. Test calculations have shown, that the influence of the 10 μm carbon dioxide bands and the 6.3 μm water vapor band on the cooling rates can be neglected. This is not the case concerning the rotational band of water vapor below 400 cm^{-1}, because this band is a very broad one. At wavelengths near the centre of the 15 μm band, on the other hand, the transmittance rapidly goes to zero and at the chosen altitude and below neither clouds nor carbon dioxide contribute to the cooling of the atmosphere.

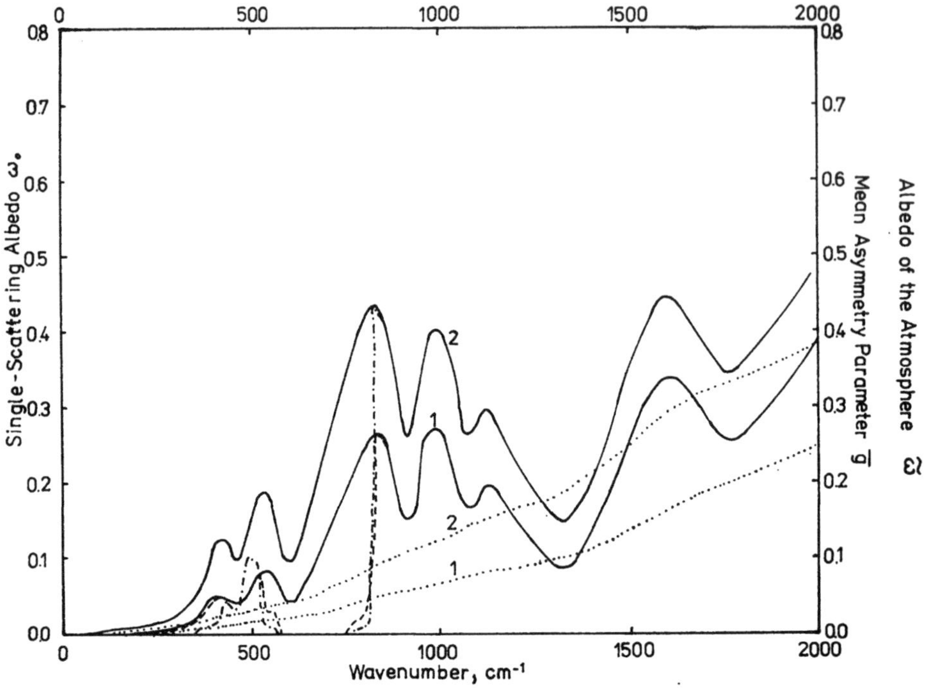

Fig. 5. Spectral dependence of single-scattering albedo ω_0 (—), albedo of the atmosphere $\tilde{\omega}$ (--- 70 km, ··· 59 km), and mean asymmetry parameter $\bar{g}$ (...) for cloud model 2 and standard temperature model T_0; 1: 70 km, 2: 59 km

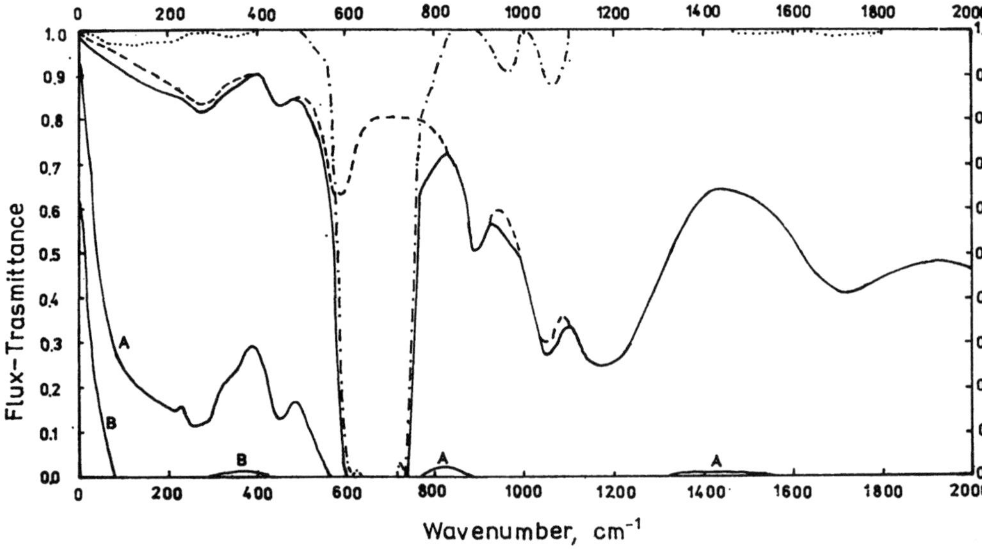

Fig. 6. Spectral dependence of atmospheric transmittance above the 35 mb or hPa pressure level for cloud (---), H_2O (...), CO_2 (-·-), and total (—). The solid lines A and B show the total transmittance above the 250 mb or hPa, and 650 mb or hPa pressure level, respectively

Models: Cloud 2, H_2O 1, standard temperature T_0

Figure 7 shows the spectral dependence of the top emission of Venus for the cloud model 2 and H_2O model 1, but only for the component F^E of the radiative flux. The solid line is the total emission, the dot-dashed line the emission of the clouds alone. The influence of the 15 μm band in the region from 570 to 850 cm^{-1} is clearly perceptible in this figure, too. The dashed line is the Planck curve for an effective temperature of 230 K.

Figure 8 illustrates in an obvious form the contribution of single absorbers to the emission to space in dependence on altitude. Above 82 km the emission and consequently the cooling rate, too, are exclusively determined by carbon dioxide, below 75 km the clouds dominate. Below 60 km water vapor has a not negligible influence on the emission.

In Figure 9 infrared cooling rates of the Venus atmosphere at altitudes between 55 and 80 km are illustrated on the basis of cloud model 2 and H_2O model 1. From this figure a plenty of informations can be won:

1. At altitudes between 80 and 65 km clouds enhance cooling up to 3.5 deg/day in contrast to a pure gaseous atmosphere. This is mainly due to aerosol absorption as well as aerosol scattering in the component Q^E as can be seen from a comparison of the three lines for Q^E. Below 65 km clouds tend to deminish cooling and give a heating below 63 km. For the conditions selected in this figure, this is mainly due to the influence of component Q^S. For other (lower) temperature profiles cloud absorption can also produce a significant heating of the atmosphere.

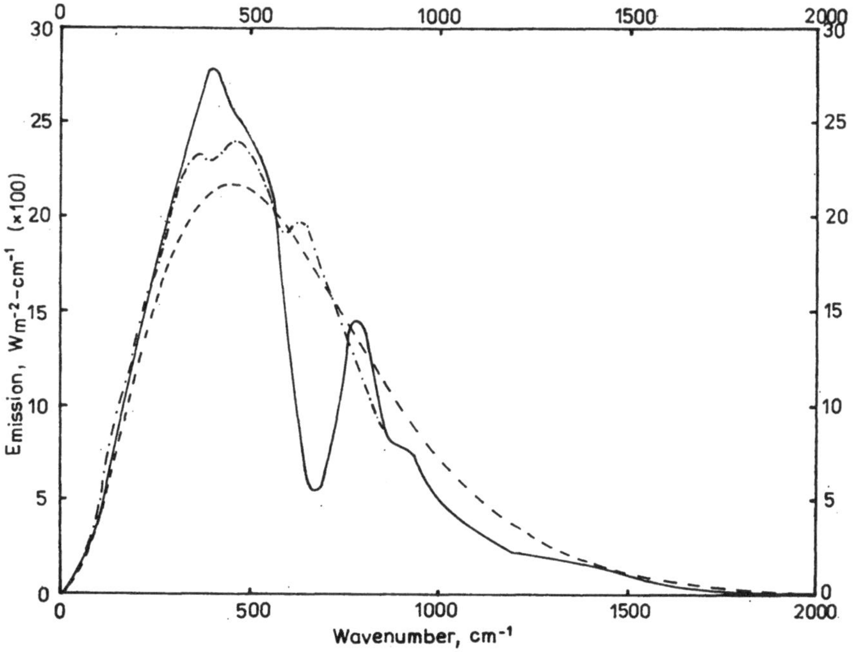

Fig. 7. Spectral dependence of Venus flux-component F^E emission (—) total, (-·-) cloud alone (- - -) Planck function for 230 K
Models: Cloud 2, H_2O 1, standard temperature T_0

2. The energetical role of cloud scattering is small above 60 km because the scattering effects of the components Q^E and Q^S compensate each other. The assumption that scattering can be neglected (i.e. $Q^S = 0$, and $\tau^S = 1$, $\bar{\gamma} = 1$ in Q^E: line (...)) yields maximum cooling errors of 1 deg/day at these altitudes with absolute values being smaller above and larger below 71.5 km, respectively. Taking into consideration the uncertainties of input data, especially of aerosol densities and size-distribution functions, these cooling rate errors can thouroughly be tolerated.
3. Multiple-scattering (line --^{1}) yields only a small modification of single-scattering (line --^{2}) results above 60 km and can be neglected. This is of great practical importance, because the expensive calculation of the scattering source function J^S turns out to be not necessary.

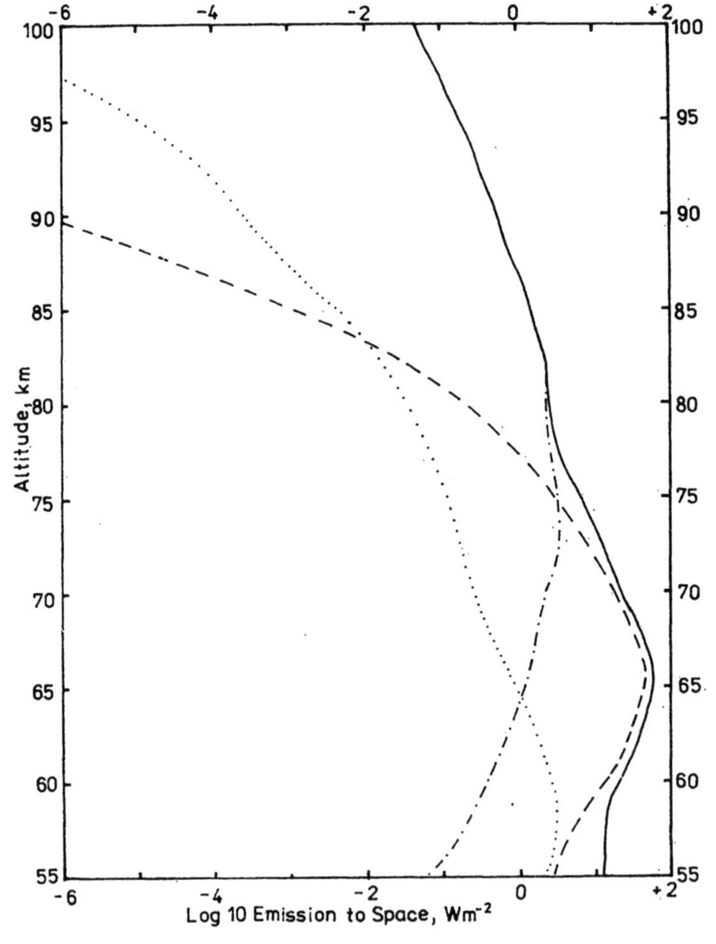

Fig. 8. Altitude dependence of Venus flux component F^E emission to space below 2000 m^{-1} (---) cloud, (---) CO_2, (---) H_2O, (—) total
Models: Cloud 2, H_2O 1, standard temperature T_0

4. With the assumption of full absorption, i.e. additional absorption of the scattered part of radiation at the place of scattering ($\tilde{\gamma} = 1$ in Q^E, $Q^S = 0$: line (---)) a stronger cooling results than for real conditions. This approximation gives cooling errors being larger than 1 deg/day, but owing to it's simplicity it can be used for an estimation of the cooling rates, especially in the region above 68 km.
5. Different water vapor profiles have only a small influence on the resulting cooling rates.

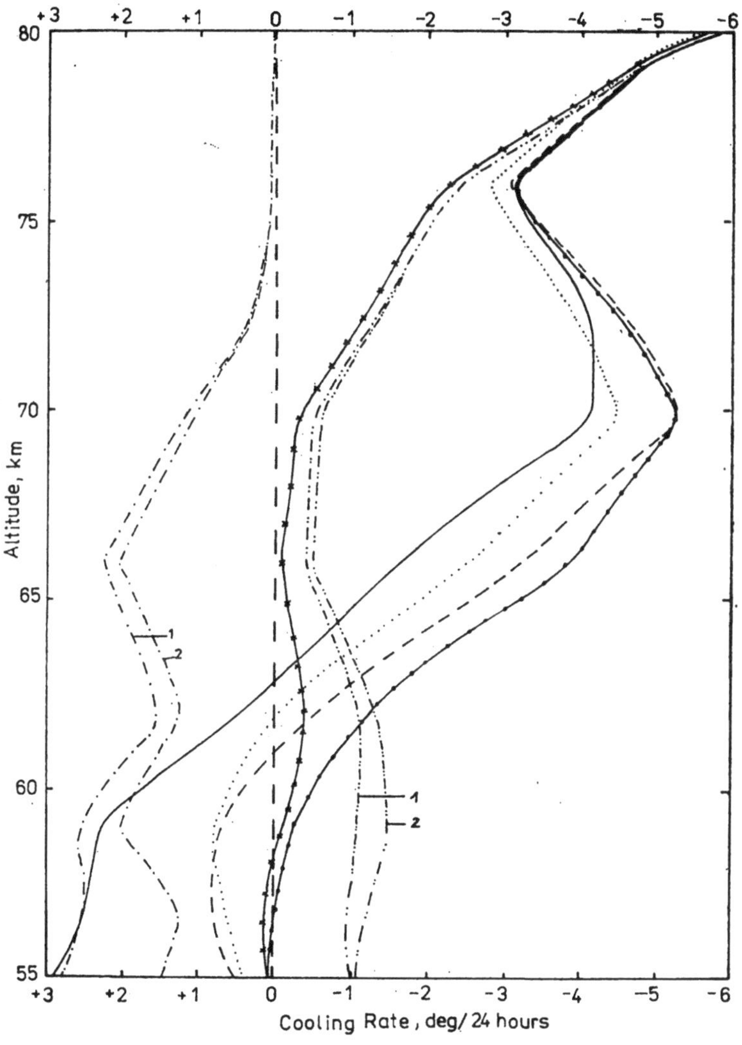

Fig. 9. Infrared cooling rates of Venus middle atmosphere below 2000 cm^{-1} for different physical conditions

Models: Cloud 2, H_2O 1, standard temperature T_0

(—) $Q^{Total} = Q^E + Q^S$ (ms), (–·–) $Q^E(\gamma \neq 1)$, (---) Q^E ($\gamma = 1$, full absorption), (...) Q^E ($\gamma = 1$, without scattering), (-··) Q^S (1: multiple-scattering, 2: single-scattering), (-×-) pure CO_2, (-···) $CO_2 + H_2O$ (1: H_2O model 1, 2: H_2O model 3)

Figure 10 shows cooling rates in the region from 100 to 55 km for the three cloud models being under consideration (standard temperature model T_0 water vapor model 1). It is clear from this figure, that different cloud models can produce very different results for the cooling rates (up to 5 deg/day). This fact supports the above mentioned statement that an expensive treatment of scattering algorithms at wavelengths above 5 µm is not necessary at present.

Figure 11 illustrates the two-dimensional (altitude-latitude dependent) cooling rates for the VIRA temperature model neglecting the latitudinal dependence of cloud distribution. No reliable data have been available for the latter.

While in the region above 95 km a strong latitudinal dependence of cooling occurs with maximum values at 100 km of about 40 deg/day at the equator and minimum values at 97 km of about 10 deg/day at the pole, there are only relative small latitudinal variations between 95 and 80 km. In the region below 80 km down to 67 km polar latitudes are characterized by strong cooling rates with maximum values of more than 20 deg/day between 75 and 70 km, while the cooling at equatorial latitudes is relative small with 4 to 6 deg/day. Below 65 km the latitudinal dependence of cooling is small. Below 62 km there is a heating for all latitudes up to 6 deg/day at the pole.

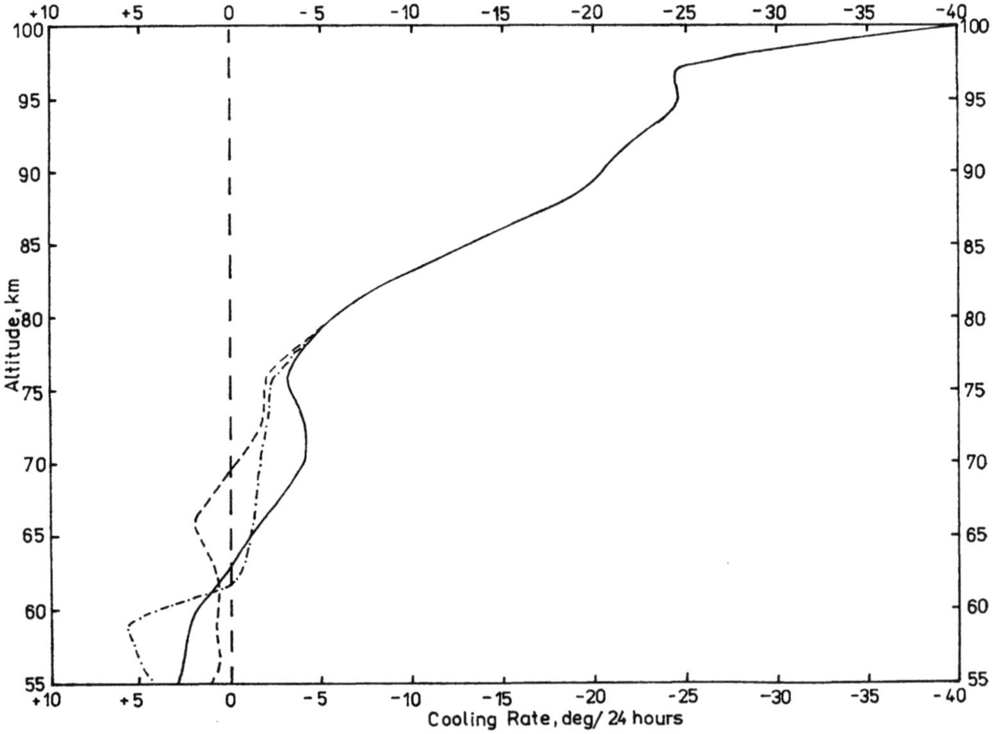

Fig. 10. Infrared cooling rates of Venus middle atmosphere below 2000 cm^{-1} for different cloud models, H_2O model 1, and temperature model T_0
(—) cloud model 2, (- - -) cloud model 1, (-·-) cloud model 3

These two-dimensional cooling or heating results closely follow the temperature model described in Figure 1. The cooling maximum at 100 km for equatorial latitudes corresponds to the temperature maximum there. The cooling maximum at around 72 km for polar latitudes is due to the large polar temperatures at those levels. The strong temperature inversion for latitudes above 60 deg is responsible for the heating maximum in the region between 60 and 55 km.

In Figure 12 the corresponding two-dimensional cooling rates for the retrieved temperatures from the Fourier spectrometer experiment aboard Venera 15 are shown. The most striking feature of this figure in contrast to VIRA is a stronger latitudinal variation between 95 and 80 km as well as smaller cooling rates at polar latitudes between 75 and 70 km.

Figure 13 shows the deviations δQ of the local cooling rates from Figure 12 from the hemispherical mean H,

$$\delta Q(\varrho, z) = Q(\varrho, z) - H(z), \tag{29}$$

$$H(z) = \int_0^{\pi/2} Q(\varrho, z) \times \cos \varrho \times d\varrho . \tag{30}$$

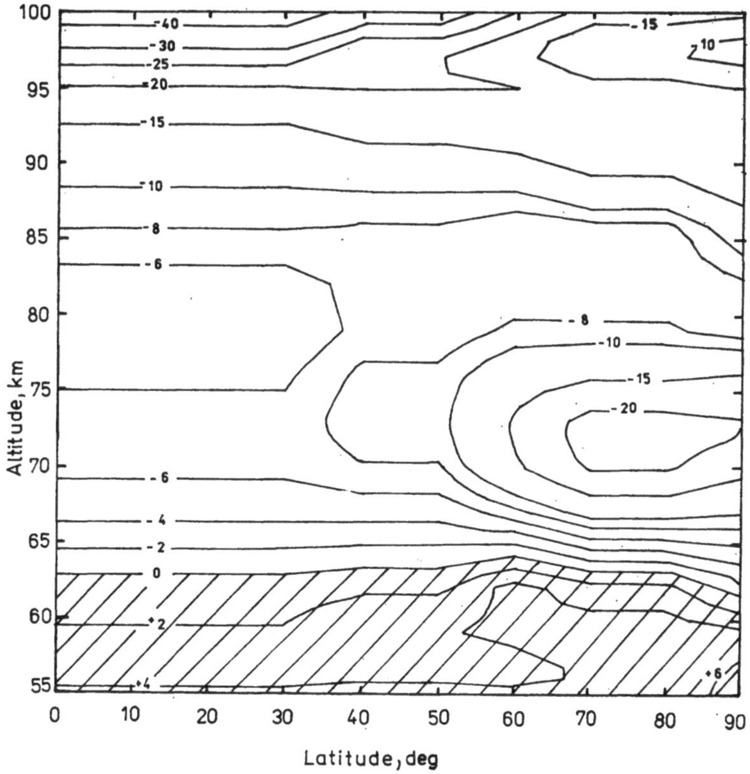

Fig. 11. Infrared cooling rates of Venus middle atmosphere below 2000 cm^{-1} for the VIRA temperature model considering cloud absorption as well as scattering Models: Cloud 2, H_2O 1

In the shaded regions the local cooling is less than the hemispherical mean. This holds true for equatorial latitudes above 65 km except a small region around 85 km, for polar latitudes above 93 km, and for middle/polar latitudes below 65/60 km, respectively.

Such a picture for the cooling rates alone has a limited significance only, but for the net heating (or cooling) rates is it very interesting, since from this a lot of informations can be found on circulation mechanisms within the atmosphere. Calculations of solar heating rates are underway.

5. Concluding remarks

It has been shown, that for the spectral region below 2000 cm^{-1} clouds have an influence up to 3.5 deg/day on the radiative cooling rates of Venus. Cloud scattering, particularly multiple-scattering is of minor importance above 60 km, because uncertainties regarding cloud composition do have a stronger influence. Making available reliable latitude-dependent cloud distributions, therefore, seems to be one of the most urgent tasks for the next future.

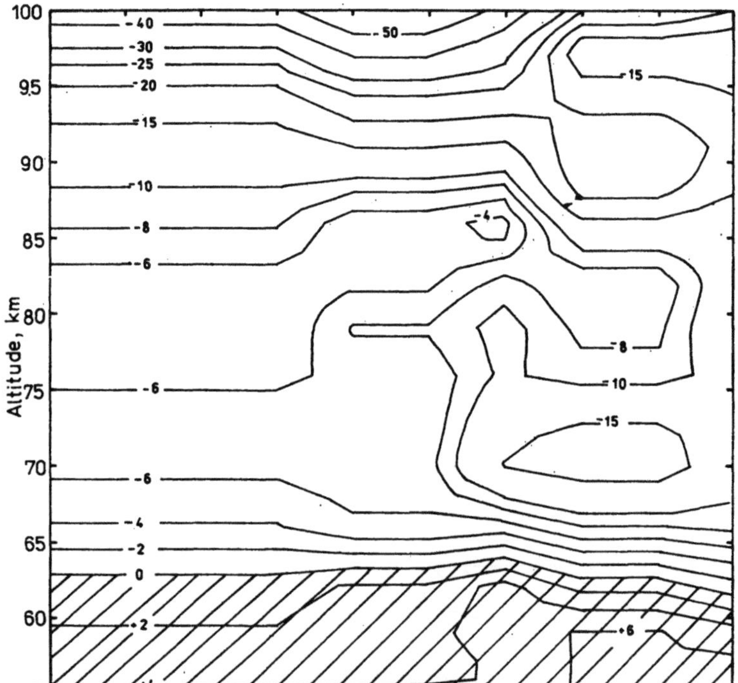

Fig. 12. Infrared cooling rates of Venus middle atmosphere below 2000 cm^{-1} for the VENERA 15 temperature model considering cloud absorption as well as scattering
Models: Cloud 2, H_2O 1

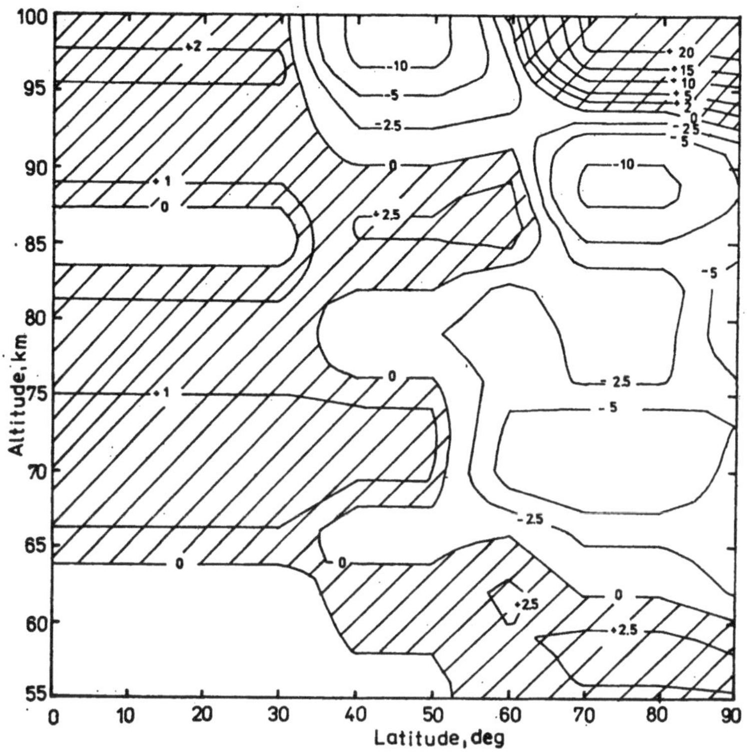

Fig. 13. Deviations of the local cooling rates from the hemispherical mean for the VENERA 15 temperature model

References

FELS, S. B. and SCHWARZKOPF, M. D.: An efficient, accurate algorithm for calculating CO_2 15 µm band cooling rates. J. Geophys. Res. **86** (1981), 1205.

HAUS, R.: Spectral highly resolved transmittion functions for CO_2 and infrared cooling rates of the Earth's atmosphere at altitudes between 30 and 110 km (in German). Doctoral thesis, Berlin, 1985.

HAUS, R.: Accurate cooling rates of the 15 µm CO_2 band-comparision with recent parameterizations. J. Atm. Terr. Physics **48** (1986), 559.

JONES, A. D.: Optical constants of sulfuric acid in the far infrared. J. Quant. Radiat. Transfer **16** (1976), 1017.

KNOLLENBERG, R. G. and HUNTEN, D. M.: The microphysics of the clouds of Venus: Results of the Pioneer Venus particle size spectrometer experiment. J. Geophys. Res. **85** (1986), 8039.

PALMER, K. F. and WILLIAMS, D.: Optical constants of sulfuric acid: Application to the clouds of Venus? Appl. Opt. **14** (1975), 208.

SCHOFIELD, J. T., TAYLOR, F. W. and MCCLEESE, D. J.: The global distribution of water vapor in the middle atmosphere of Venus. Icarus **52** (1982), 263.

SCHOFIELD, J. T. and TAYLOR, F. W.: The solar and thermal radiation balance of the middle atmosphere. VIRA workshop of the International Union of Geophysics and Geodesy, Hamburg, 1983.

Sekera, Z. and Stowe, L. L.: Effects of particulate matter on the radiances of the terrestrial infrared radiation. Beitr. Phys. d. Atm. **46** (1973), 101.

Zasova, L. V., Spänkuch, D., Moroz, V. I. et al.: Infrared experiment aboard the automatic interplanetary stations Venera 15 and Venera 16. 3. Some conclusions on the structure of clouds, based on the analysis of spectra. Veröffentlichungen des Forschungsbereichs Geo- und Kosmoswissenschaften, Heft 18, 1990.

13. Sulfur Dioxide and its Variations

LARRY W. ESPOSITO

Abstract

Ultraviolet spectroscopy from the Pioneer Venus Orbiter shows a decline in the cloud top abundance of SO_2 from about 100 ppb to about 10 ppb in the period 1978—1986. A consistent decline in polar haze has occurred over the same period with the correlation coefficient between these two observables of $r = 0.8$. Star calibrations determine the instrument sensitivity to within 10%, which rules out the possibility that this is an instrumental effect. Systematic errors could increase the SO_2 abundance to twice the inferred values in later orbits. Tracking of SO_2 features and power spectral analysis give rotation periods for the longer lived features of 3.6—5.2 days, consistent with cloud tracked winds observed at other wavelengths. The behavior of SO_2 and polar haze can be plausibly explained by episodic injection of SO_2 into the cloud top regions, for example, by active volcanism.

Introduction

The discovery of sulfur dioxide at the Venus cloud tops in 1978 (Barker 1979, Conway et al. 1979, Stewart et al. 1979) has profoundly affected our understanding of the Venus clouds and their chemistry (e.g. Winick and Stewart 1980; Young and DeMore 1982). Additionally, the variability of the SO_2 has led Esposito (1984) to infer active volcanoes on Venus. In this paper we describe the data and model procedures which establish the long and short term variability of sulfur dioxide on Venus as observed by the ultraviolet spectrometer on the Pioneer Venus Orbiter.

Experiment description

The Pioneer Venus Orbiter (PVO) began observations of Venus in 1978 (Colin, 1980). The Orbiter payload includes an Ultraviolet Spectrometer (UVS), which regularly carries out remote observations of the cloud-top region of the atmosphere (Stewart et al., 1979, Stewart, 1980, Esposito et al., 1979, Esposito, 1980). For the purpose of observing SO_2, two modes are employed. In *spectral mode*, the spectrometer grating is advanced as the spacecraft spin carries the instrument optical axis across the planet (the PVO is a spin-stabilized spacecraft). In *wavelength mode* the grating position is fixed during each scan: this yields a limb to limb profile of planetary brightness. By alternating among several wavelengths in this mode, the UVS observations create a multi-color

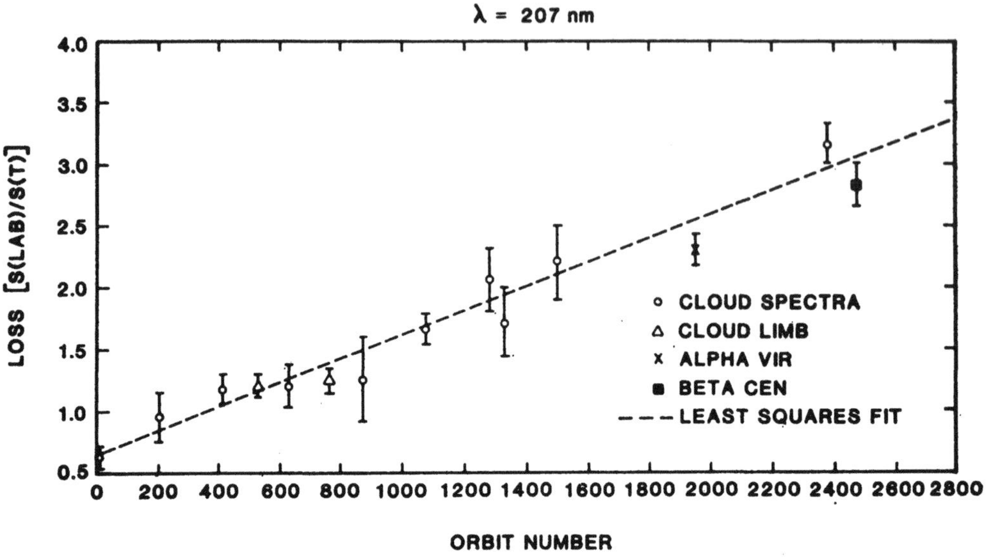

Fig. 1 a

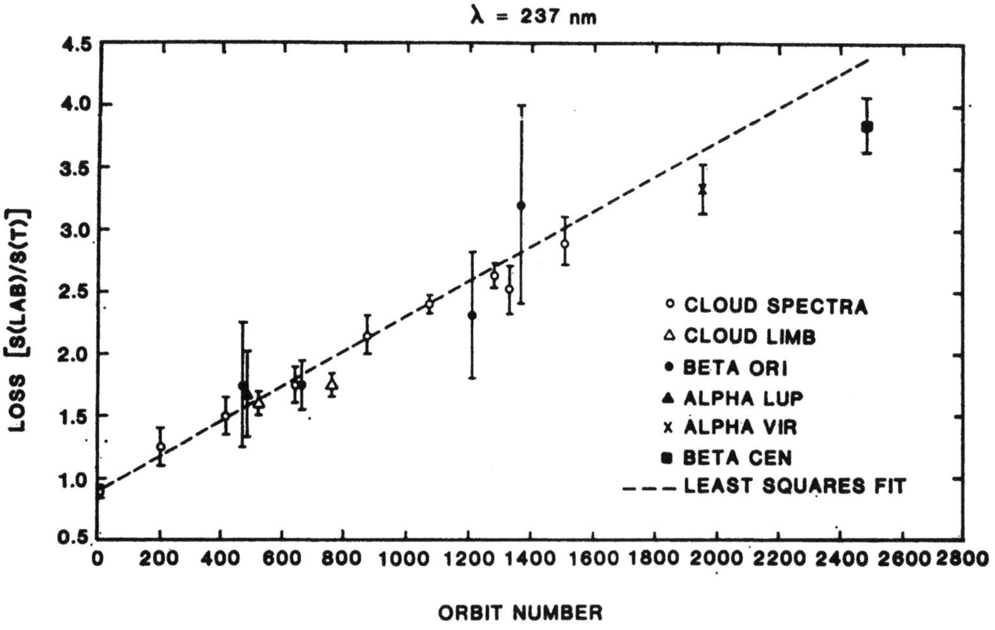

Fig. 1 b

Fig. 1. Sensitivity loss as a function of time. Instrumental sensitivity is compared to pre-flight lab calibration, S(LAB). Cloud spectra: observations of 60° N polar haze, as described in text. Cloud limb: observations of clear Rayleigh scattering above limb. Star calibrations with individual stars noted. Dashed line is least squares fit to calibrating cloud spectra. a (top) 207 nm. b (bottom) 237 nm

image of Venus. The basic data for this study of SO_2 are the brightness measured at wavelengths $\lambda = 207$ nm and $\lambda = 237$ nm from these two modes of operation, along with the associated observing geometry.

SO_2 ABUNDANCE MODEL PROCEDURE

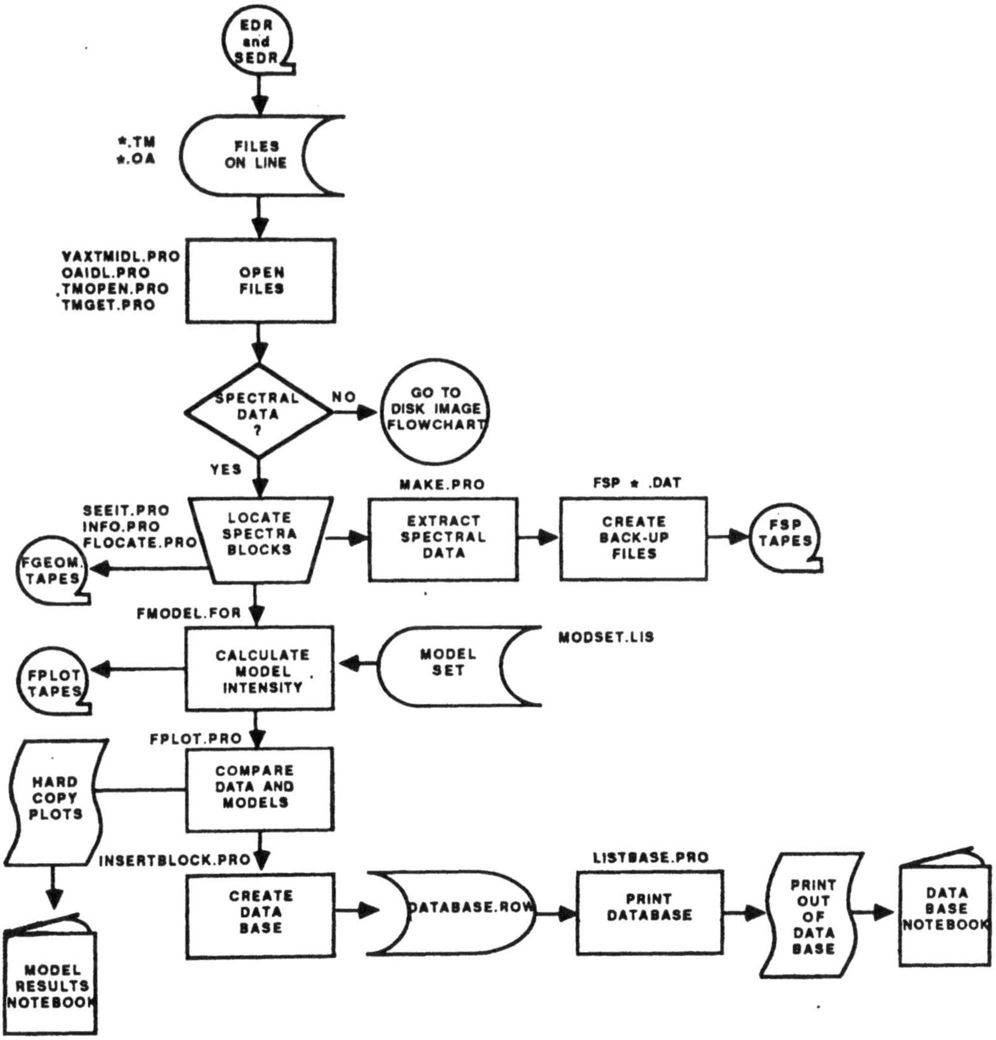

Fig. 2. Flowchart of procedure to determine SO_2 abundance. Processing begins with magnetic tapes (top) and ends with archival or results (bottom left and right). For multi-color images (middle diamond) processing continues as shown in Fig. 3

DISKIMAGE PROCESSING FLOWCHART

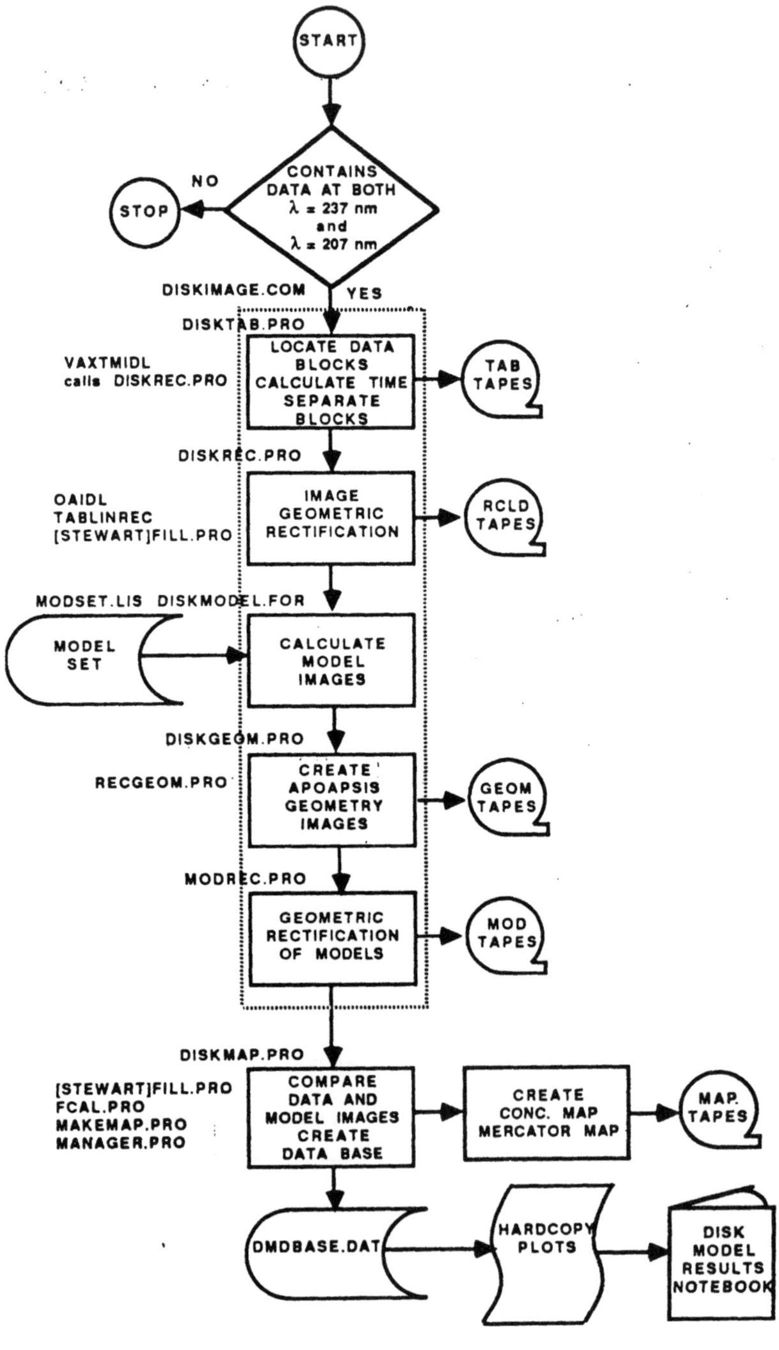

Calibration

The standard method we have used to track instrument sensitivity is to take spectra of the brightest points in the polar haze (Travis et al., 1979). This method has the advantage that the calibrating data are collected on the same orbits as the spectral mode data. To reduce the scatter, a least squares fit to these data gives the sensitivity loss as a function of time. These calibrating observations, denoted "cloud spectra" are shown in Fig. 1 along with the linear fit. During 1984 and 1985 the PVO spin axis was precessed on two occasions to observe the bright UV stars Spica (αVir) and Hadar (βCen) and to accurately measure the response variability across the FOV.

The stellar calibration results at the two wavelengths used for SO_2 analysis show excellent agreement with the planetary observations of the bright polar haze. At $\lambda = 207$ nm, the decline calculated from Venus is a factor of 4.7. For the star observations, the decline is measured as 4.34 (7.7% smaller). At $\lambda = 237$ nm, the Venus observations give a decline of 5.0. The stellar observations give 4.81 (3.8% smaller). Given the accuracy of each of the calibrating methods ($\lesssim 10\%$), we conclude that the instrument sensitivity as a function of time is determined within 10% by two sets of independent measurements over the course of the mission, and that the brightness of the polar haze as not changed appreciably over the eight years of observation. The actual sensitivity used is given by the dashed line in Fig. 1.

Model procedure

Venus spectral and multi-color data arrive on a regular schedule in the form of magnetic tapes from the Pioneer Venus Project at the NASA Ames Research Center. Once the data are in hand at the University of Colorado, the data are handled by a sequence of procedures that result in multi-color rectified images of Venus; SO_2 abundance and scale height for each spectrum; and SO_2 abundance for every picture element in the rectified images. These results are stored in various databases by orbit, latitude, Venus phase, etc. The complete procedure is sketched in the flowchart shown in Figures 2 and 3.

For the spectral data the planetary brightness is compared to a set of 10 brightness models, each calculated using the appropriate observing geometry at two wavelengths. The model parameters are the SO_2 abundance at 40 hPa (essentially the cloud top, see Esposito and Travis, 1982) and the SO_2 scale height (SO_2 is assumed to have an exponential distribution).

For the multi-color images, a subset of seven models are compared to the data. These all have the SO_2 scale height fixed at 2.5 km (the most common value derived from the analyses of the spectral data). The SO_2 abundance has the parameter values 0, 3.125, 6.25, 12.5, 25, 20, 100 ppb at 40 hPa. Fixing the value of the scale height at 2.5 km substantially reduces the computational effort required in reducing this data set. As can be seen below, the derived SO_2 abundance is in good agreement between the multi-color images and the spectra.

Fig. 3. Flowchart of processing of multi-color images. Data enter (top) from Fig. 2 and procedures end with archival of results (bottom)

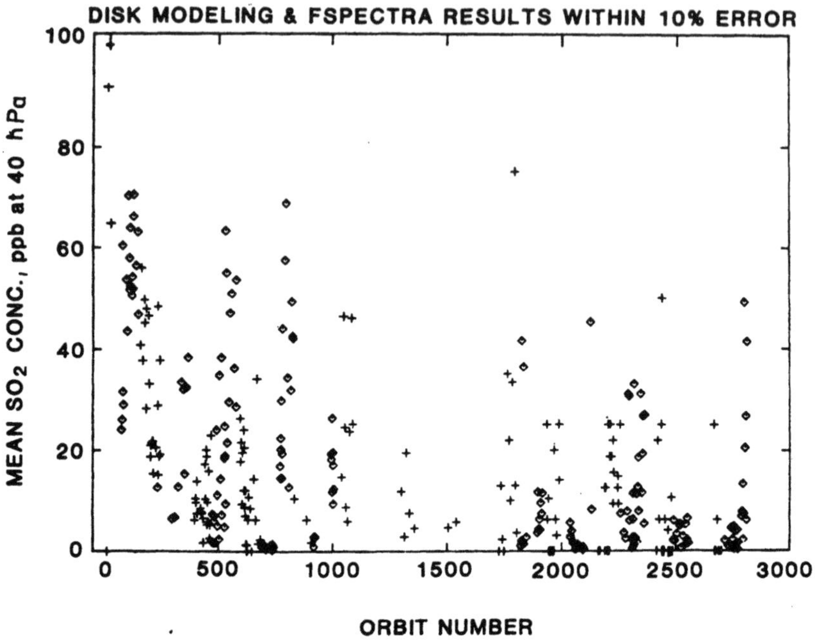

Fig. 4. Mean SO_2 amount for orbits containing spectra (sunlit periapsis, crosses) and multi-color images (sunlit apoapsis, diamonds)

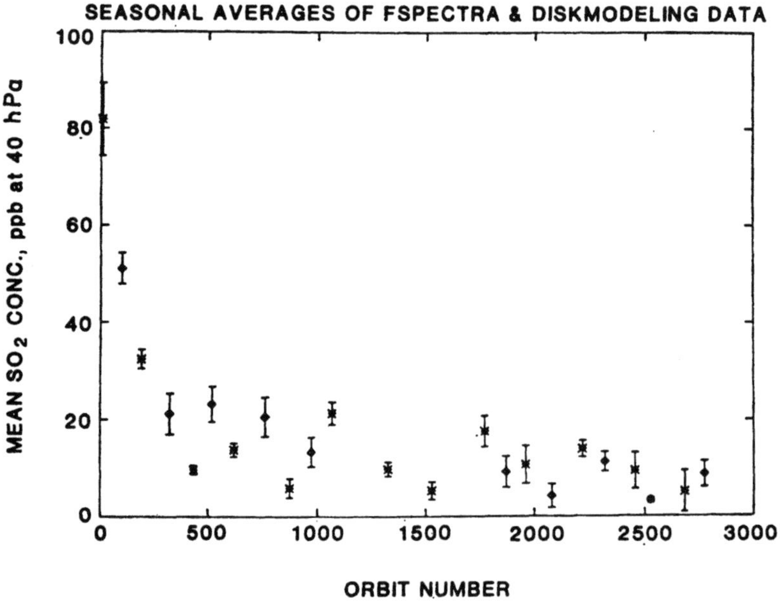

Fig. 5. Mean SO_2 amount averaged over each observing season. Stars: sunlit periapsis. Diamonds: sunlit apoapsis. Error bars show standard error of the mean

The modeling procedure is similar to that used by Esposito (1980). The reflected brightness is calculated (including multiple scattering, vertical inhomogenity and the reported observing geometry) by the method of Markov chains (Esposito, 1979). The best fit model is the one whose combination of abundance and scale height minimizes the sum of the squares of the difference between model and data at the two wavelengths. If the root-mean-square difference is greater than 10% of the observed brightness, no model fit is deemed satisfactory. The possibile causes of unacceptable fits are variability of SO_2 in the observed region which is too small to be resolved, random data errors and the limited number of models to be compared. Of 1545 spectra taken by the UVS in the first 2814 orbits (8 years) a satisfactory fit is found for 1110 (72%). For each of the 157 orbits containing spectra we average the derived values (which are the SO_2 abundances from the best fit models) to give global means for SO_2 abundance. Global means are also calculated by averaging all the picture elements from a multi-color image that are successfully modeled (i.e., within 10% error). Combing the spectral data with the 195 multi-color images gives a total of 352 orbits for which we derive individual global mean amounts of SO_2. These means are plotted in Fig. 4.

Figure 4 shows substantial scatter. We believe this scatter is real and indicates natural variability of Venus cloud top conditions (similar to 'weather' on Earth). The scatter can be reduced by averaging over Pioneer Venus observational seasons (about 100 days). See Fig. 5.

Results

Abundance of SO_2. From Figures 4 and 5 it is obvious that the amount of sulfur dioxide visible at the Venus cloud tops has declined by about one order of magnitude since the start of the mission in 1978. This decline has two parts. A rapid decline in the first 400 days of the mission has a characteristic decay time of about 200 days. Following that is a slower decline, apparently still in progress, with a decay time of about 2000 days (Fig. 6). Given the uncertainties in the instrument calibration, this latter decay rate is marginally consistent with a constant SO_2 abundance and a 10% overestimate in the instrument sensitivity decline. We note that the current abundance of SO_2 is consistent with upper limits set before the Pioneer Venus encounter (e.g., Owen and Sagan, 1972, Cruikshank and Kuiper, 1967).

Correlation with polar haze. Travis et al. (1979) have noted several other unusual aspects of the Venus atmosphere at the time of PV encounter. The cloud images from the PV Cloud Photopolarimeter (CPP) clearly show bright polar regions. Polarimetry shows that a haze of submicron particles covers the entire planet and is especially evident over the poles which can account for the bright polar regions. The last major occurrence of bright polar caps in 1959 was also associated with sub-micron particles at the cloud tops (Dollfus et al., 1979). It is possible to understand these repeated occurrences in a straightforward way. A simple explanation is as follows.

1) SO_2 is episodically injected above the Venus cloud tops;
2) this SO_2 is rapidly converted to new, small aerosols of H_2SO_4 seen as a haze and especially bright polar regions;
3) these aerosols grow and fall out into the main cloud deck, which would give rise to decade-long periods during which neither small aerosols nor SO_2 is seen.

Esposito and Travis (1982) showed that there is a negative correlation between the submicron haze at any given instant and the amount of SO_2. The natural interpretation of this is a conservation of sulfur atoms as they are exchanged between the gas (SO_2) and the haze aerosol (H_2SO_4). Since the particles are produced from the SO_2 gas, this would naturally explain the observed long term positive correlation; as the global amount of

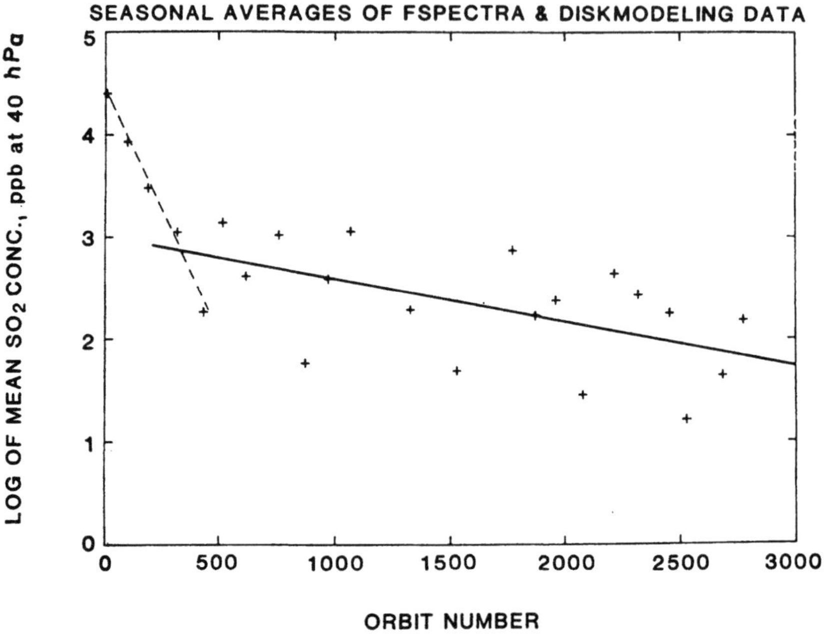

Fig. 6. Exponential decay models for seasonal averages from Fig. 5. Solid line: least squares fit. Dashed line: rapid initial decline

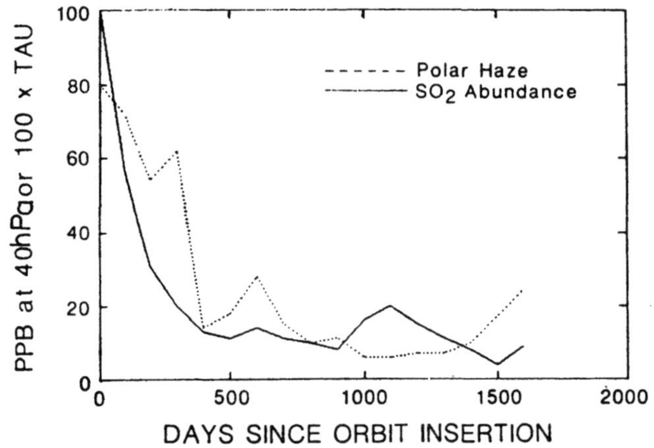

Fig. 7. Intercomparison of derived SO_2 (this work) and derived polar haze optical depth from PV Cloud Photopolarimeter (L. D. Travis, personal communication)

SO_2 declines, so should the production of haze aerosols. Indeed, Kawabata et al. (1986) show the amount of sub-micron haze decreased by an order of magnitude over the first 1000 days of the Pioneer Venus mission.

Figure 7 compares our Pioneer Venus UVS data from the first 1600 orbits to measurements of polar haze optical depth from the Pioneer Venus UVS cloud photopolarimeter (CPP) [Kawabata et al., 1986]. The correlation is striking: both measures show a strong decline since the beginning of mission, leveling off at more historically typical values after 5 years. The correlation coefficient between these two measures is 0.8. This strongly supports the hypothesis that the polar haze is formed of small, recently created aerosols of H_2SO_4.

Cloud top rotation rates. Tracking of cloud features has provided detailed measurements of the wind velocity at cloud top altitudes (e.g., Travis et al., 1979, Rossow et al., 1980, Del Genio and Rossow, 1982). We have examined the rotation rates of SO_2 features to compare with these wind measurements.

These analyses show rotation rates for SO_2 features comparable to Venus mean cloud top rotation (~4.8 days) and rotation of the dark "Y" feature (4—5 days) as determined by Rossow et al. 1980.

Discussion

Sources of error and uncertainty in SO_2 abundance. The data analyzed here represent the longest uniform spectroscopic data set for the planet Venus. Identical Venus observations have been made on a weekly basis for more than 8 years and those observations have been reduced and analyzed using a single established protocol. This analysis, although optimized to measure SO_2 abundance, has a number of imperfections, which we now discuss.

The total decline of instrument sensitivity over the length of the mission is now known within 10%, because of careful star calibrations. However, we have modeled this as a purely linear decline. Some evidence shows that the rate of decline is less in recent years. A second characteristic of the analysis is the use of a simple two-parameter model (SO_2 abundance at 40 hPa, and scale height at 40 hPa). Model results are only calculated for discrete pairs of these parameters. Venus is certainly more complicated than this! We guard against some problems by not using data which do not lie within 10% r.m.s. of one of the model spectra.

The models include a number of assumptions about the Venus atmosphere. For example, the amount of the second, unknown UV absorber is assumed constant and to have a flat absorption spectrum over the wavelengths of interest. The sub-micron haze varies only in the polar region model in a slow way that accounts for the broad long term trends measured by the CPP. As the SO_2 abundance declines, these other scattering components have become more important, thus decreasing the reliability of our SO_2 abundance determinations.

Altogether, however, we believe that observations over a season (see Table 1) tend to approximate the actual global means, and that the data quality does not justify a model with substantially more parameters or allowance for unresolved horizontal variability. In the interests of uniformity, we have not tinkered with the model set as a function of time. Because of these considerations, the systematic errors in the later determinations

Table 1.
Seasonal Mean SO_2 Abundance

Sunlit periapsis seasons	Mean PV orbit	Date	Derived SO_2 abundance at 40 hPa
1	8	12 Dec 78	82 ppb
2	191	13 Jun 79	32 ppb
3	433	11 Feb 80	10 ppb
4	619	15 Aug 80	14 ppb
5	871	25 Apr 81	6 ppb
6	1067	7 Nov 81	21 ppb
7	1326	24 Jul 82	10 ppb
8	1527	10 Feb 83	5 ppb
9	1771	12 Oct 83	18 ppb
10	1961	19 Apr 84	11 ppb
11	2216	29 Dec 84	14 ppb
12	2455	25 Aug 85	9 ppb
13	2681	8 Apr 86	5 ppb
Sunlight apoapsis seasons	**Mean PV orbit**	**Date**	**Derived SO_2 abundance at 40 hPa**
1	100	14 Mar 79	51 ppb
2	320	21 Oct 79	21 ppb
3	519	7 May 80	23 ppb
4	759	2 Jan 81	20 ppb
5	971	3 Aug 81	13 ppb
9	1871	20 Jan 84	9 ppb
10	2077	12 Aug 84	4 ppb
11	2318	10 Apr 85	11 ppb
12	2527	5 Nov 85	3 ppb
13	2771	7 Jul 86	9 ppb

are probably larger than those in the first 400 orbits. A number of these systematic errors go in the same direction so that SO_2 abundance may be underestimated by 50 to 100% in the last 1000 orbits.

Active Venus volcanoes. Esposito (1984) has interpreted the observed SO_2 injection and resulting haze formation as due to episodic volcanism. Radar studies by the Pioneer Venus Orbiter (Masursky et al., 1980) have detected topographic features on Venus that resemble terrestrial volcanic landforms. Prinn (1985) argues that the observed amount of SO_2 in the lower atmosphere is out of equilibrium with the surface composition measured by the latest Venera landers (Moroz, 1983). This current imbalance requires a geologically recent source of sulfur compounds at the surface.

Phillips and Malin (1983) have suggested volcanic activity may be more vigorous on Venus than on Earth because of the lack of organized plate tectonics on Venus. The more vigorous volcanic activity on Venus might be due to larger volcanoes, a greater number of volcanoes, or more frequent eruptions. The last two of these possibilities could account for a higher frequency of SO_2 injections into the visible Venus atmosphere.

However, we will not know for certain that volcanoes are erupting until we have a direct sighting or observe before-and-after topography. The best opportunity for this is from the radar imaging on the Magellan mission, to be launched to Venus later in this decade.

Conclusions

Pioneer Venus spectroscopy of the Venus cloud tops shows a decline by about a factor of ten in the amount of SO_2 since 1978. A consistent decline is also seen in the amount of polar haze, which was also unusually abundant in 1978. Star calibrations determine the instrument sensitivity well enough to rule out the decline as an instrumental effect. Systematic errors are probably larger in the later orbits and may be as large as 50–100% for the SO_2 abundance. A plausible explanation for the observed behavior is injection of SO_2 into the Venus middle atmosphere by a volcanic explosion (Esposito, 1984).

The assumed injection can account for the current geochemical imbalance between the surface and atmosphere, noted by Prinn (1985) and also the failure of steady-state photochemical models to match both the small scale height of SO_2 and the upper limits on O_2. Since SO_2 is the major precursor of the Venus clouds and is an important absorber of solar radiation, steady-state models of the chemistry and dynamics of the cloud-top region must be used cautiously. The abundances of other chemically active constituents such as Cl_2, CO, and polymorphic sulfur must also experience coupled episodic variations.

Acknowledgments

Larry Travis supplied the information in polar haze optical depths. We have had helpful discussions with Y. L. Yung and H. Masursky. The Pioneer Project Office allowed us to turn the spacecraft to observe stars. This work has been supported by the Pioneer Venus Project Office and the NASA Planetary Atmospheres Program (NAGW-389).

References

BARKER, E. S.: Detection of SO_2 in the UV spectrum of Venus. Geophys. Res. Lett. **6** (1979), 117–120.

COLIN, L.: The Pioneer Venus program. J. Geophys. Res. **85** (1980), 7575–7598.

CONWAY, R. R., R. P. McCOY and C. A. BARTH: IUE detection of sulfur dioxide in the atmosphere of Venus. Geophys. Res. Lett. **6** (1979), 629–631.

CRUIKSHANK, D. P. and G. P. KUIPER: Communications Lunar Planetary Lab. **6** (1967), 196.

DEEMING, T. J.: Fourier Analysis with Unequally-Spaced Data. Astrophys. Space Sci. **36** (1975), 137–158.

DELGENIO, A. D. and W. B. ROSSOW: Temporal variability of ultraviolet cloud features in the Venus stratosphere. Icarus **51** (1982), 391–415.

DOLLFUS, A., M. AURIERE and R. SANTER: Astron. J. **84** (1979), 1419.

ESPOSITO, L. W.: An adding algorithm for the Markov chain formalism for radiative transfer. Astrophys. J. **233** (1979), 661.

ESPOSITO, L. W.: Ultraviolet contrasts and the absorbers near the Venus cloud tops. J. Geophys. Res. **85** (1980), 8151–8157.

Esposito, L. W.: Sulfur dioxide: Episodic injection shows evidence for active Venus volcanism. Science **223** (1984), 1072–1074.

Esposito, L. W. and L. D. Travis: Polarization studies of the Venus UV contrasts: Cloud height and haze variability. Icarus **51** (1982), 374–390.

Esposito, L. W., J. R. Winick and A. I. Stewart: Sulfur dioxide in the Venus atmosphere: Distribution and implications. Geophys. Res. Lett. **6** (1979), 601–604.

Hansen, J. E. and J. W. Hovenier: Interpretation of the polarization of Venus. J. Atmos. Sci. **31** (1974), 1137–1160.

Kawabata, K., D. Coffeen, J. Hansen, W. Lane, M. Sato and L. Travis: Cloud and haze properties from Pioneer Venus polarimetry. J. Geophys. Res. **85** (1980), 8129–8140.

Kawabata, K. and J. E. Hansen: Interpretation of the variation of polarization over the disk of Venus. J. Atmos. Sci. **32** (1975), 1133–1139.

Kawabata, K., M. Sato and L. D. Travis (in preparation), 1986.

Moroz, V. I.: Summary of preliminary results of the Venera 13 and 14 missions in Venus (D. M. Hunter, Ed.), Arizona, 1983.

Masursky, H., E. Eliason, P. G. Ford, G. E. McGill, G. H. Pettengill, G. C. Schaber and G. Schubert: Pioneer Venus radar results: Geology from images and altimetry. J. Geophys. Res. **85** (1980), 8232–8260.

Owen, T. and C. Sagan: Icarus **16** (1972), 557.

Phillips, R. J. and M. C. Malin: The interior of Venus and tectonic implications, in Venus, D. M. Hunten, ed., Arizona, 1983.

Ponman, Trevor: The analysis of periodicities in irregularly sampled data. Mon. Not. R. Astr. Soc. **196** (1981), 583–596.

Prinn, R. G.: The photochemistry of the atmosphere of Venus, in Photochemistry of Atmospheres, J. Levine, ed., Academic, 1985.

Rossow, W. B., A. D. DelGenio, S. S. Limaye, L. D. Travis and P. H. Stone: Cloud morphology and motions from Pioneer Venus images. J. Geophys. Res. **85**, (1980), 8107–9128.

Stewart, A. I.: Design and operation of the Pioneer Venus Orbiter ultraviolet spectrometer. IEEE, Trans. GE-18, **65** (1980).

Stewart, A. I., D. E. Anderson, L. W. Esposito and C. A. Barth: UV spectroscopy of Venus: Initial results from the Pioneer Venus Orbiter. Science **203** (1979), 23.

Taylor, H. A. and P. A. Cloutier: Venus: Dead or Alive? Science **234** (1986), 1087.

Thomas, G. E., B. M. Jakosky, R. A. West and R. W. Sanders: Satellite limb scanning thermal infrared observations of the El Chichon stratospheric aerosols: First results. Geophys. Res. Lett. **10** (1983), 997.

Thompson, B. A., P. Harteck and R. V. Reeves, Jr.: Ultraviolet absorption coefficients of CO_2, CO, O_2, H_2O, N_2O, NH_3, NO, SO_2, and CH_4 from 1850 to 4000 Å. J. Geophys. Res. **68** (1963), 6431.

Travis, L. D., D. L. Coffeen, J. E. Hansen, K. Kawabata, A. Lacis, W. A. Lane, S. S. Limaye and P. H. Stone: Orbiter cloud photopolarimeter investigation. Science **203** (1979), 781 to 785.

Winick, J. R. and A. I. E. Stewart: Photochemistry of SO_2 in Venus upper cloud layers. J. Geophys. Res. **85** (1980), 7849–7860.

14. Water vapor and sulfur dioxide in the Venus' atmosphere from the Venera-15 data

V. I. Moroz, D. Spänkuch, D. V. Titov, W. Döhler, L. V. Zasova, K. Schäfer, A. V. D'jachkov, D. Oertel, V. M. Linkin, J. Nopirakowski

Abstract

About 1500 spectra of Venus' thermal radiation in the spectral range from 250 to 1600 cm^{-1} were obtained in the infrared experiment on board Venera-15 orbiter [1, 2]. A lot of details are seen in these spectra: absorption bands of CO_2, H_2O and SO_2, and wide features of aerosol H_2SO_4 (Fig. 1). The equivalent width and shape of spectral details depend not only on the gas abundances but also on the temperature and aerosol structure of the atmosphere at the given site. First results of spectra analysis inside H_2O and SO_2 absorption bands can be found in earlier publications [3]: a large group of H_2O and SO_2 features was identified and the mixing ratios of these gases were determined for the middle latitudes. These results were obtained by comparison of synthetic spectra with measured for SO_2 and by method of the curves of growth for water vapor. The main goal of the present work is to continue these investigations and to obtain the gas abundances for various latitudes and local times.

Variations of SO_2 and H_2O spectral details

The equivalent widths of spectral details vary significantly from one spectra to another. There are both systematic and occasional effects connected with latitude and local time. The latter can be partially explained by noise and partially by the real variations of abundances. For presentation we selected two spectral details: ν_2 band of sulfur dioxide (507 cm^{-1}) and water vapor absorption feature near 352 cm^{-1}. Instead of equivalent width the depth of detail (in K) was used to simplify the analysis. For $\nu_2 SO_2$ band it was determined as follows

$$T_{SO_2} = T_{517.4} - T_{507.0}$$

and for H_2O

$$T_{H_2O} = \frac{T_{343} + T_{346}}{2} - \frac{T_{350.7} + T_{354}}{2}.$$

The maps of distribution of these values calculated on the basis of all the data obtained are presented in Fig. 2, 3. (Empty regions in these figures correspond to the sites where no measurements were made.) The SO_2 band has its maximal depth at high latitudes (60—80° N) and disappears in the spectra at low latitudes. The region of maximal value

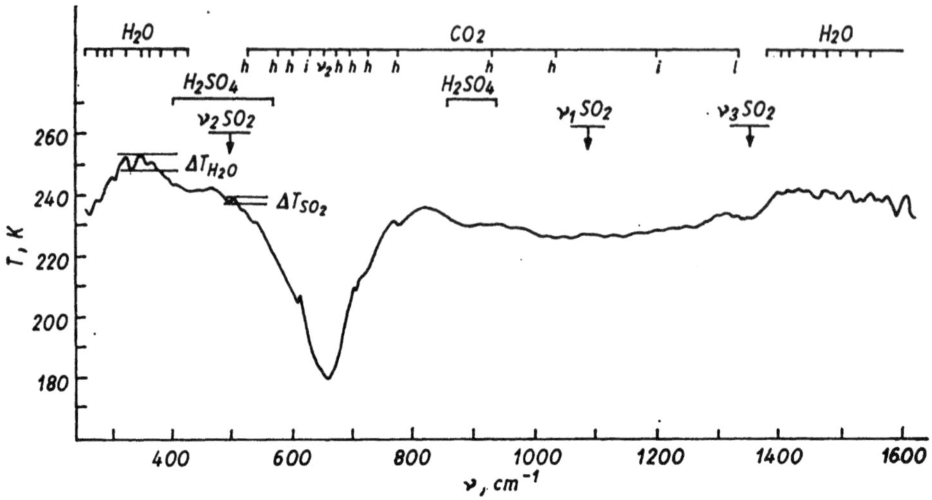

Fig. 1. An example of Venus thermal radiation spectra obtained in the infrared experiment on board Venera-15 orbiter

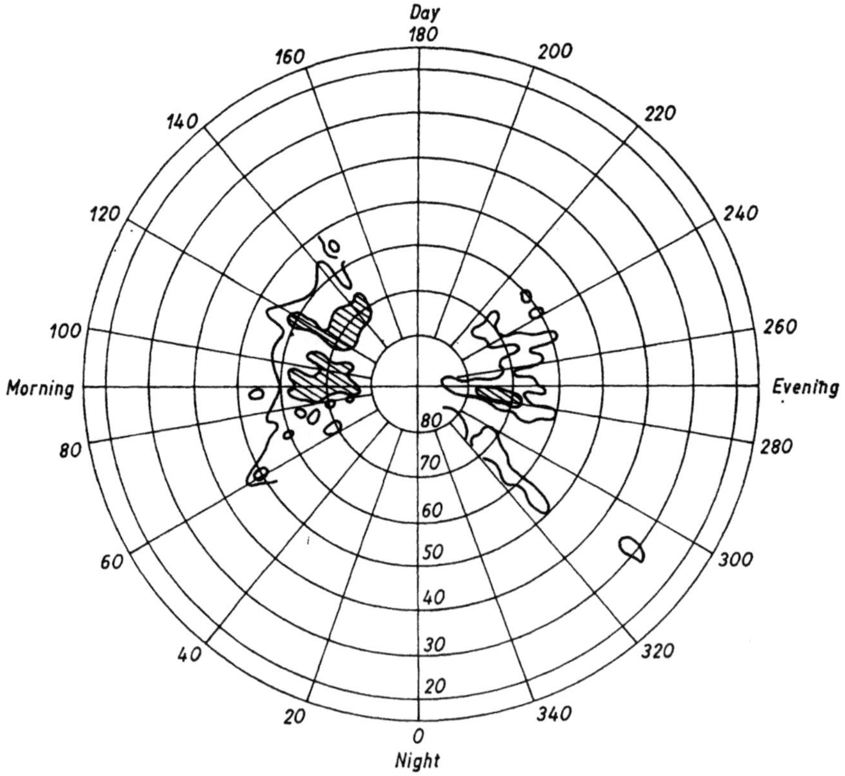

Fig. 2. The distribution of depth of ν_2SO_2 absorption band. The shaded areas represent the regions with $\Delta T_{SO_2} > 2.5$ K. The curves show the boundaries inside which $\Delta T_{SO_2} > 1.5$ K. No measurements were made in the ranges 60–320 and 150–220 degrees local solar time

of H_2O absorption feature ($T_{H_2O} > 5$ K) is located between 45 and 60° N (Fig. 3). At lower latitudes the depth of water vapor detail decreases. However the depth of spectral detail is the function not only of the amount of absorbing molecules but also of the vertical profiles of temperature and aerosol extinction. That's why a-priori it is not necessary that the maps in Fig. 2, 3 represent the spatial variations of H_2O and SO_2 amounts in the atmosphere. Nevertheless we'll show further that the spatial distribution of gas amounts saves in general the main features of the maps.

Method of determination of the gas abundance

To solve this problem one must have vertical profiles of temperature and aerosol extinction. So the determination of the minor constituents' abundances must be preceded by the retrieval of temperature and aerosol structure of the atmosphere. Temperature profiles obtained from the measurements inside the 15 μm CO_2 band providing gas atmosphere [4] and extrapolated down on the basis of existing lander data were used in our previous paper [3]. Aerosol profiles were determined from the same spectra [5], within the a-priori given model (for example, constant or exponential). Later [6] a self-consistent

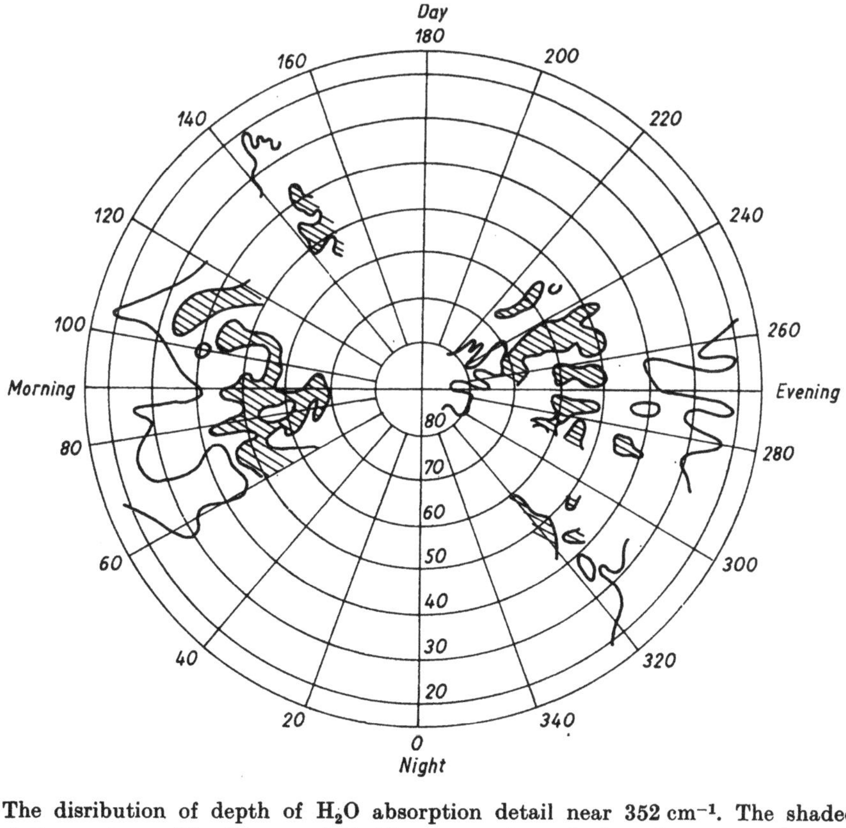

Fig. 3. The disribution of depth of H_2O absorption detail near 352 cm^{-1}. The shaded areas represent the regions with $\Delta T_{H_2O} > 5$ K. The curves show the boundaries inside which $\Delta T_{H_2O} > 2.5$ K. No measurements were made in the ranges 60–320 and 150–220 degrees local solar time

retrieval of temperature and aerosol profiles was fulfilled and these results are used here. It's extremely important that to determine gas amounts temperature and aerosol properties obtained from the same spectrum are used. In this paper we apply a modification of method of curves of growth slightly different from that used in [3]. Here we shall define the curve of growth as an equivalent width dependence on the integral gas amount above the level in the atmosphere where aerosol optical depth equals 1 for the given wavenumber. We applied this definition because gas optical thickness is very small for all spectral details of H_2O and SO_2 and it is aerosol structure that determines the position of the level where the measured radiation comes from. The curve of growth is obtained as follows. A set of synthetic spectra is calculated for the known temperature and aerosol structure and different gas profiles. The dependence of equivalent width W on the column density above the level $\tau_a = 1$ is obtained from these spectra. Some examples of curves of growth are shown in Fig. 4. An approximate method [7] was used to calculate the monochromatic absorption coefficient of water vapor in this procedure. Absorption between observed spectral details of H_2O was taken into account. One needs such a correction to determine the level of continuum which is important for calculation of equivalent width of absorption feature. An approximation of strong lines' overlapping in Elsasser model was employed for SO_2 band. Besides this an overlapping with CO_2 band was also taken into account. So each absorption feature allows one to determine column density of the gas above the altitude where $\tau_a = 1$. But if there are measurements in two spectral details of the gas that gives column densities above two levels $N_1 = N(z_1)$ and $N_2 = N(z_2)$ than there is a chance to estimate the profile of absorbing molecules. For the exponential gas model the scale heigth is given by the expression

$$H = -\frac{z_1 - z_2}{\ln (N_2/N_1)}$$

and the number density at the altitude z_1 is defined as

$$n_1 = N_1/H .$$

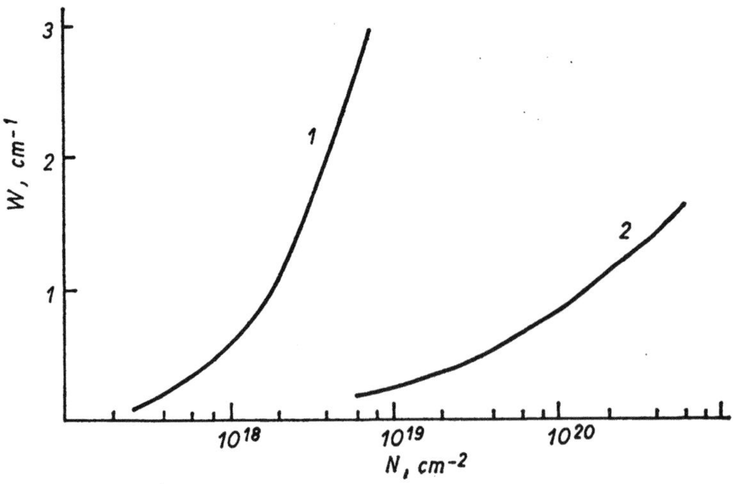

Fig. 4. The examples of the curves of growth for ν_2SO_2 band (1) and water vapor spectral feature near 352 cm^{-1} (2)

Sulfur dioxide

The described method was applied to determine the SO_2 abundance from about 100 spectra of seven revolutions. For this purpose two bands were considered: ν_2 517 cm^{-1} and ν_3 1380 cm^{-1}. Measurements in ν_1 1050 cm^{-1} band usually give only the upper limit. Results of several retrievals are shown in Fig. 5. The data in two bands permit us (though with a significant error) to determine the parameters of gas profile. The following values can be cited: for the middle latitudes (50—60° N) SO_2 scale height is 2 ± 1 km with $f = 6 \pm 2$ ppm at 62—63 km, for the 'collar' region ($\approx 70°$ N) the gas scale height increases to 5 km with mixing ratio ≈ 10 ppm at 60 km. But to use measurements in $\nu_3 SO_2$ band is not always admittable due to the large errors.

The most reliable results are obtained for the measurements in $\nu_2 SO_2$ band (517 cm^{-1}). We used them for the determination of SO_2 latitude dependence (Fig. 6). Then the inte-

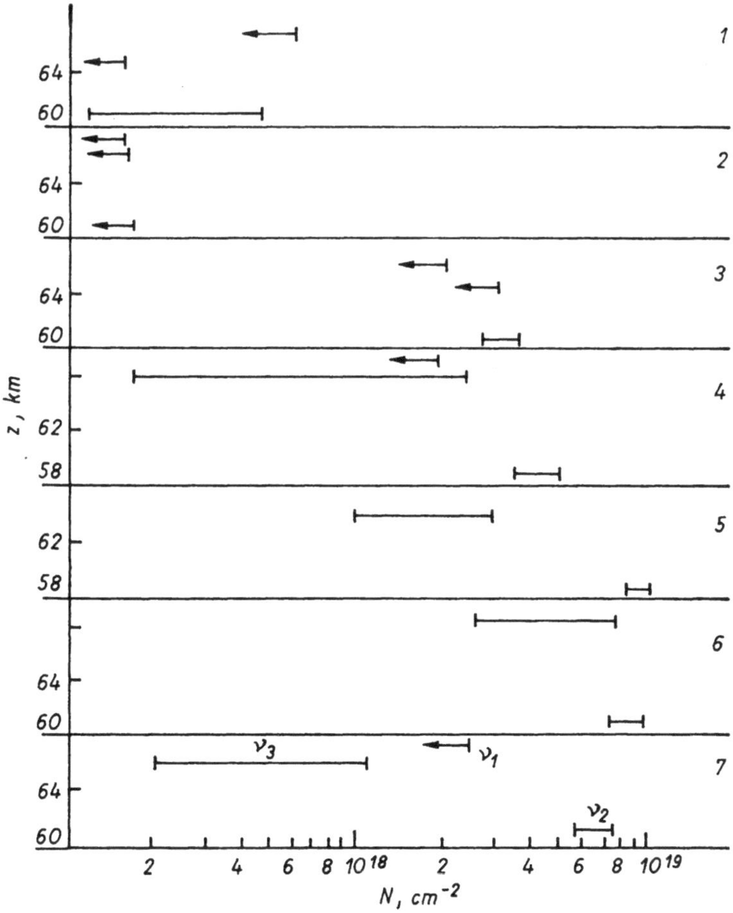

Fig. 5. The examples of determination of SO_2 abundances from seven individual spectra using the measurements in three absorption bands. The corresponding latitudes are: 30° N (1), 50° N (2), 59° N (3), 64° N (4), 74° N (5), 73° N (6), 53° N (7)

gral mixing ratio can be obtained:

$$\bar{f}_{SO_2} = N_{S_2O}/N_{C_2O}.$$

The value of $\bar{f}_{S_2O}$ gives one the lower limit for the local gas mixing ratio providing $f_{SO_2}(z)$ decreasing with height

$$f_{S_2O}(z_{\tau_a=1}) \geqq \bar{f}_{SO_2}.$$

Its latitude dependence is also shown in Fig. 6. Both column density and integral mixing ratio increase 1.5—2 orders of magnitude from 40° N to 60° N. The interpretation of such strong latitude dependence is ambiguous. The first possible reason is the increase of SO_2 abundance with latitude at all levels in the atmosphere. The other one is the coming down of the cloud tops (or the level $\tau_a = 1$) at the high latitudes while the SO_2 profile remains independent of latitude. In this case deeper and more abundant with SO_2 regions of atmosphere are observed at high latitudes. In the frame of the last assumption latitude behavior of column density can be used to derive the SO_2 vertical profile. The detailed analysis of this opportunity is not done yet, only one estimation was made for one session of measurements (14.10.83). Linear regression for 14 points gave the scale height $H \approx 1.3$ km and mixing ratios for different latitude zones.

Latitude	Altitude of the level	Mixing ratio
deg	$\tau_a = 1$	ppm
50	64	4.5
73	62	15

We must note however that the scale height of ≈ 1 km seems to be too small because the extrapolation of mixing ratio down and up with such a scale will provide too large and too low mixing ratios inconsistent with the results of other experiments. The estimation obtaines is likely to be the lower limit of SO_2 scale height and both effects are responsible for the dependence in Fig. 6; i.e. the sinking of cloud tops at high latitudes and the increase in SO_2 abundance with latitude. Probably it will be possible to separate these effects further for example by exact modelling of SO_2 band shape.

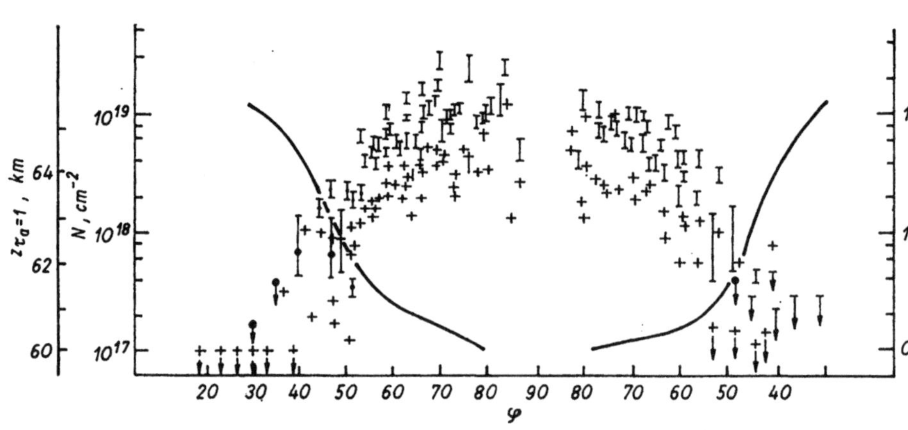

Fig. 6. The latitude dependence of SO_2 column density N (vertical line segments), SO_2 integral mixing ratio $\bar{f}$ (crosses) and the height of the $\tau_a = 1$ level (solid curve)

One additional comment can be made about the latitude variations of SO_2 abundance. Sulfur dioxide photolysis by the UV sunlight ($\lambda = 0.2-0.22\ \mu m$) mainly takes place near the level where $\tau_{UV} = \sigma_{UV} \times N_{SO_2} = 1$. SO_2 absorption cross-section for this spectral region equals $3 \cdot 10^{-18}$ cm^2 [8]. And hence for the obtained latitude dependence of column density (Fig. 6) the level of photolysis is below the level of infrared observations for $\varphi \leqq 50°$ N and for $\varphi > 50°$ N — above it.

These results allow one to make some conclusions on the dependence of SO_2 column density on the local time. In the sessions considered the Venera-15 orbiter moved from the nightside (near the morning terminator) to the dayside (evening terminator). This provides a chance to compare the gas column density for day and night conditions. The results are shown in Fig. 7. The integral mixing ratio for the dayside is about 2 ppm less than for the nightside. And the level $\tau_a = 1$ at night is located a little bit higher than at the daytime (Fig. 7). That's why it seems likely that the night atmosphere is more abundant with SO_2 than the day. But we draw this conclusion with care because it needs further confirmation.

The results of the infrared experiment on board Venera-15 are compared in Fig. 8 with the data on SO_2 abundance obtained from the other measurements. Our results correspond to the altitude 65—60 km, i.e. they fill in the gap between the data of remote measurements in UV spectral range and the results of in-situ experiments. The results of infrared spectrometry confirm rather well the trend for SO_2 abundance to decrease rapidly with height near the cloud tops. However the latitude dependence of SO_2 mixing ratio obtained in Pioneer-Venus Orbiter UV-experiment [14] is opposite to that observed in our IR-experiment: SO_2 mixing ratio is less at high latitudes than at

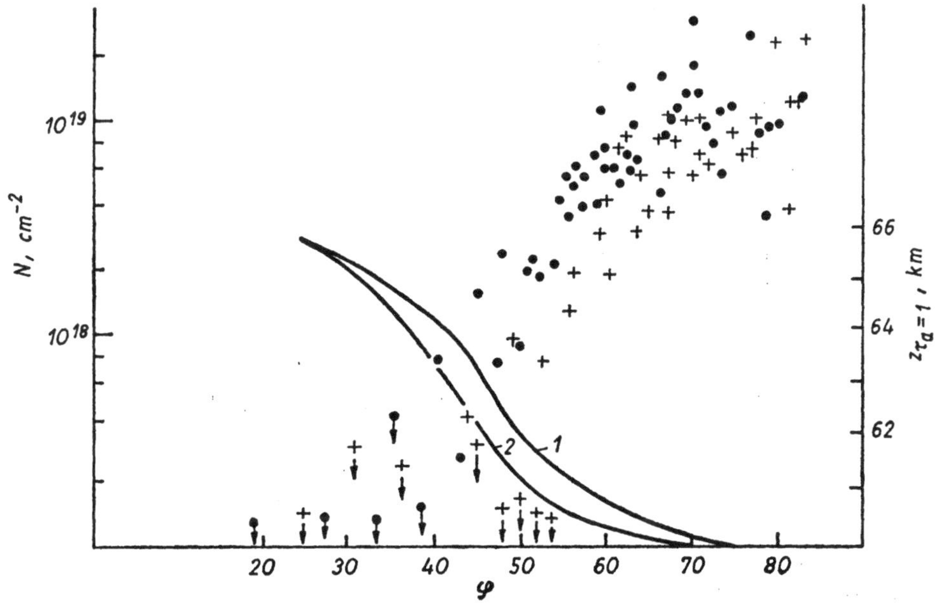

Fig. 7. The comparison of SO_2 column densities N for the day (crosses) and night (dots) conditions. The solid curves represent the latitude dependence of the altitude of the level $\tau_a = 1$ for the night (1) and day (2) atmosphere

middle. Nevertheless it can't be excluded that both sets of measurements could be agreed in principle by taking into account the polar haze at high latitudes [15], which has large extinction coefficient in UV and small in IR-spectral range.

Water vapor

Several absorption details of water vapor rotation band are clearly identified in the longwavelength part of the spectra (250—420 cm^{-1}) (Fig. 1). In principle the described method could be applied to each of them of obtain several values of column density N_{H_2O} corresponding to different levels. This procedure was used for processing of a number of spectra. But most of these features usually give too small difference in the altitudes of the level $\tau_a = 1$, and this fact practically excluded the chance to obtain reliable data about the H_2O profile by using more than one absorption feature. That's why we decided to restrict the analysis to the strongest feature near 352 cm^{-1} and to determine only the column density above the level $\tau_a = 1$ (approximately 58—60 km). The $\nu_2 H_2O$ band is difficult to use because of too large errors in measured spectra. The latitude dependence of water vapor column density obtained from the processing of data for seven sessions is shown in Fig. 9. The height of the level $\tau_a = 1$ for 352 cm^{-1} depends weakly on the latitude hence the values in Fig. 9 correspond to the altitudes of 58—60 km. For further interpretation beyond the data in Fig. 9 one may use the fact that Venus' aerosol is composed mainly of sulfuric acid. This can provide us some new data on the concentration of acid in droplets and on the H_2O partial pressure in the atmosphere. The procedure can be done as follows. Water vapor must be in equilibrium

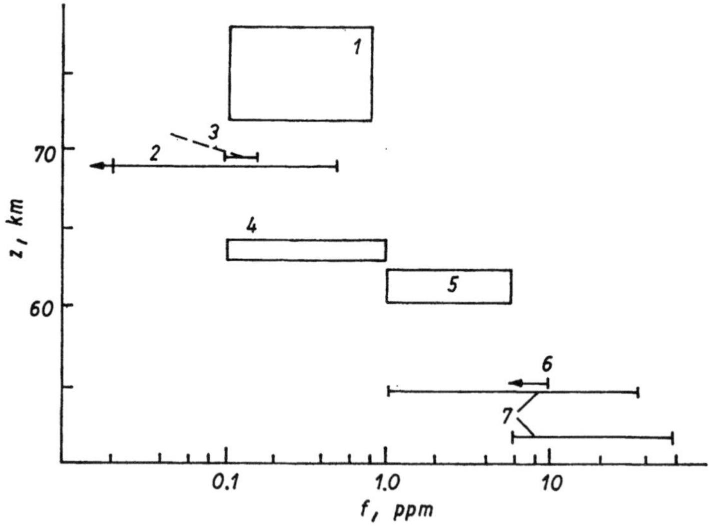

Fig. 8. The comparison of SO_2 abundances obtained from different experiments: 1 — UV spectroscopy on board of the International Ultraviolet Explorer [9], 2 — groundbased observations [10], 3 — UV spectroscopy on board Pioneer-Venus orbiter [11], 4, 5 — the results of this work for 50—50° N and 60—80° N latitude ranges respectively, 6 — Pioneer-Venus Large Probe mass-spectrometry [12], 7 — UV photometry on board Venera-13 lander [13]

with aerosol sulfuric acid. In this case water vapor pressure $p^s_{H_2O}$ above the H_2SO_4 solution is a function of acid concentration in the droplet and temperature. The data on the thermodynamic equilibrium of sulfuric acid solutions were taken from [16]. Owing to our knowledge of temperature profile for each spectrum H_2SO_4 concentration C in the solution remains the only free parameter that determines H_2O column density in equilibrium. Applying the data from Fig. 9 one can fix the droplet composition so that $N^s_{H_2O}$ is equal to the measured value of N_{H_2O}. This method gives the acid concentration in the range 70—85% which is in good agreement with the other data.

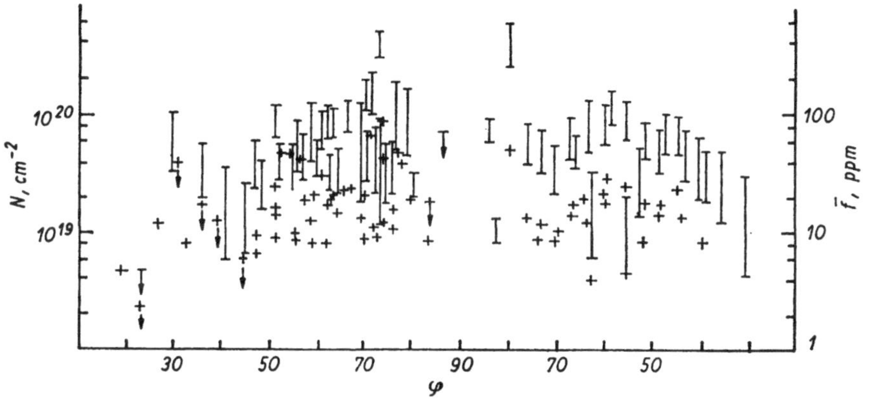

Fig. 9. The latitude dependence of H_2O column density N (vertical line segments) and integral mixing ratio $\bar{f}$ (crosses)

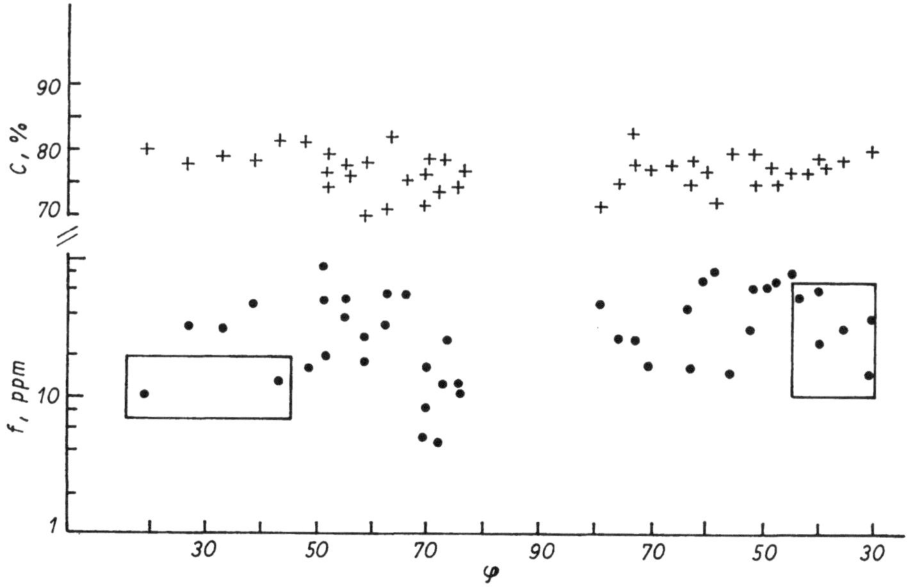

Fig. 10. Sulfuric acid concentration in the droplets C (crosses) and H_2O mixing ratio (dots) in the case of the equilibrium with aerosol H_2SO_4 solution. The results of Pioneer-Venus IR radiometry [17] are represented with rectangles for comparison

The results of various experiments on water vapor abundance in the Venus atmosphere are compared in Fig. 11. As in the case of sulfur dioxide results of Venera-15 correspond to the intermediate range of altitudes (58—60 km). Our results may be compared to the results of Pioneer-Venus radiometer experiment [17] (Fig. 10), that were obtained with the filter that picked out the longwavelength part of the H_2O rotation band. Pioneer-Venus results correspond to the low latitudes (0—45° N) and Venera-15 measurements were made at the latitudes higher than 30° N: that's why we can compare the data in a rather narrow latitude range. Taking into account the errors there is no contradiction between two sets of measurements. Unfortunately nothing can be said about the subsolar maximum of H_2O mixing ratio fixed by the PV Orbiter because this solar longitude range was not explored by Venera-15. But this feature is possibly an artefact because the retrieval procedure applied in [17] didn't take into account the variations of aerosol optical depth in the water vapor (220 cm^{-1}) and aerosol (872 cm^{-1}) channels. The authors [17] showed that the results of H_2O retrievals are very sensitive to the value of this ratio. And the processing of Venera-15 data has shown [6] that spectral dependence of aerosol extinction coefficient varies from one spectrum to another and there is no reason to consider it constant for the whole cloud deck. So the subsolar maximum of water vapor mixing ratio [17] may result from not adequate cloud layer model.

Conclusions

1. Column density of sulfur dioxide and water vapor above the level $\tau_a = 1$ was determined from the measured spectra for seven revolutions of Venera-15 orbiter. The

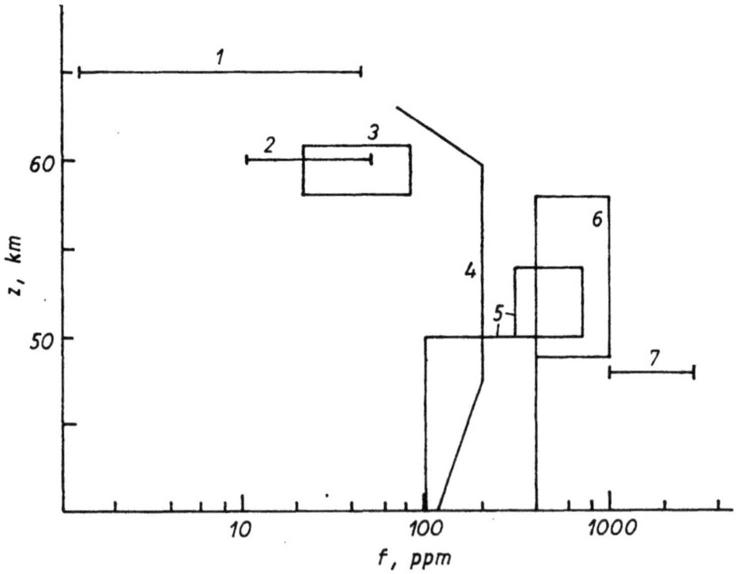

Fig. 11. The comparison of H_2O abundances obtained from different experiments: 1 — ground-based spectroscopy [18], 2 — Pioneer Venus IR radiometry [17], 3 — this work, 4 — spectrophotometry on board Venera-11 lander [19], 5 — spectrophotometry on board Venera-13, -14, landers [20], 6 — Venera-13, -14 gas chromatography [21], 7 — Venera-14 humidity sensor [22]

altitude of $\tau_a = 1$ level varies with latitude and local time but is known for each spectrum. The attempts to determine the scale height of SO_2 and H_2O distributions didn't give reliable results yet.

2. The integral mixing ratio (i.e. the ratio of the number of SO_2 molecules to that of CO_2 above the level $\tau_a = 1$) varies from 1 ppm at 50° N to 10–20 ppm at 70–80° N (corresponding altitudes are 64 and 61 km respectively). For the low latitudes ($\varphi < 40°$ N) only the upper limit of ≈ 0.2 ppm can be estimated.
3. The observed strong latitude dependence of N_{SO_2} is likely result from two effects: the coming down of the cloud tops with latitude and the increase of SO_2 abundance in polar regions as compared to the low latitudes.
4. The latitude dependence of SO_2 abundance obtained from the infrared observations has the opposite sign as that determined from the ultraviolet experiment on board Pioneer-Venus orbiter.
5. Sulfur dioxide integral mixing ratio for the night conditions is ≈ 2 ppm greater than for the dayside and it seems to be the result of the enrichment with SO_2 of the night atmosphere as compared to the day.
6. Water vapor integral mixing ratio at the height 58–60 km is 20 ± 10 ppm.
7. The results of temperature and water vapor column density retrievals fix the value of sulfuric acid concentration in aerosol droplets at 70–85%.

References

[1] Oertel, D., Moroz, V. I., Linkin, V. M. et al.: Venera-15 and Venera-16: first results of infrared spectroscopy. Pis'ma v Astron. Zh. **10** (1984) 4, 243 (Russian).

[2] Oertel, D., Moroz, V. I., Nopirakowski, J. et al.: Infrared experiment aboard the automatic interplanetary stations Venera 15 and Venera 16. 1. Methodics and first results. Veröffentlichungen des Forschungsbereichs Geo- und Kosmoswissenschaften, Heft 18, 1990.

[3] Moroz, V. I., Döhler, W., Ustinov, E. A. et al.: Infrared experiment aboard the automatic interplanetary stations Venera 15 and Venera 16. 4. Preliminary results of spectral analysis in the region of H_2O and SO_2 absorption bands. Veröffentlichungen des Forschungsbereichs Geo- und Kosmoswissenschaften, Heft 18, 1990.

[4] Spänkuch, D., Zasova, L. V., Schäfer, K. et al.: Infrared experiment aboard the automatic interplanetary stations Venera 15 and Venera 16. 2. Preliminary results of temperature profile retrieval. Veröffentlichungen des Forschungsbereichs Geo- und Kosmoswissenschaften, Heft 18, 1990.

[5] Zasova, L. V., Spänkuch, D., Moroz, V. I. et al.: Infrared experiment aboard the automatic interplanetary stations Venera 15 and Venera 16. 3. Some conclusions on the structure of clouds, based on the analysis of spectra. Veröffentlichungen des Forschungsbereichs Geo- und Kosmoswissenschaften, Heft 18, 1990.

[6] Zasova, L. V., Spänkuch, D., Moroz, V. I. et al.: Venusian clouds from Venera 15 data. Veröffentlichungen des Forschungsbereichs Geo- und Kosmoswissenschaften, Heft 18, 1990.

[7] Moroz, V. I., Zasova, L. V.: Venus' infrared radiation: approximate methods of calculation of the spectrum inside the absorption bands of atmospheric gases. Kosmich. Issled. **23** (1985) 2, 259 (Russian).

[8] Yung, Y. L., Demore, W. B.: Photochemistry of the stratosphere of Venus: Implications for atmospheric evolution. Icarus **51** (1982) 2, 198.

[9] Conway, R. R. McCoy, R. P., Barth, C. A., Lane, A. L.: IUE detection of sulfur dioxide in the atmosphere of Venus. Geophys. Res. Lett. **6** (1979) 7, 629.

[10] Barker, E. S.: Detection of SO_2 in the UV spectrum of Venus. Geophys. Res. Lett. **6** (1979) 2, 117.

[11] Esposito, L. W., Winnick, J. R., Stewart, A. I.: Sulfur Dioxide in the Venus atmosphere: distribution and implications. Geophys. Res. Lett. **6** (1979) 7, 601.

[12] Hoffman, J. H., Hodges, R. R., Donahue, T. M., McElroy, M. B.: Composition of the Venus lower atmosphere from the Pioneer-Venus mass-spectrometer. J. Geophys. Res. **85** (1980) A 13, 7882.

[13] Ehkonomov, A. P., Moshkin, B. E., Moroz, V. I. et al.: Experiment on UV-photometry on board Venera-13 and Venera-14 landers. Kosmich Issled. **21** (1983) 2, 254 (Russian).

[14] Esposito, L. W., Gates, L. J.: Horizontal and vertical distributions of sulfur dioxide on Venus. **62** EOS (1981), 17, 321.

[15] Kawabata, K. et al.: Cloud and haze properties from Pioneer Venus polarimetry. J. Geophys. Res. **85** (1980) A 13, 8129.

[16] Gmitro, J. I., Vermuelen, T.: Vapor-liquid equilibria for aqueous sulfuric acid. A.I.Ch.E. Journal **10** (1964) 5, 740.

[17] Schofield, J. T., Taylor, F. W., McCleese, D. J.: The global distribution of water vapor in the middle atmosphere of Venus. Icarus **52** (1982) 2, 263.

[18] Barker, E. S.: Observation of Venus water vapor over the disk of Venus. The 1972—74 data using the H_2O lines at 8197 and 8176 Å. Icarus **25** (1975) 2, 268.

[19] Moroz, V. I., Parfent'ev, N. A., San'ko, N. F.: Spectrophotometric experiment on board Venera-11 and Venera-12 landers. 2. Analysis of Venera-11 spectral data by layer adding method. Kosmich. Issled. **17** (1979) 5, 727 (Russian).

[20] Moroz, V. I., Moshkin, B. E., Ehkonomov, A. P. et al.: Spectrophotometric experiment on board Venera-13 and Venera-14 landers. — Preliminary results of spectra analysis in the region of H_2O absorption bands. Kosmich. Issled. **21** (1983) 2, 524 (Russian).

[21] Mukhin, L. M., Gel'man, B. G., Lamonov, N. I. et al.: Gas chromatographic analysis of chemical composition of the Venus atmosphere on board Venera-13 and Venera-14 landers. Kosmich. Issled. **21** (1983) 2, 225 (Russian).

[22] Surkov, Yu. A., Ivanova, V. F., Pudov, A. N. et al.: The water vapor abundances in the Venus atmosphere from Venera-13 and Venera-14 data. Kosmich. Issled. **21** (1983) 2, 231 (Russian).

15. Venusian clouds from Venera 15 data

L. V. Zasova, D. Spänkuch, V. I. Moroz, R. Dubois, V. M. Linkin, K. Schäfer, I. A. Macygorin, W. Döhler, D. V. Titov, D. Oertel

Abstract

Spectra of infrared thermal emission of Venus contain an important information concerning composition and structure (altitude and horizontal) of upper clouds. Inside the latitude range between approximately 60° S and 60° N they demonstrate H_2SO_4 absorption bands near 11 and 22 μm. At the higher latitudes these bands are less pronounced and even disappear. Quantitative analysis shows that this could be explained by the change of structure of clouds and temperature profiles. Self-consistent retrieval of altitude profiles of cloud volume absorption coefficient and temperature profiles was fulfilled for the main bulk of Venera-15 infrared spectra. Examples of these profiles in the range from 55—60 km to 90 km are presented and general properties of upper clouds and haze (including latitude and time variations) derived from these data are described.

The Venusian spectra obtained by Fourier-spectrometer aboard the Venera-15 in the wavenumber range 250—1600 cm^{-1} were used to retrieve the vertical temperature profiles and aerosol parameters of the clouds of Venus [1, 2]. The spectral region contains CO_2 15 μm band which can be convenient to obtain a temperature profile in a large interval of altitudes of the Venusian atmosphere.

Both wings of CO_2 15 μm band are free of absorption of other gases. The aerosol absorption becomes noticeable here only at levels lower than 50 hPa. On the other hand there are spectral regions free of any gases absorption and these regions may be used for aerosol soundings. For example these are the region $\nu = 365\ cm^{-1}$, where the coefficient of aerosol extinction of sulfuric acid is minimal and the region $\nu = 1152\ cm^{-1}$ of the maximal coefficient of aerosol extinction. The values of the extinction coefficient for these wavenumbers differ by a factor of 10—30 (depending on the partical sizes distribution). This makes it possible to sound the cloud layer in the height range 57 to 72 km (at low and middle latitudes).

The intensity of the radiation as a function of the wavenumber ν for a semi-infinite purely absorbing medium is determined by the well known solution of the transfer equation

$$I_\nu = \sec\theta \times \int_0^\infty B_\nu[T(z)] \times \varkappa_\nu[p(z),\ T(z)] \times \exp\left\{-\sec\theta \times \int_z^\infty \varkappa_\nu[p(z'),\ T(z')] \times \mathrm{d}z'\right\} \times \mathrm{d}z, \quad (1$$

where z is the altitude, $T(z)$, $p(z)$ are the temperature and pressure at the altitude z, respectively, $B(T)$ is the Planck function, $\varkappa_\nu(p, T)$ is the bulk absorption coefficient, and θ is the angle between the normal and the direction of the radiation.

It is convenient to use the pressure p rather than altitude z, which are related by the hydrostatic equilibrium equation

$$\mathrm{d}p/p = -\mathrm{d}z/H. \tag{2}$$

Here H is the scale height. We introduce then the concept of the transmission function of an atmosphere from a given level to $p = 0$:

$$t_\nu = \exp\left(\sec\theta \times \int_{\ln p}^{\infty} H \times \varkappa_\nu(p', T)\right) \times \mathrm{d}\ln p'. \tag{3}$$

Then equation (1) may be rewritten as

$$I_\nu = \int_{-\infty}^{\infty} B_\nu[T(p)] \times K_\nu(p, T) \times \mathrm{d}\lg p, \tag{4}$$

where

$$K_\nu = -\frac{\partial t_\nu}{\partial \lg p}. \tag{5}$$

Here K_ν is the weighting function. It determines the contribution of some layer of the atmosphere to the intensity of the escaping radiation at a given wavenumber.

We are to determine the vertical profile $T(P)$ or $T(z)$, from the measured spectrum I_ν^{mes} inside the 15 μm CO_2 band, and to retrieve $N(p)$ or $N(z)$ — the vertical profile of particle number density from the measured intensity in continuum spectral channels.

Generally the absorption coefficient is the sum of two quantities

$$\varkappa_\nu = \varkappa_{\nu,\mathrm{g}} + \varkappa_{\nu,\mathrm{a}} \tag{6}$$

where $\varkappa_{\nu,\mathrm{g}}$ is the coefficient of CO_2 gas absorption, $\varkappa_{\nu,\mathrm{a}}$ is the aerosol absorption. Method of calculating $\varkappa_{\nu,\mathrm{g}}$, the transmission, and the weighting functions is presented in [1].

In the aerosol medium outside of the gas absorption bands, we have

$$\varkappa_{\nu,\mathrm{a}} = 10^{-3} \times \pi \times \sum_{i=1}^{N} n_i(z) \times \langle Q \times r^2 \rangle_i, \qquad \mathrm{km}^{-1}, \tag{7}$$

where

$$n_i(z) = \int_0^\infty n_i(z, r) \times \mathrm{d}r, \qquad \mathrm{cm}^{-3}. \tag{8}$$

Here $n_i(z, r)$ is a number density of mode-i-particles with a radius ranging from r to $r + \mathrm{d}r$ at an altitude z.

The parameter $\langle Q \times r^2 \rangle$ is the averaged extinction cross-section (within a factor of π) for mode i.

$$\langle Q \times r^2 \rangle_i = \frac{1}{n_i(z)} \times \int_0^\infty n_i(z, r) \times Q_i\left(\frac{2 \times \pi \times r}{\lambda}\right) \times r^2 \times \mathrm{d}r, \tag{9}$$

where $Q_i\left(\frac{2 \times \pi \times r}{\lambda}\right)$ is extinction efficiency factor.

Table 1.
The channels inside the CO_2 15 μm band used for temperature profiles retrieval

Wavenumber cm^{-1}	Pressure at the level of weighting function maximum hPa	Wavenumber cm^{-1}	Pressure at the level of weighting function maximum hPa
666.75	0.5	666.75	0.5
656.33	1	677.17	1
642.44	2	691.06	2
632.02	5	701.48	5
614.66	12	711.90	10
600.77	20	732.73	16
593.82	25	746.62	30
590.35	40	750.09	35
586.88	45	753.57	45
579.93	50	757.04	50
576.46	80	760.51	80
572.99	100	770.93	90
562.57	150	781.35	150
555.63	200	802.18	200

To retrieve the temperature profile, 27 channels inside the 15 μm CO_2 band were chosen in both wings. The number of channels was increased as compared to [1] to improve the retrieval of temperature profiles, having an inversion. The channels are listed in Table 1. Table 2 contains the spectral channels used for aerosol sounding, and the altitudes at the levels where the weighting functions are maximal for the exponential aerosol model

$$n(z) \sim \exp\left[(H_1 - z)/H_a\right] \tag{10}$$

with $H_1 = 68$ km and $H_a = 3.5.$ km. Here H_1 is the altitude where $\tau_{(1152)} = 1$ and H_a is the aerosol scale height.

The CO_2 transmission functions as well as the weighting functions were calculated for three models specified a-priori [1, Fig. 2]. This makes it possible to obtain interpolation coefficients which were used for calculating the transmission function for real temperature profiles.

The highest level for which the transmission functions were calculated is 10^{-2} hPa level. However, for higher channels near the CO_2 band center the transmission functions are not equal to unity at the 10^{-2} hPa level and accordingly the weighting functions are not equal to zero here. To account for the contribution of the upper tails of the weighting functions we suggested that all transmission functions are equal to unity and all weighting functions equal to zero at levels higher than 10^{-4} hPa.

To solve the inverse problem we resort to relaxation method [3, 4, 5]. To retrieve the aerosol profiles ten channels free of gas absorption were used (Tab. 2). In principle not all of these channels are independent. Whether some channels are dependent or independent may be determined by aerosol model. The altitudes corresponding to the maxima of the weighting functions are presented in Table 2 for two aerosol models. As

Table 2.
The channels used for aerosol sounding

Wavenumber cm^{-1}	Altitude of the weighting function maximum km*)	Wavenumber cm^{-1}	Altitude of the weighting function maximum km*)
364.63	58	1000.12	62.8
406.30	58.5	1107.78	63.2
451.45	59	1152.92	70
871.64	61.5	1201.54	64
899.42	62.5	1312.66	62.5

*) The altitude are for exponential aerosol model with $H_1 = 68$ km and $H_a = 3.5$ km.

one can see from Tab. 1 maximum of the lowest weighting function for the 802 cm^{-1} channel is at the 200 hPa level ($\sim$60 km). However the radiation in the 365 cm^{-1} spectral region — the lowest aerosol channel, is formed somewhat lower in the atmosphere, the level at which $\tau_{(365\,\mathrm{cm}^{-1})} = 1$ is situated at 57—60 km depending on latitude. Thus to calculate the radiation near 365 cm^{-1} it is necessary to know the temperature at levels which are lower than we can obtain from our temperature sounding.

It is known from radio occultation data and from in situ measurements (Venera-9—14, 15, 16 and Pioneer-Venus) which were summarized in VIRA [6] that the temperature at 50 km level is stable enough and only depends on altitude. From the radio occultation data the altitude of tropopause for different latitudes was defined. So we shall use the retrieved temperature profiles above the tropopause to extrapolate them at lower altitudes using an adiabatic lapse rate to get the true value of temperature at the 50 km level.

The temperature vs. latitude dependence at the 50 km level was taken from the VIRA model [6] and is shown in Fig. 3.

To retrieve the aerosol number density we should accept some hypothesis concerning the wave length dependence of extinction cross section, which is governed by both the absorption coefficient of particle material and the particle size distribution. Particle size distribution are obtained from Earth-based [7] and onboard polarization observations [8] as well as from the in-situ measurements by Venera 9, 10, 11 [9] and Pioneer-Venus probes [10]. The possible particle size distributions ever suggested for different altitudes in Venusian clouds are summarized in VIRA [6]. The spectral extinction coefficients for some of them are presented in Fig. 2. We have used two kinds of particle size distribution: the gamma distribution

$$n(r) \sim r^{-(1-3\times\nu_{\mathrm{eff}})/\nu_{\mathrm{eff}}} \times \exp\left(-r/r_{\mathrm{eff}} \times \nu_{\mathrm{eff}}\right) \tag{11}$$

and the log-normal one

$$n(r) \sim \frac{1}{r \times \ln \sigma} \times \exp\left\{-\frac{1}{2} \times \left[\frac{\ln (r/r_0)}{\ln \sigma}\right]^2\right\}, \tag{12}$$

with different values of parameters ν_{eff}, r_{eff}, σ and r_0.

For the calculations within the framework of the Mie theory we have used the sulfuric acid optical constants obtained by Palmer and Williams [11] and Jones [12]. We proceeded from the generally accepted hypothesis that Venusian clouds consist of concentrated (about 80%) sulfuric acid. The values of the parameter $\langle Q \times r^2 \rangle$ for 75% and 84.5% H_2SO_4 are shown in Fig. 1 for the particle size distribution corresponding to curve 2, Fig. 2. The values of this parameter for the two different acid concentrations differ noticeably in the spectral range $\nu < 550\ cm^{-1}$. For the different mode 2 particle size distributions (with the same sulfuric acid concentration) the extinction coefficient may change several times in our spectral region. So the different behavior of $\langle Q \times r^2 \rangle$ for different particle size distributions as well as for the different sulfuric acid concentration requires its updating. As the initial extinction cross section wavelength dependence we take one corresponding to mode 2, curve 2, Fig. 2. The sulfuric acid concentration and the size of aerosol particles may vary with altitude in the clouds. So using our spectral data we can only retrieve the altitude profile of the extinction cross section or $\tau_a(h)$. However τ_a depends on wavenumber so it is more convenient to use the concept of the 'equivalent number density', that is the particle number density which is obtained under the assumption that throughout the Venusian clouds only mode 2 particles are present with the parameter $\langle Q \times r^2 \rangle$ determined by curve 2, in Fig. 2. Sharp increase in the retrieved particle number density ($h < 60$ km) may be attributed to an increase of their effective size, for example, to the presence of large mode 3 particles [10]. For hypothetical mode 3 particles ($r_0 = 3.8\ \mu m$) the extinction cross section differs (for $\nu = 365\ cm^{-1}$) from one for mode 2 particles ($r_0 = 1.05\ \mu m$) by a factor of 150. We shall not consider the haze because of its negligible optical depth in our spectral region.

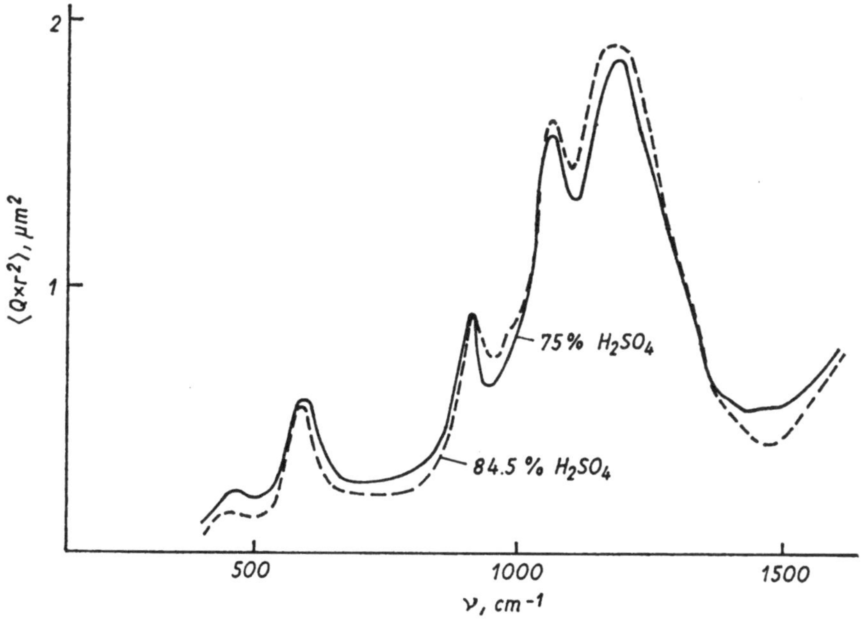

Fig. 1. Parameter $\langle Q \times r^2 \rangle$ for 75% and 84.5% H_2SO_4 for log-normal particle sizes distribution with $r_0 = 1.05\ \mu m$, $\sigma = 1.21$

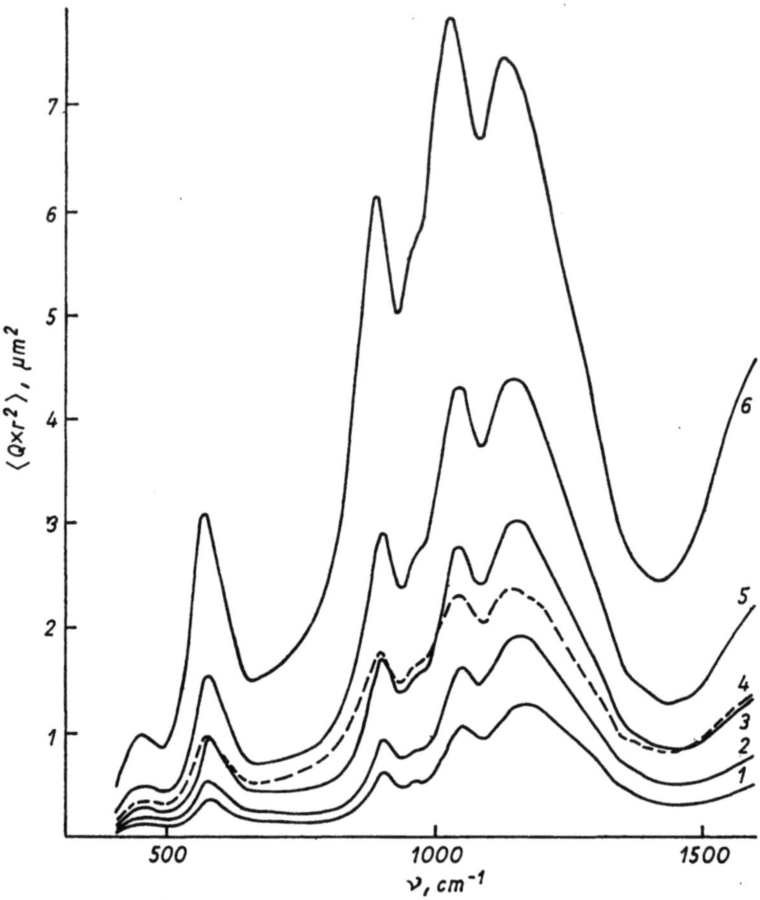

Fig. 2. $\langle Q \times r^2 \rangle$ vs. ν for different particle sizes distribution for mode 2: 1 — gamma-distribution (Γ); $r_{eff} = 1.05$ µm, $\nu_{eff} = 0.07$; 2 — log-normal ($l - n$), $r_0 = 1.05$ µm, $\sigma = 1.21$; 3 — Γ, $r_{eff} = 1.35$, $\nu_{eff} = 0.04$; 4 — Γ, $r_{eff} = 1.6$ µm, $\nu_{eff} = 0.2$, 5 — $l - n$, $r_0 = 1.4$ µm, $\sigma = 1.23$, 6 — Γ, $r_{eff} = 1.9$ µm, $\nu_{eff} = 0.04$

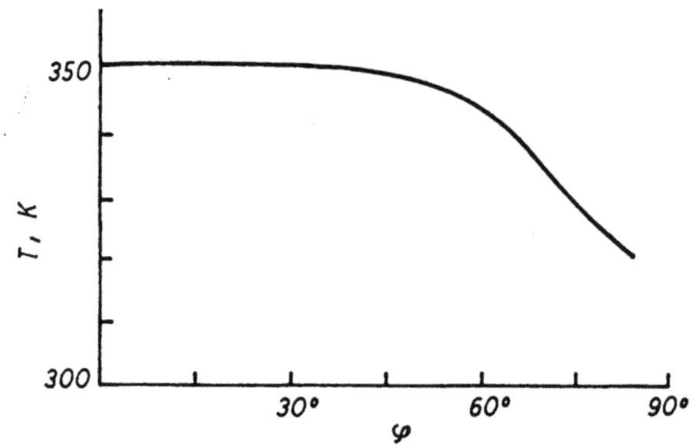

Fig. 3. Temperature at altitude 50 km in the Venus atmosphere as a function of latitude

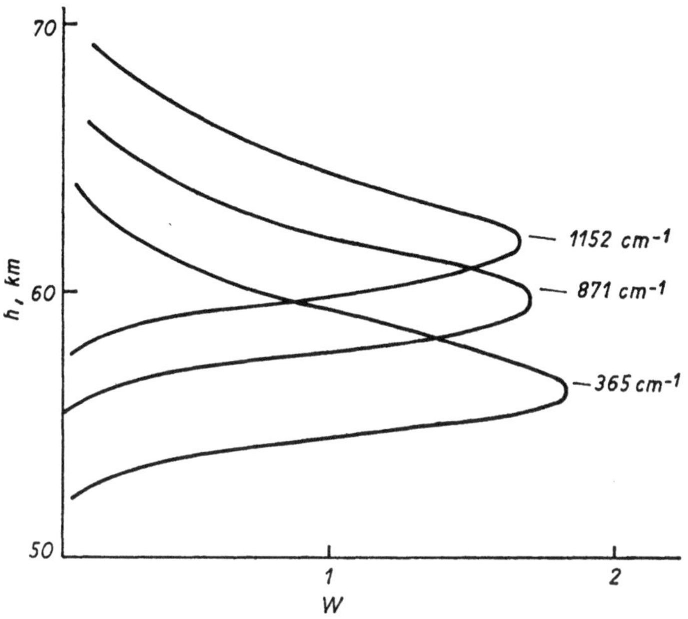

Fig. 4. Weighting functions for exponential aerosol model with $H_1 = 64$ km, $H_a = 2$ km

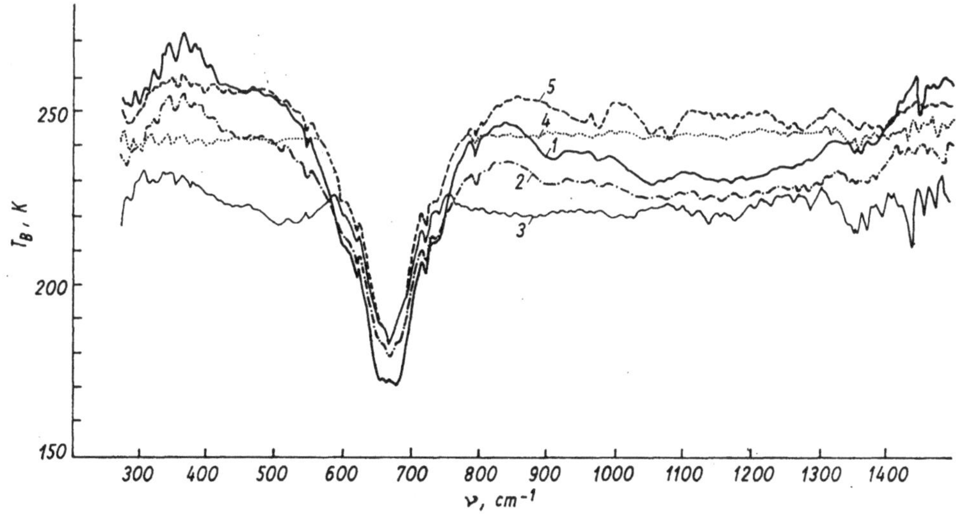

Fig. 5. Spectra averaged for homogeneous spectral groups corresponding to five basic latitudinal zones of Venus:

1 — group 1, $\varphi = 0.5-9°$-equatorial; 2 — group 2, $\varphi = 36-71°$-mid-latitudal; 3 — group 3, $\varphi = 66-73°$-'cold collar'; 4 — group 4, $\varphi = 85-87°$ — polar; 5 — group 5, $\varphi = 77-80°$-'hot dipole'

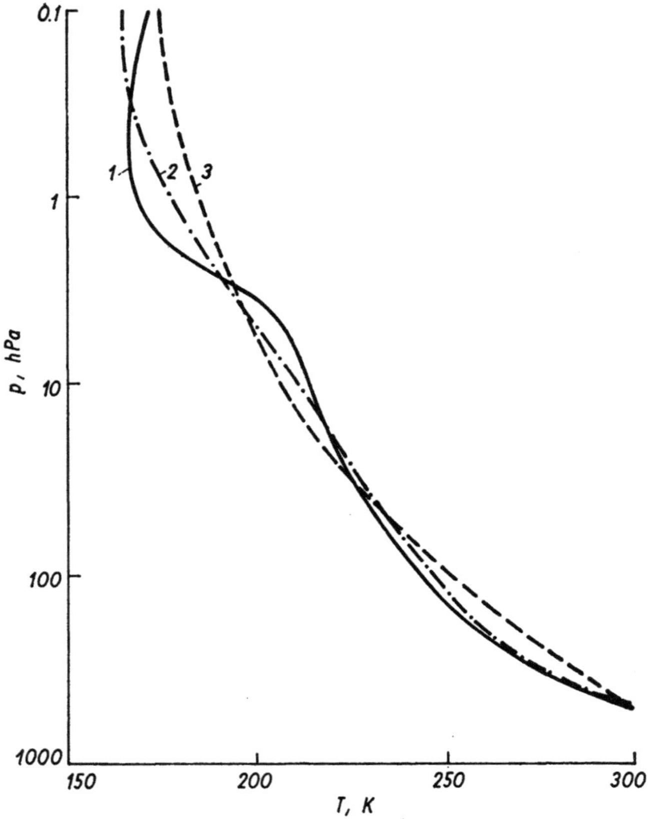

Fig. 6a)

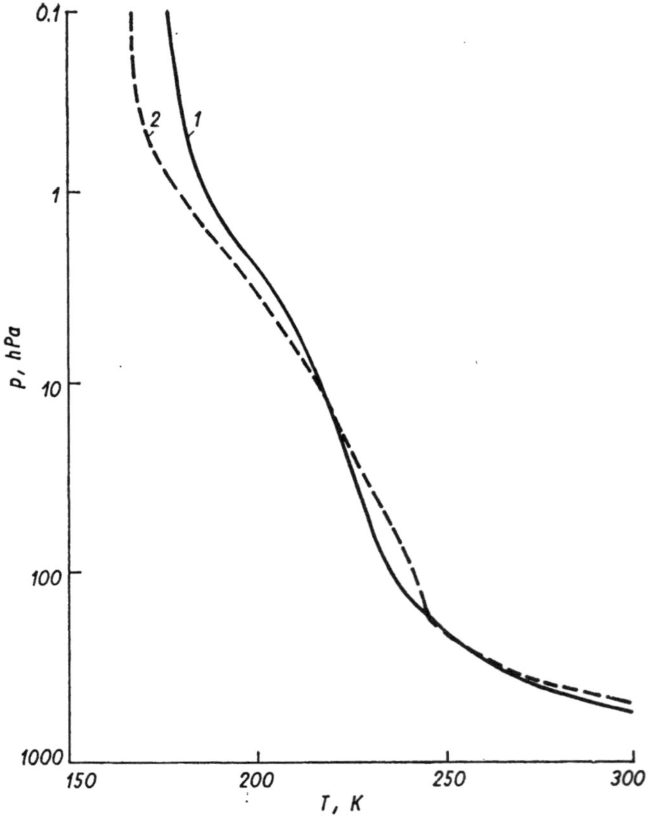

Fig. 6b)

Fig. 6. Temperature profiles for five typical groups of spectra:

a) 1 — gr. 1, 2 — low latitudal model profile [1]; 3 — $\varphi < 30°$ (VIRA);

b) 1 — gr. 2, 2 — $\varphi = 45°$ (VIRA)

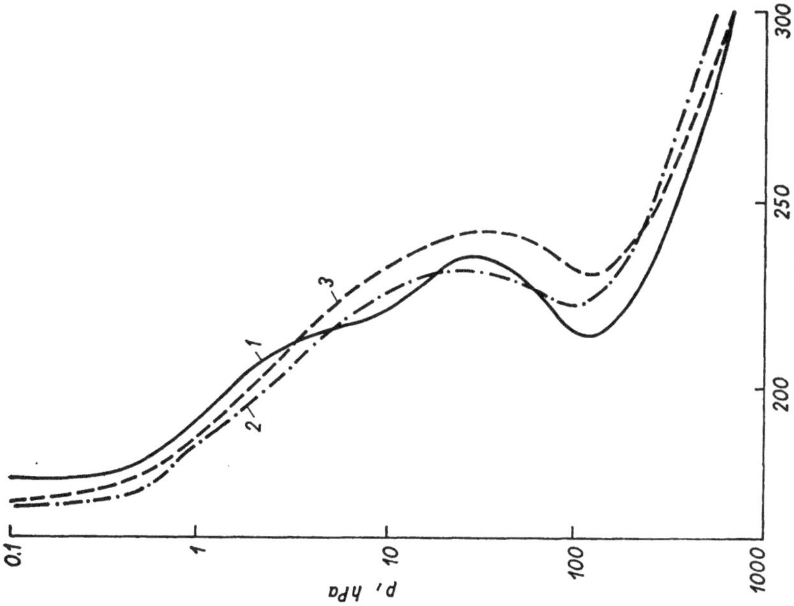

Fig. 6c)

c) 1 — gr. 3, 2 — $\varphi = 60°$, 3 — $\varphi = 75°$ (VIRA)

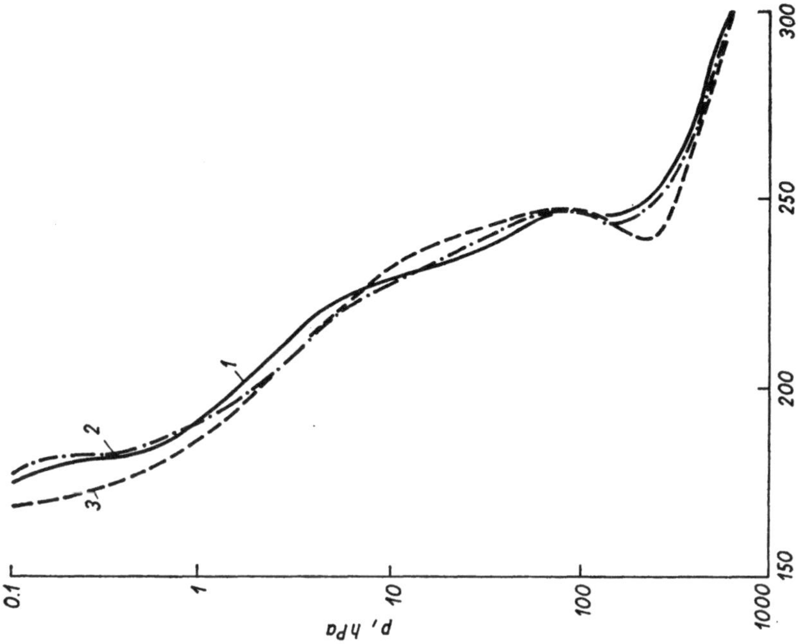

Fig. 6d)

d) 1 — gr. 4, 2 — gr. 5, 3 — $\varphi = 85°$ (VIRA)

The relaxation method was used for the temperature and aerosol profiles retrieval. The method consists of the following steps:

1. The temperature profile $T(p)$ was retrieved from intensity $I\nu_i$, measured in channels ν_i, given in Tab. 1.
2. The aerosol profile $N(p)$ or $N(z)$ was retrieved for aerosol channels listed in Tab. 2.
3. We retrieved $T(p)$ taking into account the presence of aerosol. Aerosol particles have the values of $\langle Q \times r^2 \rangle$ for mode 2 ($r_0 = 1.05$ µm, $\sigma = 1.21$) and $N(p)$ obtained at the second step.

Steps 2 and 3 are repeated until the residual is less than the measurement errors: $\sigma_i < \sigma_{imes}$ (a) in each channel, or when the convergence is reached and the condition $\partial\sigma/\partial n < 10^{-6}$ (b) is fullfilled (n is the iteration number). If the condition (b) is met and condition (a) is not it may indicate that real spectral dependence of $\langle Q \times r^2 \rangle$ values differ from the one accepted as an initial approximation. In this case we made some iteration to update $\langle Q \times r^2 \rangle$. Then we again returned to step 2 until both conditions (a) and (b) were met.

As a first-guess for $T_0(p)$ we have taken the temperature profile obtained for low latitudes [1] and presented in Fig. 6a, curve 2. It is the same for all latitudes. As aerosol profile first-guess $N_0(p)$ we have taken the aerosol models obtained during a preliminary analysis of our data [2], the profiles being different for different latitudinal zones. As an initial approximation $\langle Q \times r^2 \rangle$ we used curve 2, Fig. 2 for 75% H_2SO_4.

Figure 5 presents several typical spectra, obtained by averaging over the homogeneous groups of spectra [2]. In Fig. 6 the temperature profiles retrieved for the spectral groups are compared with the model ones (VIRA) for corresponding latitudinal zones. One can

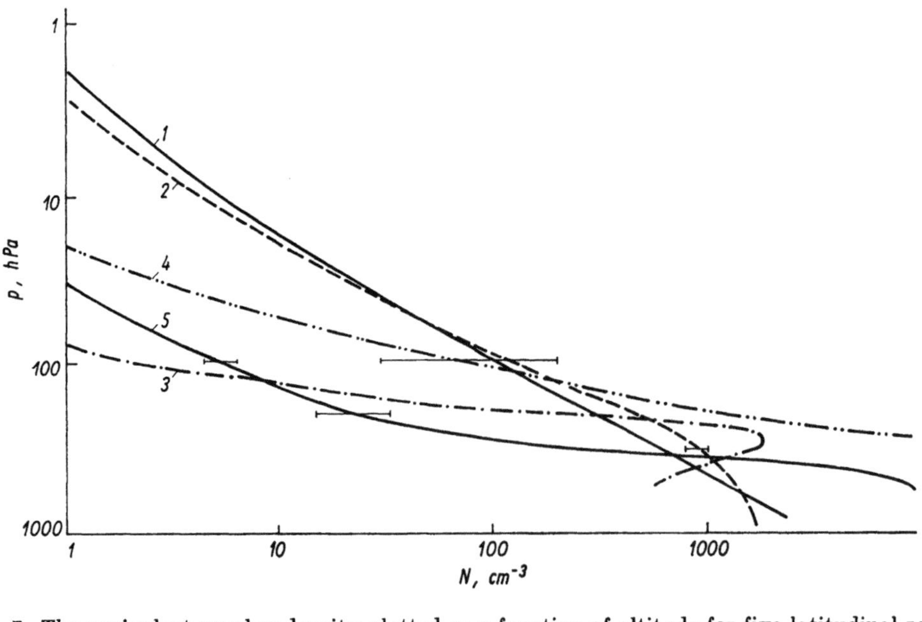

Fig. 7. The equivalent number density plotted as a function of altitude for five latitudinal zones of planet. The figures show the spectral group number

see that the temperature profiles obtained for equatorial region differ from the models at $p < 10$ hPa, where the inversion is observed in retrieved temperature profiles. Obtained mid-latitudinal profiles are similar to the model ones, but the temperature at levels where $p < 10$ hPa is higher than the model predicts. Retrieved high-latitudinal temperature profiles behave in the same manner as the corresponding model profiles. Throughout we used profile 2 in Fig. 6a as a first-guess.

The vertical number density profiles are presented in Fig. 7. In equatorial and mid-latitudinal zones (spectral groups 1 and 2), the cloud layer is rarefied enough in its upper part and has a scale height of about 4 km. Spectral group 3 characterizes the region of the cold collar. The inversion in the wings of the 15 μm CO_2 band is well defined. Here the upper boundary of the clouds is situated lower than in mid-latitudes and the scale height being less or equal to 1 km. The lowest position of the upper boundary is observed in hot spots (dipole). At pole the cloud layer has sharp upper boundary and is situated higher than in the dipole.

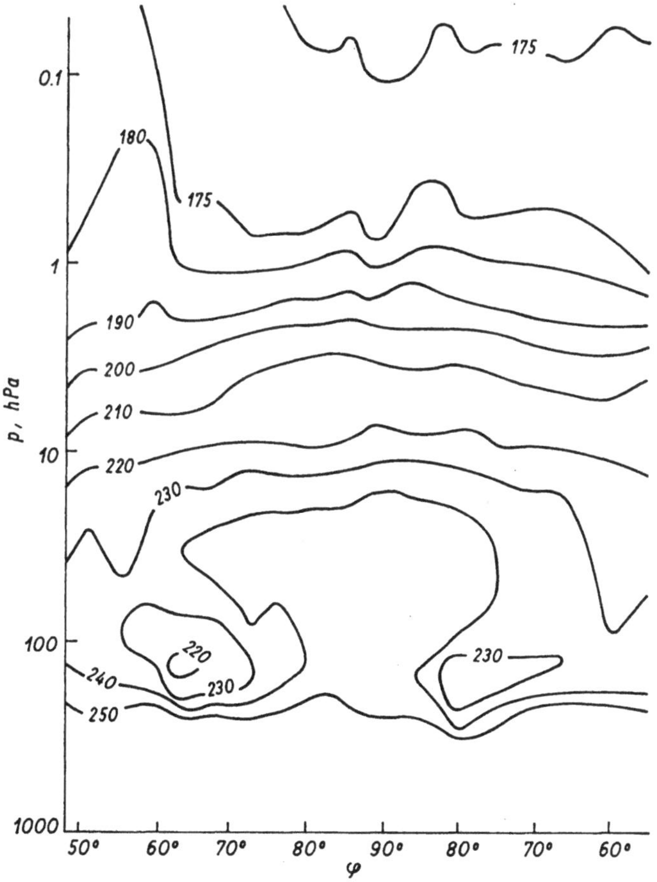

Fig. 8. The temperature field for session 12 Oct. 1983 contoured as a function of pressure and latitude

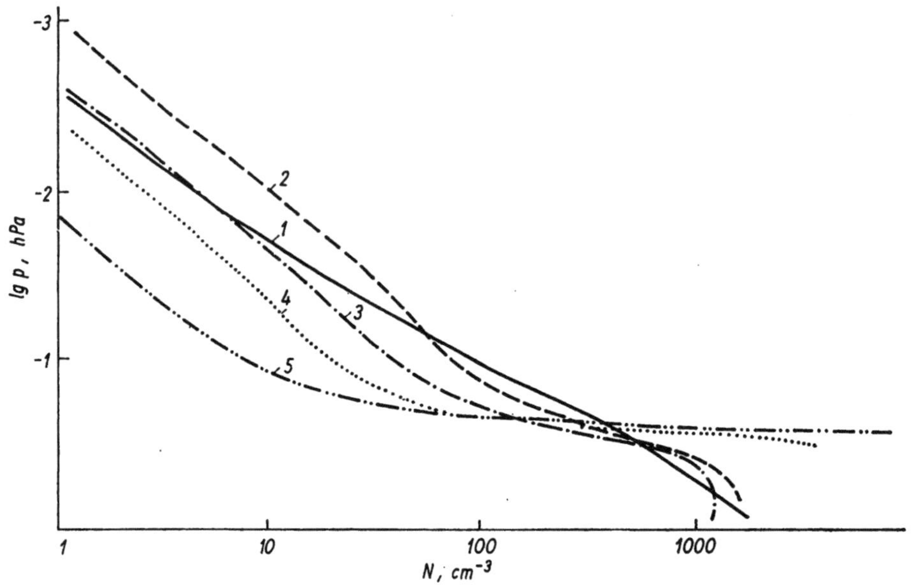

Fig. 9a)

Fig. 9. Vertical aerosol profiles for different latitudes for session Oct. 12

a) ascending part of trace, 1 — $\varphi = 43°$, 2 — $\varphi = 66°$, 3 — $\varphi = 73°$, 4 — $\varphi = 80°$, 5 — $\varphi = 83°$

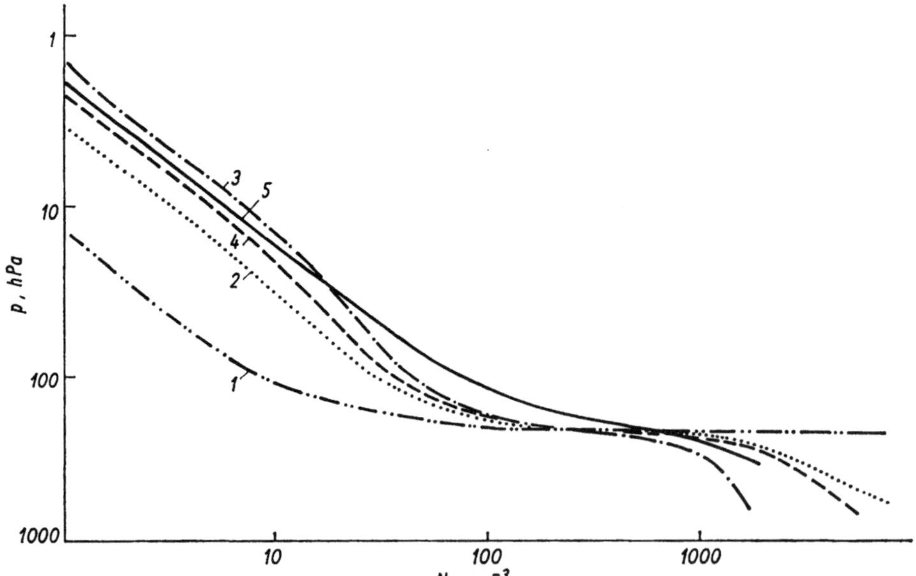

Fig. 9b)

b) descending part, 1 — $\varphi = 87°$, 2 — $\varphi = 80°$, 3 — $\varphi = 73°$, 4 — $\varphi = 63°$, 5 — $\varphi = 44°$

The measurements by the Fourier-spectrometer were organized in the following manner: the measurements began near the equator (first limb crossing), passed through a point of 2.7° from the north pole (the line of sight is directed almost along the nadir here) and ended in the equatorial region on the other side from the pole (second limb crossing). The part of the session before passing through the point closest to the pole will be called the ascending part, and the other part will be called the descending part. During the first measurements the ascending part of the traces was on the night side at about 30° from the morning terminator, while the discending part was on the day side at about 30° from the evening terminator (October 12). The trace of Nov. 5 passed almost along the terminator and afterwards the ascending part was on the day side of the planet. During the Nov. 11. session the equatorial region and south latitudes were observed: the trace began at $\varphi = 66°$ in the southern hemisphere and ended at $\varphi = 20°$ in the northern hemisphere.

The spectra change gradually during every session changing from the spectra with strong variations of the brightness temperature, so that the differences between T_B (365 cm^{-1}) and T_B (1152 cm^{-1}) reaches ≈ 40 K at equator and mid-latitudes, to isothermal ones in the polar region (the 15 μm CO_2 band is not considered), and vice versa after passing through the pole. The session (exception is Nov. 11) is symmetric with respect to the pole, the asymmetry is observed in the latitudinal zone $\varphi > 65°$ only which is characterized by such specific features as the cold collar and the hot dipole.

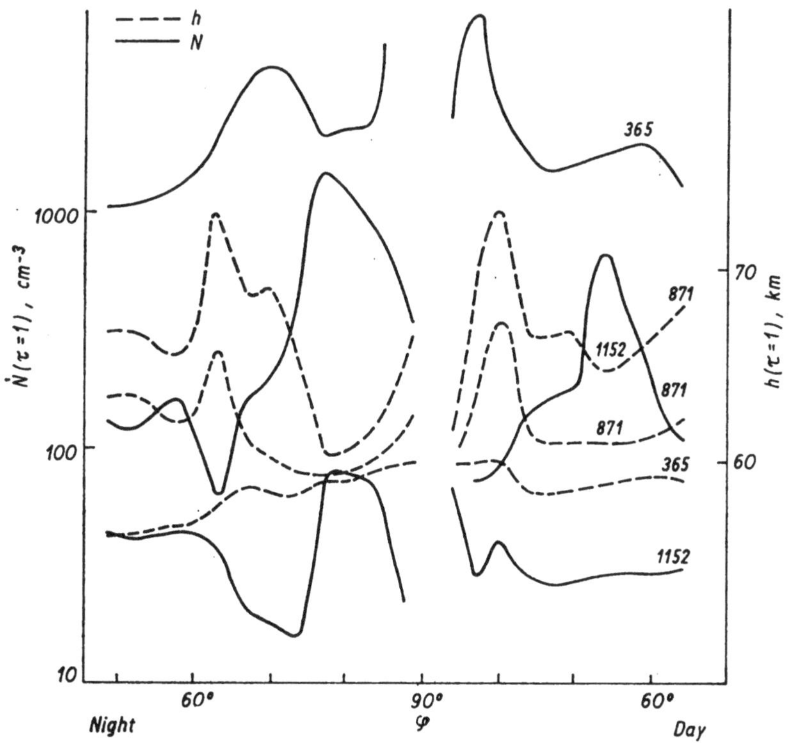

Fig. 10. The altitude of $\tau_a = 1$ level plotted as a function of latitude and equivalent number density at this level for three wavenumber: 365, 871, 1152 cm^{-1} for Oct. 12 session

Because the spectra vary gradually during the session it is possible to use the temperature and aerosol profiles obtained for one spectrum as initial approximations to retrieve such profiles for the next spectrum. As a rule, the first spectra in the session were measured at low latitudes, where the temperature and aerosol profiles are of the most simple monotonous nature.

Our aim was to obtain the difference between calculated and measured spectra as small as 1 K (which may be smaller than measurement errors in the aerosol channels). Generally not all aerosol channels are independent. Excess of channels improve accuracy of retrieval vertical profiles. So the measurement errors may influence the retrieving $\langle Q \times r^2 \rangle$ values more than vertical number density profile. The obtained spectral dependence of $\langle Q \times r^2 \rangle$ differs from the initial one because of the differences which exist in reality as well as of the measurement errors.

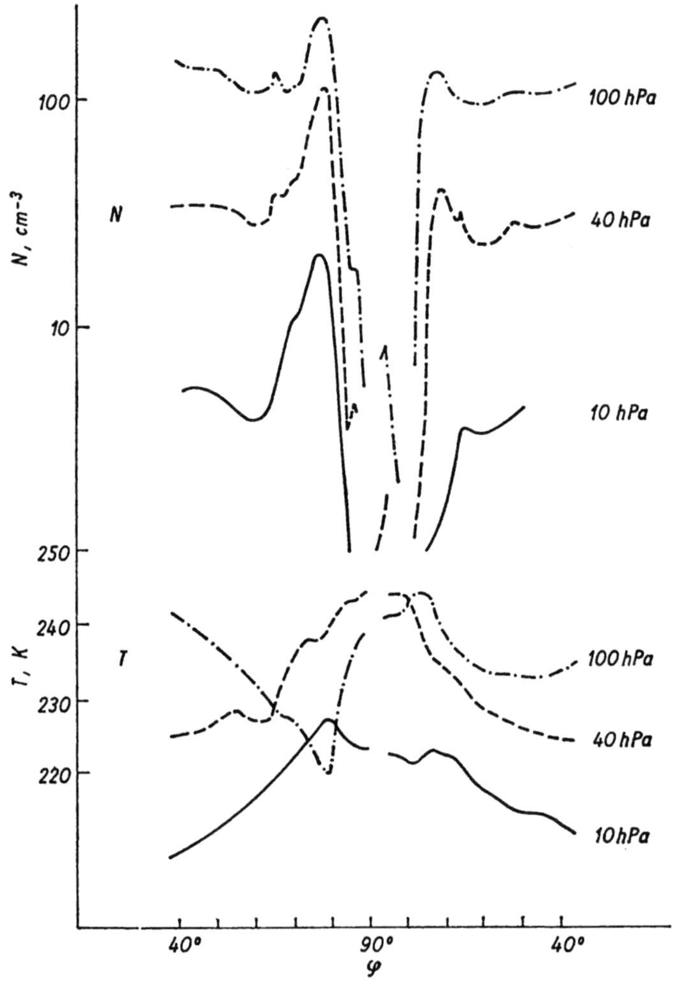

Fig. 11. Latitude dependence of equivalent number density at three isobaric levels for Oct. 12 session

The temperature field for the first session Oct. 12 is presented in Fig. 8. The warm zone above the pole ($p < 100$ hPa) and the features of the cold collar at a level where $p = 100$ hPa are well pronouced. The vertical number density profiles for this session are given in Fig. 9a for ascending part of trace and in Fig. 9b for the descending part. If we go over from one spectrum to another the cloud scale height at first slightly increases from ≈ 4 km at low latitudes to 5—6 km in the direction of the cold collar then it drops in the cold collar and is going over from the cold collar to the pole. After the pole the changes at the descending part of trace occur in the reverse order (Fig. 9b).

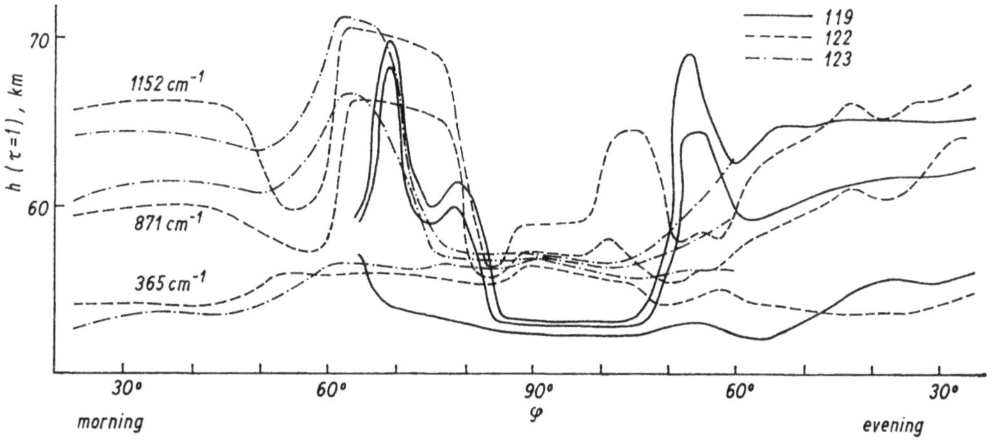

Fig. 12. Altitude of $\tau_a = 1$ level for three sessions, closest to the terminator

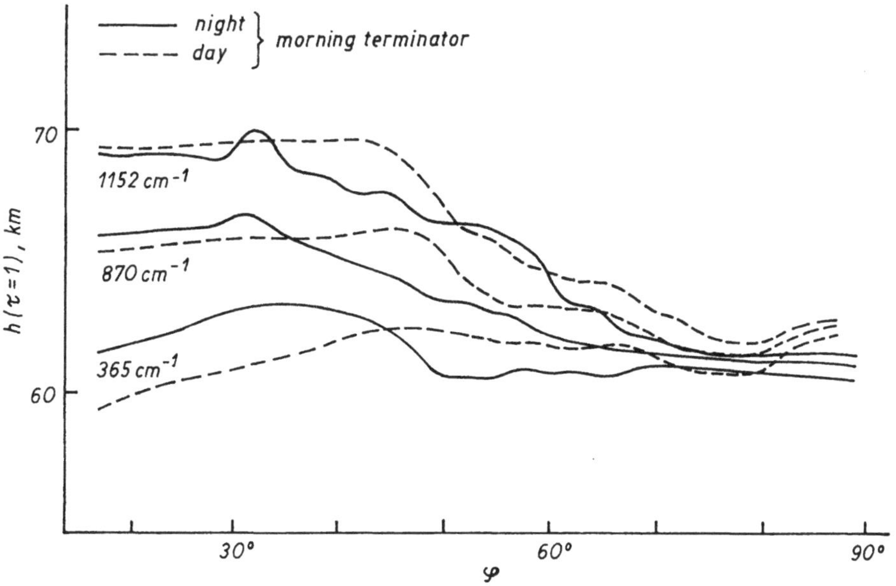

Fig. 13. Altitude of $\tau_a = 1$ level near the terminator at the day side (solid line) and at the night side (dashed line)

Fig. 10 shows the height of the $\tau_a = 1$ level for three chosen wavenumbers and number density vs. latitude at these altitudes.

The temperature and number density at three isobaric levels for session Nov. 14 are given in Fig. 11. One can see that the temperature increases with increasing latitude except for the 100 hPa level in the ascending part of the trace where the cold collar is observed. There is a sharp minimum in the number density at the polar region which indicates that the upper boundary of the cloud layer is lower than 100 hPa level.

The altitudes of the level $\tau_a = 1$ for three sessions, passing along the terminator are presented in Fig. 12. Session Nov. 10 is the nearest of them to the terminator. For this session the cold collar is observed at ascending and descending part of trace and has an equal intensity. Fig. 13 allows to compare the regions of morning terminator at day and

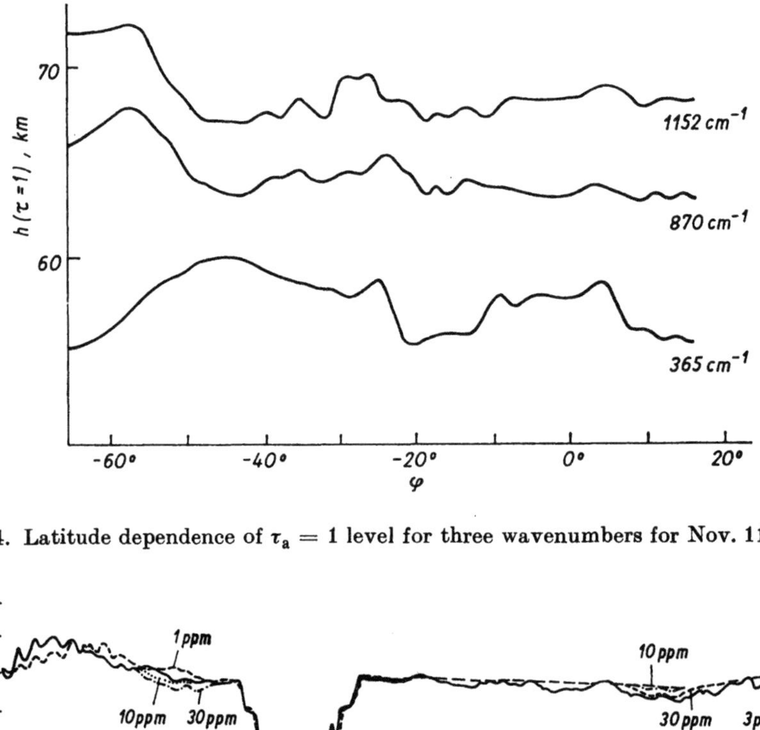

Fig. 14. Latitude dependence of $\tau_a = 1$ level for three wavenumbers for Nov. 11 session

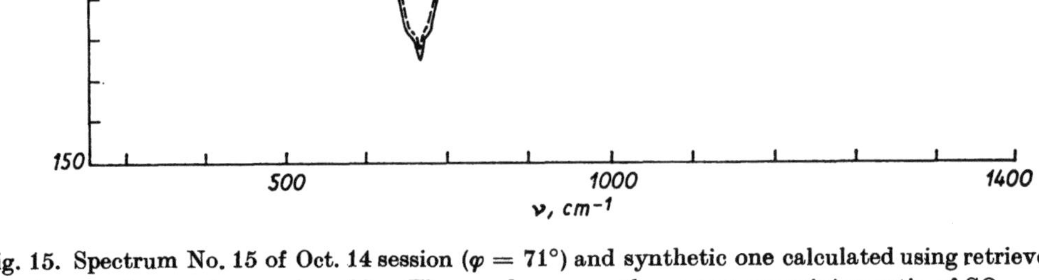

Fig. 15. Spectrum No. 15 of Oct. 14 session ($\varphi = 71°$) and synthetic one calculated using retrieved temperature and aerosol profiles. The numbers near the curves are mixing ratio of SO_2

night sides. One can see some differences between temperatures at the level $\tau_a = 1$ for the day and night sides but in the same manner it may be spatial variations. The existence of spatial inhomogeneities in equatorial region one can see in Fig. 14, which are especially noticeable at altitudes where $\tau_{(365 cm^{-1})} = 1$.

We compare the synthetic spectrum calculated for retrieved $T(p)$ and $N(p)$ with the measured one (spectrum No. 15 in session Oct. 14). This spectrum relates to the cold collar. We have taken into account the presence of SO_2 with constant mixing ratio which is indicated at the curves in Fig. 15 and H_2O with definite model concentration. A good coincidence between measured spectrum and calculated one exists except for the spectral regions where absorption bands of SO_2 and H_2O are situated.

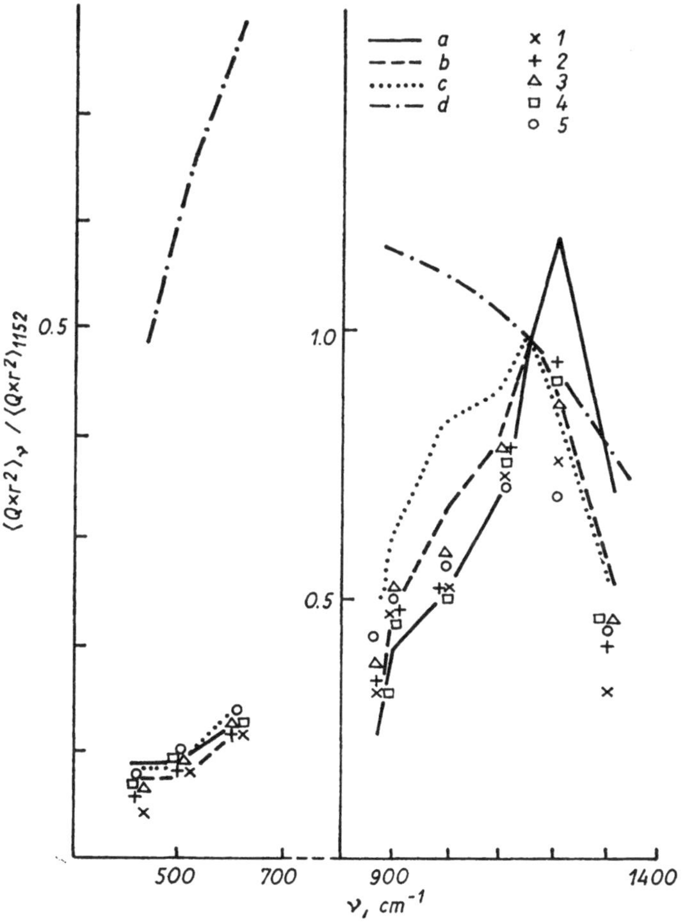

Fig. 16. The normalized $\langle Q \times r^2\rangle_\nu/\langle Q \times r^2\rangle_{1152}$ extinction cross-section for ten aerosol channels. The values obtained were averaged by ten spectra inside of several latitudinal zones:

1 — $\varphi < 30°$, Nov. 11; 2 — $30° < \varphi < 45°$, Nov. 11; 3 — $45° < \varphi < 60°$, Dec. 3; 4 — $60° < \varphi < 80°$, Dec. 3; 5 — $80° < \varphi < 87°$, Dec. 3. The calculated values are given (for 75% H_2SO_4): a — Γ, $r_{eff} = 0.23$ µm, $v_{eff} = 0.18$, b — $l - n$, $r_0 = 1.05$ µm, $\sigma = 1.21$, c — $l - n$, $r_0 = 1.4$ µm, $\sigma = 1.23$, d — $l - n$, $r = 3.85$ µm, $\sigma = 1.30$

We compared the retrieved $\langle Q \times r^2 \rangle$ values with calculated ones (Fig. 16). It is clear that the hypothesis about H_2SO_4 aerosol nature is good for all investigated latitudinal zones.

Summary

The upper clouds of Venus consist of the concentrated sulfuric acid at all latitudes involved. There are H_2SO_4 absorption bands near ≈ 400 and $900\ cm^{-1}$ in spectra, obtained inside the 60° S to 60° N latitude range. These bands disappear in high latitudinal spectra because of the isothermal temperature profiles and sharp upper boundary of clouds. The structure of the clouds change with latitude. In low and mid-latitudes the clouds are rarified in their upper part, and they have sharp boundary at high latitudes. The temperature in cold collar may be 30—40 K less than in the polar region. The hot dipole is characterized by lower position of upper cloud boundary. Aerosol scale height is about 4 km at low and mid-latitudes, increases to cold collar to 5—6 km, and then drops to $\leqq 1$ km in polar region. The $\tau_a = 1$ level is at different altitudes depending on latitude. At low and mid-latitudes H_1 (365) $\approx$ 57 km; H_1 (871) $\approx$ 61—64 km; H_1 (1152) $\approx$ 67—69 km. In polar region the level $\tau_a = 1$ is at the altitude range 56—64 km for all three wavenumbers.

References

[1] Spänkuch, D., Zasova, L. V., Schäfer, K. et al.: Infrared experiment aboard the automatic interplanetary stations Venera 15 and Venera 16. 2. Preliminary results of temperature profile retrieval. Veröffentlichungen des Forschungsbereichs Geo- und Kosmoswissenschaften, Heft 18, 1990.

[2] Zasova, L. V., Spänkuch, D., Moroz, V. I. et al.: Infrared experiment aboard the automatic interplanetary stations Venera 15 and Venera 16. 3. Some conclusions on the structure of clouds, based on the analysis of spectra. Veröffentlichungen des Forschungsbereichs Geo- und Kosmoswissenschaften, Heft 18, 1990.

[3] Chahine, M. T.: Inverse problem in radiative transfer: determination of atmospheric parameters. J. Atm. Sci. **27** (1970), 960.

[4] Twomay, S.: Comparison of constrained linear inversion and iterative nonlinear algorithm. J. Compt. Phys. **18** (1975), 188.

[5] Schofield, J. T., Taylor, F. W.: Measurements of the mean solarfixed temperature and cloud structure of the middle atmosphere of Venus. Quart. J. R. Met. Soc. **109** (1983), 57.

[6] The Venus International Reference Atmosphere. Adv. in Space Res. **5** (1985) 11.

[7] Hansen, J. E. and Hovenier, J. W.: Interpretation of the polarisation of Venus. J. Atm. Sci. **31** (1974), 1137.

[8] Kawabata, K. et al.: Cloud and haze properties from Pioneer Venus polarimetry. J. Geophys. Res. **85** (1980), 8129.

[9] Marov, M. Ya. et al.: The structure and microphysical properties of the Venus clouds: Venera-9, 10 and 11 data. Icarus **44** (1980), 608.

[10] Knollenberg, R. G. and Hunten, D. M.: The microphysics of the clouds of Venus results of the Pioneer Venus particle size spectrometer experiment. J. Geophys. Res. **85** (1980), 8039.

[11] Palmer, K. F. and Williams, D.: Optical constants of sulfuric acid: appplication to the clouds of Venus? Applied Optics **14** (1975), 208.

[12] Jones, A. D.: Optical constants of sulfuric acid in the far infrared. J. Q. Spec. Rad. Transf. **16** (1976), 1017.

16. The structure of the aerosol medium in the Venusian atmosphere from the data of ISAV-A experiment on Vega descent probes

B. E. Moshkin, V. I. Moroz, V. I. Gnedykh, A. V. Grigor'ev, L. V. Zasova and A. P. Ehkonomov

Abstract

The aerosol particle size distribution and number density profiles have been measured by means of the aerosol spectrometer at 55—33 km altitude. Approximately the same structure as obtained in previous in-situ aerosol studies on Venus is found: 3-layers main cloud cover, intermediate region and undercloud haze. However there are large differences in number densities and size distributions. The number density of particles with diameter $D > 0.4$ μm is only in a few times less in the haze than in main clouds. The size distribution is satisfactory described by Junge law. No sign of the earlier by Knollenberg and Hunten proposed three modal distribution was found.

The backscattering coefficient was measured. It shows only a little variation between 63 and 32 km altitude. Tropospheric nightglow was discovered at about 1 μm wavelength. A large amount of submicron particles in the atmosphere above the landing sites seems to be present.

The main objective of the experiment was to determine the aerosol particle size distribution, the profile of particle number density, and particle shape and refractive index.

The experiment has been fostered to a certain extent by the uncertainty in the mode-2 existence (Knollenberg, 1980). Soon after the hypothesis on the three mode composition of Venus clouds has been published, the authentity of detecting the third mode became questinable (Toon, 1984).

Each VEGA lander carried an ISAV-A instrument to study the aerosols of Venusian clouds.

The ISAV-A instrument comprises a photoelectric aerosol spectrometer and the backscattering sensor. The spectrometer measured the light fluxes scattered in four directions as individual aerosol particles passed through the instrument. The backscattering sensor measured the light scattered by the aerosol medium near the instrument, and/or the atmosphere glow. The block-diagram is given in Fig. 1. The light from a halogen filament lamp 1 (5 W) with a mirror system 2 is directed to the so-called counting volume (0.5 mm^3) where the light scattering by an aerosol particle is measured. The fluxes of light scattered in forward, backward and lateral directions (the appropriate scattering angles are 7°—17°, 165°—175°, 25°—65°) are received by four optical-electronic units 3, 4, 5, 6, where the solicon avalanche photodiodes are used as photodetectors. Two lateral detectors operate with similar scattering angles but with different azimuths.

The back-scattering sensor and the aerosol spectrometer are mounted in the same case. The light from lamp 1 is directed by reflector 7 via window 8 to the atmosphere. The back-scattered light (160° — 175°) enters the instrument via window 9 and is received by the optical-electronic unit 10. The effective operating volume of the backscattered sensor is equal to 0.5 cm^3 and its center is separated from the window by about 3 cm. Periodically each 2 min. lamp 1 is switched off and during 1.5 s the backscattering sensor records only the external radiation, that is it operates as a photometer.

The output pulses of the four photodiodes of the aerosol spectrometer corresponding to the passage of a particle are registered, their logarithms are taken, and the maximum value of each pulse is stored in the memory till the next particle passes. The instrument is interrogated twice per 0.43 s, and the amplitude values of the four pulses which characterize the phase function of each particle are transmitted to the Earth.

The instrument includes a 64-channel amplitude analyzer which receives all pulses of the photodiode which measures scattering in the lateral direction. The analyzer is designed for determining the size spectrum of all particles passing through the instrument during 27 s. The analyzer's data are transmitted to the Earth at the same intervals.

When the lander descends particles are introduced into the counting volume with an incoming gas flux through an aspirator (see Fig. 1). It consists of the coaxially arranged input nozzle 11, oriented downward in the descent direction, tube 12 connected with tube 13, and capillar 14. To produce aerodynamic focusing the purified atmospheric gas is let into the instrument through the branch pipe 15 and the aerosol filter 16.

The minimum diameters of the particles registered by the VEGA-1 and VEGA-2 instruments are 0.9 and 0.5 μm, respectively, with the upper limit being about 25 μm.

The backscattering sensor measured the baskscattering coefficients ranging from 3×10^{-6} to $10^{-3}\,m^{-1} \times sr^{-1}$ and the upward radiation intensity from 0.1 to 50 $mW/m^2 \times sr$.

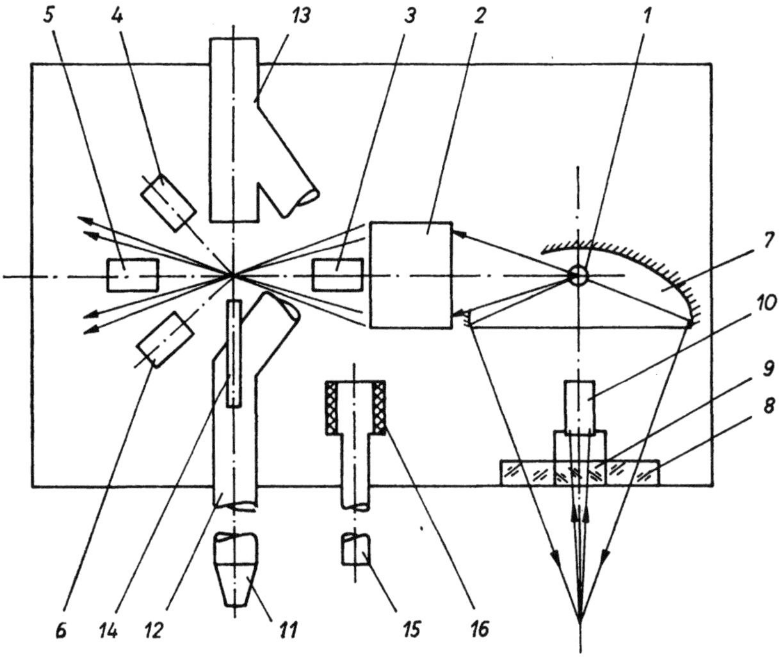

Fig. 1. The ISAV-A set-up

The effective wavelength of the spectrometer and backscattering sensor was 0.85 μm. Gas expenditure in the counting volume was about 0.3 cm^3/s.

The instruments operated reliably over the altitude range 55—33 km: each experiment yielded 30 particle size spectra with the help of the amplitude analyzer, about 3700 data on the phase functions of individual particle scattering, about 450 measurements of number densities and more than 1800 values of the atmosphere's backscattering coefficient.

The particle size distribution is characterized by two modes. Mode 1 consists of small particles described by the Yung distribution with a power of 5 ± 1 for VEGA-1 and 4 ∓ 0.5 for VEGA-2. Mode 2 consists of particles with diameters of 2—5 μm. Out of 9000 (VEGA-1) and 34000 (VEGA-2) particles that have passed through the counting

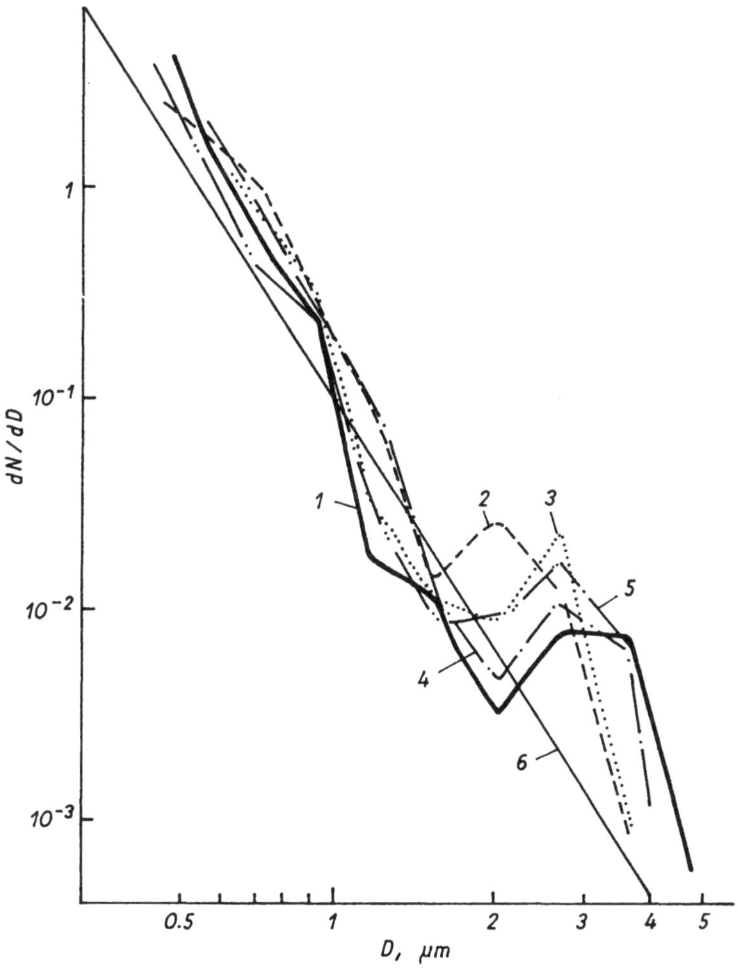

Fig. 2. Size spectrum in the altitude range 48.5 to 52 km, VEGA-2

1 — amplitude analyzer results; 2, 3, 4, 5 — over channels: 'forward', 'backward' (for $m = 1.4$), 'to the right', 'to the left'; 6 — Junge's distribution with a power of 4

volume only about 20 particles with diameters exceeding 5 μm have been recorded (modes 1 and 2). Figure 2 gives a typical example of a particle size spectrum. The third mode has not been detected.

As was mentioned above analysis of light scattering by individual particles in four directions has revealed the particles shape and their refractive index. Only the signals from large particles have been analyzed (0.8 μm, VEGA-2) to exclude noise. The ratio of signals from two 'lateral' channels slightly differs from unity, and the observed spread is explained by the finite width of the response curve. This implies that the large particles are spherical.

To determine the particle refractive index each recorded particle was put into corre-

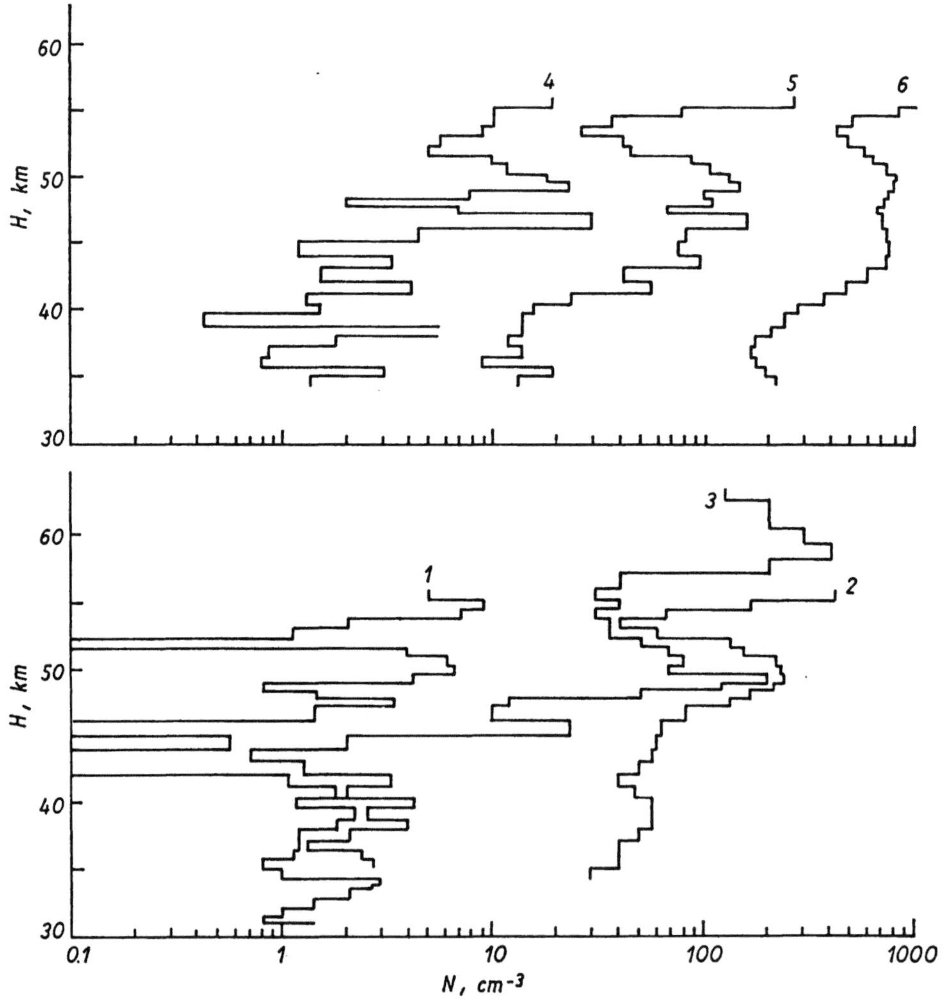

Fig. 3. Altitude profiles of aerosol's particle number density in different size ranges

1 — VEGA-1, $D \geqq 2.3$ μm; 2 — VEGA-1, $D \geqq 0.8$ μm; 3 — Pioneer-Venus, $D \geqq 0.6$ μm (Esposito, 1983); 4 — VEGA-2, $D \geqq 2.3$ μm; 5 — VEGA-2, $D \geqq 0.8$ μm; 6 — VEGA-2, $D \geqq 0.5$ μm

spondence with a point in the coordinate system: 'backward' signal vs. 'to the right' signal.

The calculated lines of constant values of the refractive index with the 0.05 step were drawn on the correlation field. The particles of 2 μm in diameter proved to have a refractive index $m = 1.4 \pm 0.05$. The particles of 0.8—2 μm in diameter are subdivided into two groups: for 80%, $m = 1.4 \mp 0.1$, for 20%, $m = 1.7 \mp 0.1$.

Figure 3 gives the number densities of particles in different size ranges. Attention is drawn to the small number density of mode-2 particles which is equal to several particles per cm^3. According to the Pioneer-Venus data (Knollenberg, 1980; Esposito, 1983), the maximum particle number densities of mode 1 and 2 are about 10^3 and 10^2 cm^{-3}, respectively.

The number density altitude profils obtained from the VEGA-1 experiments have a maximum in the 48—52 km range, the densities of both large (> 2 μm) and small particles being at their maximum in this range. Below 48 km large particles disappear and appear again at 35—38 km.

The altitude profiles at the VEGA-2 landing site are quite unusual — the sharp boundary of the main cloud layer is absent at 48 km. Approximately at the same altitude the number density of mode-2 particles decreases by about one order of magnitude, whereas the number density of small particles has a gentle maximum in the 40—50 km range smoothly decreasing at 33 km.

Using the estimated density profiles of particles with different size, we obtain that the mass density with a material density of 1.7 g/cm^3 is as high as 0.7 mg/m^3 at about 50 km, and decreases to 0.12 mg/cm^3 at 40 km.

The backscattering sensor recorded simultaneously two fluxes: the radiation from the instrument lamp scattered by the outer medium and the radiation from the Venusian atmosphere. The former was measured 8 times at the moments when the lamp was switched off. This made it possible to single out the altitude dependence of the backscattering coefficient and radiation power.

The measured values of the backscattering coefficient change with altitude in a random manner between two altitude functions given in Fig. 4, curves, 3, 4. This is most pronounced in the 56—47 km range.

The random variations are due to the fluctuations in the number of large particles of different size in the counting volume of the backscattering sensor. The recorded altitude profiles differ markedly from the results of earlier measurements (Esposito, 1983) in that we haven't found the pronounced layered structure of clouds.

The backscattering coefficient was calculated independently using the measured spectra of particles size and their number density under the assumption that $m = 1.42$ and particles are distributed according to Junge's law in the range from 0.1 to 5 μm.

The results of calculations are given in Fig. 4, curves 1, 2.

The elevated values of the coefficient as compared to the calculated ones probably indicate the presence of a great number of submicron particles, which cannot be recorded by the aerosol counter.

Assuming that the size of these particles ranges from 0.1 to 0.3 μm and the refractive index is equal to 2 (for example, sulphur), we obtain that the mass density is equal to 0.3—2 mg/m^3.

With periodic switching off the lamp, the backscattering sensor recorded a pronounced upward radiation flux. As the probes descended the value of this flux increased from

1.5 to 8.5 m W/m² × sr (VEGA-1) and from 53 to 22 mW/m² × sr (VEGA-2). The absolute values are known within a factor of 3.

The most probable assumption is that this radiation is emitted by the planet's surface and passes through the subcloud atmosphere into the spectral atmosphere window with any equivalent width of 0.03 μm at a wavelength of 1 μm. The calculated value of the radiation power at 30 km and at a surface temperature of 740 K is 12 mW/m² × sr.

Table 1 summarizes the results of measurements referred to 31 km.

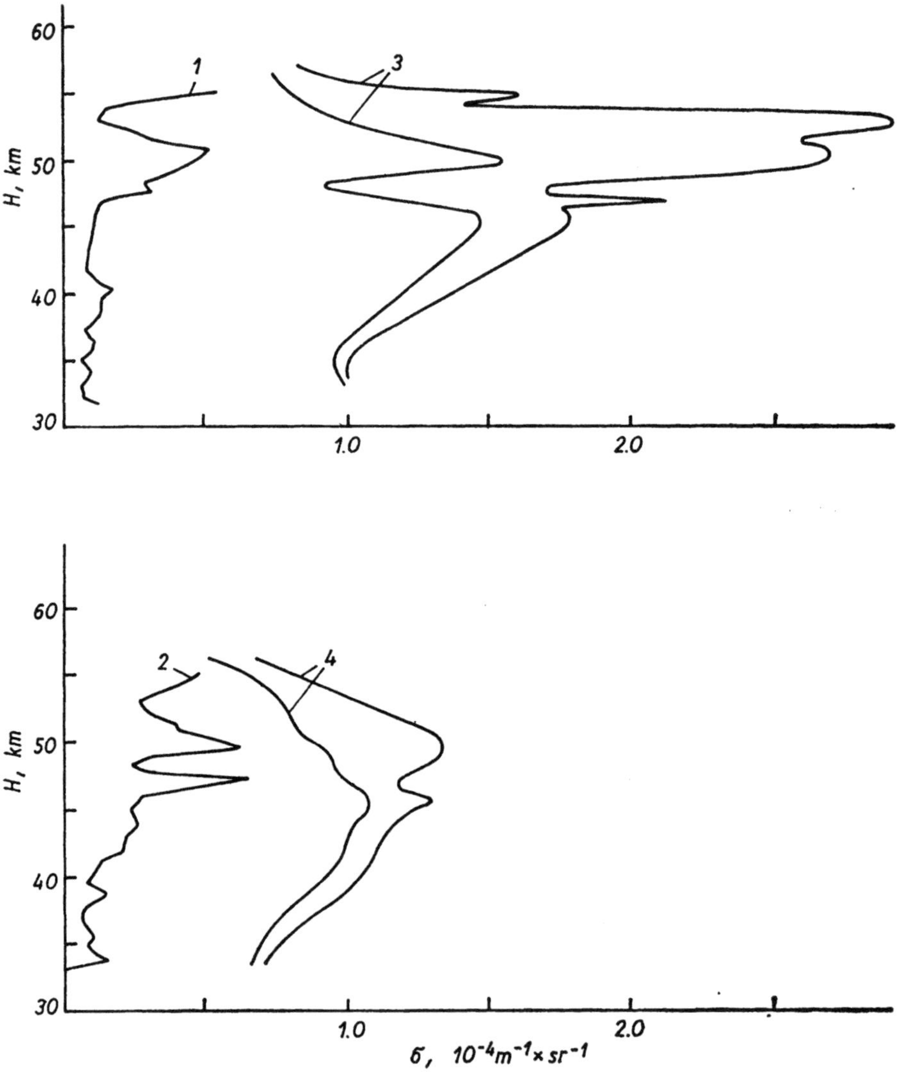

Fig. 4. Backscattering coefficient

1 — VEGA-1, calculation; 2 — VEGA-2, calculation; 3 — VEGA-1, experiment; 4 — VEGA-2, experiment

Table 1
Intensity, related units

H, km	31.0	33.8	37.2	41.3	47.0	49.5	53.0	57.5
Vega 1	1	0.93	0.75	0.65	0.50	0.37	0.25	0.18
Vega 2	1	0.96	0.81	0.62	0.49	0.40	0.31	0.24

The experimental results can be explained if it is assumed that in addition to aerosol particles recorded by the aerosol spectrometer there are many small particles which cannot be measured. With a refractive index of 2 their mass density should be 0.2 to 1.5 mg/m^3 (VEGA-1) or 0.1—0.8 mg/m^3 (VEGA-2).

An analysis of the altitude distribution of mode-1 number density the altitude profile of the backscattering coefficient and outer upward radiation lead to the surprising conclusion on the fairly thick and practically altitude-independent haze of submicron particles above the VEGA-1 and VEGA-2 landing sites. The value of the haze scattering coefficient exceeds by one-two orders of magnitudes the values obtained earlier (Esposito, 1983; Golovin, 1981).

It is worth noting that in the practice of optical investigations of the Venusian atmosphere the event was recorded in which the high values of the extinction coefficient were observed below 48 km, namely during the first VENERA-8 experiment in 1972 which measured luminosity (Avduevskijj, 1973). After these measurements it was assumed during several subsequent years that the lower boundary of the main cloud layer is located at 35 km.

The basic conclusions on the cloud porperties at the VEGA-1 and VEGA-2 landing sites are as follows:

1. The size spectrum of aerosols can be presented by two modes: mode 1 is described by Junge's distribution from, at least, 0.5 to 5 μm, mode 2 consists of 2—5 μm particles.
2. Mode-2 particles are spherical in shape, their refractive index is 1.4 ∓ 0.05.
3. The number density of mode-2 particles sharply changes at 45—50 km that of mode-1 particles smoothly changes at altitudes from 35 to 55 km, without forming pronounced cloud layers.
4. At 30—60 km submicron aerosol with a mass density of 0.1—2 mg/m^3 (refractive index $m = 2$) is present in the atmosphere.
5. At 63—30 km the Venusian cloud glow has been detected, whose possible source is the planet's hot surface.

References

Avduevskijj, V. S., Marov, M. Ya., Moshkin, B. E., Ehkonomov, A. P.: Venera-8: Measurements of Solar Illumination Through the Atmosphere of Venus. Journal of Atmospheric Science **30** (1973) 6, 1215.

Ehkonomov, A. P., Golovin, Yu. M., Moroz, V. I., Moshkin, B. E.: Solar Scattered Radiation Measurements by Venus Probes. "Venus", Edited by Hunten, D. M., Colin, L., Donahue, T. M., Moroz, V. I., Tucson, Arisona, 1983, 632.

Esposito, L. M., Knollenberg, R. G., Marov, M. Y., Toon, O. B., Turco, R. P.: The Clouds and Hazes of Venus. "Venus", Edited by Hunten, D. M., Colin, L., Donahue, T. M., Moroz, V. I., Tucson, Arisona, 1983, 484.

Knollenberg, R. G., Hunten, D. M.: The Microphysics of the Clouds of Venus: Results of the Pioneer Venus Particle Size Spectrometer Experiment. J. Geophys. Res. 85 (1980), 8039.

Toon, O. B., Ragent, B., Colburn, D., Blamont, J., Cot, C.: Large, Solid Particles in the Clouds of Venus: Do They Exist? Icarus 57 (1984), 143.

17. Microphysical processes in the Venus' cloud layer

D. V. Titov

Abstract

The main microphysical characteristics of aerosol in the Venus cloud layer were distinguished on the basis of experimental data. Size spectrum evolution of sulfuric acid aerosol is considered. The role of various microphysical processes (nucleation condensation, coagulation, sedimentation, turbulent mixing) in forming of the observed aerosol properties is analysed.

The exploration of microphysical properties of Venus aerosols allows one to understand the processes that result in the formation ob observed particle size spectra. Rather rich experimental material is collected in this field by now. It includes a set of nephelometer experiments on board Venera-9, -10, -11, -13, -14 [1, 2] and Pioneer-Venus landers [3], the results of interpretation of Venera-11 spectrophotometer measurements [4], Venera-15, -16 infrared spectrometer data [5], and experiments on direct investigation of cloud microphysics on board Pioneer-Venus Large probe [6] and Vega-1, -2 landers [7]. The following main aerosol properties may be distinguished on this experimental basis.

1. The cloud layer is located between the altitudes 75 and 30 km and can be devided in 5 sublayers: upper haze (85—70 km), upper (70—57 km), middle (57—50 km), lower (50—48 km) layers and lower haze (48—30 km).
2. Experimentally observed particle radii are in the range 0.2—3.0 μm.
3. Particle size spectra are usually unimodal or bimodal.
4. Aerosol number density doesn't exceed 10^3 cm^{-3} and usual values are 300—500 cm^{-3} for mode 1 and 50—100 cm^{-3} for mode 2.
5. Particle shape was directly investigated in the only experiment on board Vega-1, -2 [7] which results confirm droplet spherisity for the sizes $r > 0.5$ μm.

Beside the enumerated properties common for all the experiments there are some particularities, namely the detection of large non-spherical particles by the Large Probe Particle Size Spectrometer of Pioneer-Venus [6]. But these results are in doubt now and another interpretation without mode 3 particles and spherical shape was proposed [8]. Now it's difficult to make the definite choice between these two points of view.

The results of spectrometry and chemical experiments strongly confirm sulfuric acid composition of aerosol droplets though there must be some cotaminants to explain for example the phenomenon of ultraviolet absorption. So it seems important to consider the formation and growth of sulfuric acid particles in venusian atmosphere in order to understand what microphysical processes are responsible for the observed aerosol properties.

Photochemical models of Venus atmosphere [9, 10] predict the photooxidising of SO_2 by ultraviolet solar radiation in the upper portion of the cloud following by the rapid formation of sulfuric acid molecules. The model after Yung and Demore [10] gives one the rate of this process $R(z)$ that is shown in Fig. 1. Owing to the very low saturation pressure sulfuric acid codenses quickly to form aerosol particles. Let's dwell on this process. Suppose the cloud is in the beginning of its formation, the atmosphere is clear. The production of H_2SO_4 molecules with the rate $R(z)$ results in the increase of acid vapor pressure. The rate of formation of particle embryos with the radii r^* or the rate of homogeneous nucleation is given by the following expression [11]

$$J = 4 \times \pi \times r^{*2} \times \beta_{H_2SO_4} \times n_{H_2O} \times \exp(-\Delta G^*/kT), \qquad (1)$$

where $\beta_{H_2SO_4}$ is the molecule flux to the surface of growing droplet, n_{H_2O} — the number density of H_2O molecules, and ΔG^* is the energy of embryo formation. Average profiles of temperature and water vapor abundance (Fig. 1) were used in the calculations. For these average conditions one can determine the dependence of nucleation rate on the abundance of H_2SO_4 in the gas phase. The data on vapor-liquid equilibrium for sulfuric acid solutions were taken from the calculations of Gmitro and Vermeulen [12], taking into account some corrections after Ayers et al. [13].

The examples of nucleation rate dependence on the acid abundance in ambient gas is plotted in Fig. 2. We'll consider the process of aerosol formation to be important if one

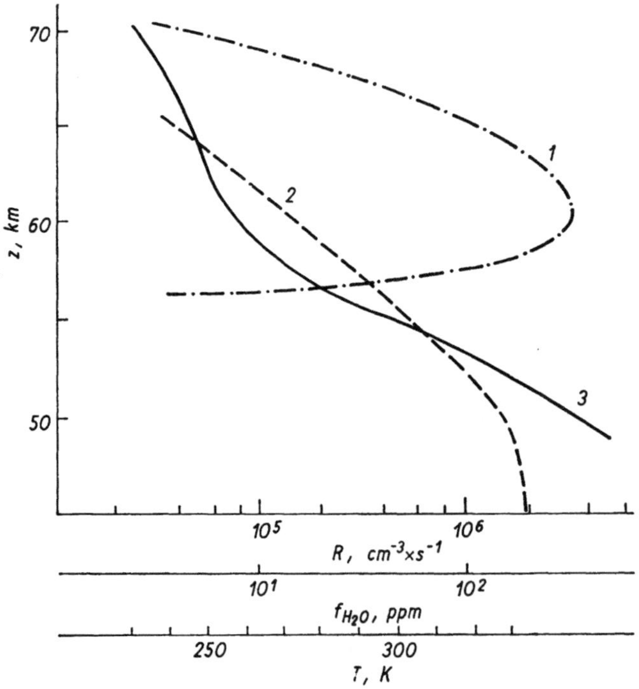

Fig. 1. Average altitude profiles in the Venus atmosphere adopted for the calculations: 1 — the rate of photoproduction of H_2SO_4 molecules $R(z)$; 2 — water vapor mixing ratio f_{H_2O}; 3 — temperature

new particle is generated in 1 cm^{-3} in the turbulent mixing time $\tau_t = H^2/K$ (about 10^7 s at 65 km). In this case the rate of particle formation is equal $1/\tau_t$. Fig. 2 helps to determine acid abundance at different levels, or the value of acid supersaturation. The results are shown in Fig. 3. So if the acid supersaturation at some height in the atmosphere is in the shaded part of this figure one may expect formation of new particles through the homogeneous nucleation.

What are the possible values of S in the Venus cloud layer? In order to answer this question let's consider the balance between photochemical source of H_2SO_4 molecules and their condensational sink onto existing aerosol particles. The balance may be discribed as follows

$$\frac{d(S-1)}{dt} = R \times \frac{k \times T}{p_s} - 4 \times \pi \times D \times (S-1) \times r \times N_a, \qquad (2)$$

where D — molecular diffusion coefficient, p_s — H_2SO_4 saturation vapor pressure over the flat surface and S — supersaturation. This equation assumes that only monodispersed particles with radii r and number density N_a are present in the atmosphere. Equation (2) has the following solution

$$S - 1 = \exp(-t/\tau) + b \times \tau,$$

where

$$\tau = 1/(4 \times \pi \times D \times r \times N_a), \quad \text{and} \quad b = R \times \frac{k \times T}{p_s}.$$

The balance between sources and sinks of acid molecules results in some sort of equilibrium with characteristic supersaturation

$$S^\infty - 1 = b \times \tau.$$

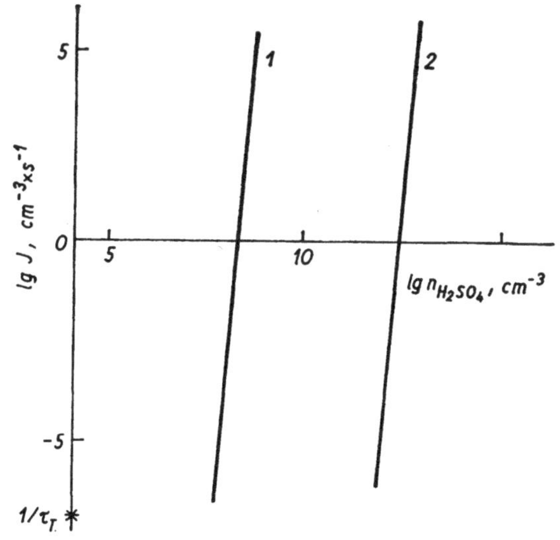

Fig. 2. The dependence of the rate of aerosol nuclei formation J on the number density of acid molecules in the ambient gas for: 1 — 62 km, 2 — 66 km

The system reaches this state in rather short time ~ 100 s when $N_a = 10^2$ cm^{-3} and $r = 10^{-4}$ cm. Using experimentally obtained particle size spectra [1—7] one can calculate the altitude dependence of S^∞. Such a profile is shown in Fig. 3 and the fact that obtained curve crosses the shaded area means that homogeneous nucleation can be an effective source of new particles at 60—65 km. At lower altitudes this process is of no significance for the sulfuric acid aerosol. It is worth mentioning that the real atmosphere isn't clear at all and condensational nuclei are always present. This helps droplets to form. So for the real atmosphere the shaded area moves to the left from its position for homogeneous nucleation.

The initial steps of cloud formation were modelled on the computer by solving the general dynamic equation for the aerosol size distribution function $n(v, t)$

$$\frac{dn(v,t)}{dt} = J_{v*} + \frac{\partial}{\partial v} \times \left[\frac{\partial v}{\partial t} \times n(v,t)\right]$$
$$+ \frac{1}{2} \times \int_0^v K(v - v', v') \times n(v - v', t) \times dv' - n(v,t) \times \int_0^\infty K(v, v') \times n(v', t) \times dv'. \quad (3)$$

The first term on the right describes the birth of aerosol nuclei due to the homogeneous nucleation. Second one is the rate of formation for the particles of volume v. The third and the fourth terms take into account the coagulation of droplets. The terms connected with sedimentation and turbulent mixing are absent in this analysis. That's why the

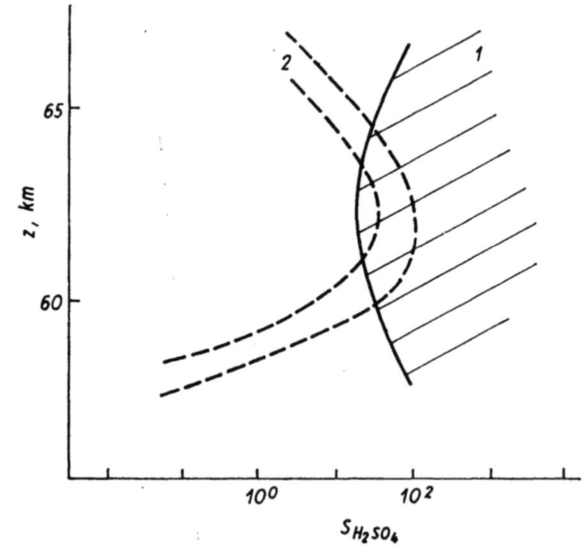

Fig. 3. The profiles of H_2SO_4 supersaturation in the Venus' atmosphere:

1 — necessary for the effective generation of new particles by homogeneous nucleation; 2 — resulted from the balance of photoproduction of H_2SO_4 molecules and their condensation on the droplets existing in the Venus atmosphere

results of these calculations are valid for the times $t < \tau_t$, i.e. untill different levels in the atmosphere may be considered separately. Fig. 4 shows that nucleation, condensation and coagulation result in the formation of unimodal size distributions with particle radii in the range 0.5—2.0 μm and number density 100—1000 cm^{-3}. This figure allows one to propose a qualitative explanation of observed size spectra bimodality. We see that droplet sizes depend on the conditions at their birth-place. So mixing of unimodal size spectra that formed at different altitudes will result in the bimodal or wide unimodal spectra that include both submicron and micron size ranges. Once appeared such a complex size distribution can be conserved in the middle cloud layer where the size spectra change very slowly with time.

Equation (3) describes only the initial phase of cloud formation. For times $t > \tau_t = 10^7$ s turbulent mixing plays a significant role and when the droplets grow up to the size 2—3 μm sedimentation must be taken into account. The influence of these processes on the aerosol size spectra can be evaluated through comparison of their time constants like is was done by Rossow [14]. One of the most important dynamical characteristics is the time of turbulent mixing $\tau_t = H^2/K$ which is 10^6—10^7 s for the upper cloud. The rate of condensational growth of a droplet is determined by the rate of production of condensable species in the gas phase, in the case of the upper cloud — by the rate of photoproduction of acid molecules $R(z)$. So it will take time of about $\tau_{\text{cond}} = \frac{N_a \times \pi \times r^3}{v_m \times R(z)}$ for N_a particles to grow to the size r, where v_m is the volume per one H_2SO_4 molecule. The time for the number density to decrease twice is usually called the time of coagulation $\tau_{\text{coag}} = 1/N_a \times C_{\text{coag}}$ where $C_{\text{coag}} \simeq 0.5 \times 10^{-9}\ cm^3 \times s^{-1}$ — coagulation constant, that determines the probability of collisions and is independent of the chemical nature of aerosol [15]. For the number densities observed in Venus clouds (100—1000 cm^{-3}) coagulation time is approximately 10^6—10^7 s. For rather large particles ($r > 3$ μm) sedimentation becomes significant and the droplets fall with velocity that is determined by the Stokes-Davies formula:

$$V_S = \frac{2}{g} \times \frac{\varrho_a \times g}{\eta} \times r^2 \times \left[1 + Kn \times \left(1.257 + 0.4 \times e^{-\frac{1.1}{Kn}}\right)\right]$$

where $\varrho_a = 1.8\ g/cm^3$ — aerosol density, g — free fall acceleration, $\eta = 1.4\ 10^{-4}\ g/cm \times s$ — CO_2 viscosity, and $Kn = \lambda/r$ — Knudsen number. Substituting values into this expression we obtain V_S [cm/s] $= 2.5 \times 10^{-2} \times r^2$ [μm]. Sedimentation time constant is determined as $\tau_S = H/V_S$.

Time constants for different microphysical processes at 62 km are compared in Fig. 5. This figure allows one to make a conclusion that the presence of particles with $r > 2$—3 μm is rather unlikely because for such sizes the time of condensational growth is larger than the time of sedimentation: large acid droplets haven't enough time to grow up. The role of coagulation seems to be restricted to that regions where intensive formation of new particles occurs and aerosol number density exceeds $10^4\ cm^{-3}$. Coagulation results in fast decrease of number density to the values of 100—1000 cm^{-3} for which $\tau_{\text{coag}} \sim \tau_t$. So aerosol systems with $N_a > 10^4\ cm^{-3}$ are rather unstable with respect to coagulation and that's why they are not observed in experiments.

In the previous part we considered the microphysical processes in the upper cloud layer ($z > 57$ km) where aerosol formation takes place. Below this level the main micro-

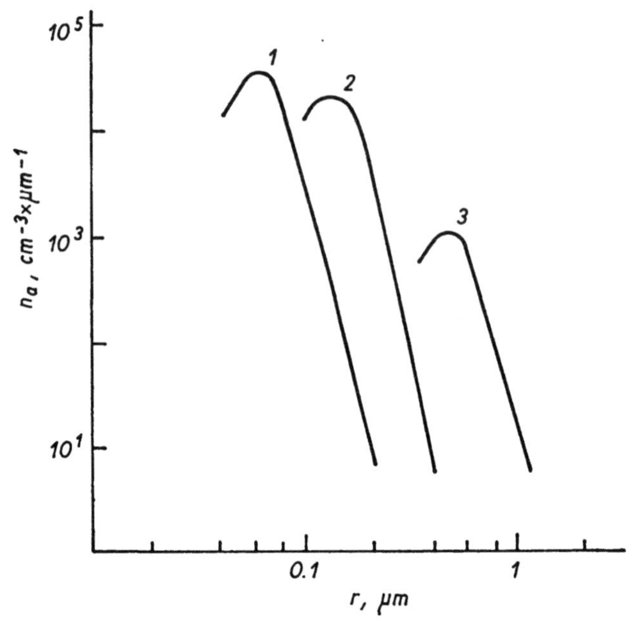

Fig. 4a. The results of computer modelling of the evolution of size spectra at the initial phase of cloud formation at the level 66 km. The curves correspond to the different times: 1 — 10^5 s, 2 — 10^6 s, 3 — 10^7 s

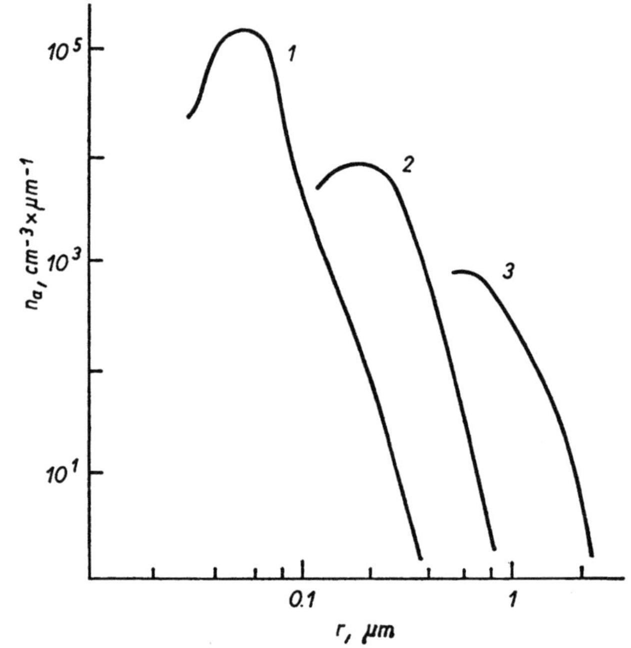

Fig. 4b. Same as fig. 4a but for the altitude 62 km

physical processes are condensation, sedimentation and turbulent mixing. The rate of particle growth is a function of supersaturation S in the gas phase. In the absence of photochemical source of H_2SO_4 molecules supersaturation results from the gas colling in the rising parcels of the atmosphere. If it is due to the turbulent mixing with coefficient $K(z)$ one can derive the expression for the rate of supersaturation changing

$$\frac{dS}{dt} = \left[V \times \left(-\frac{1}{H} + \gamma_a \times \frac{\partial \ln p_s}{\partial T}\right) - 4 \times \pi \times D \times N_a \times r\right] \times (S - 1) + V \times \left(\frac{1}{H} + \gamma_a \times \frac{\partial \ln p_s}{\partial T}\right). \tag{4}$$

This equation describes the balance of supersaturation S: its increase in upward motions and decrease due to the condensation on existing droplets of radii r and number density N_a. This equation is similar to the equation (2) and its solution shows that in time $\tau = 1 \Big/ \left[V \times \left(-\frac{1}{H} + \gamma_a \times \frac{\partial \ln p_s}{\partial T}\right) - 4 \times \pi \times D \times N_a \times r\right]$ the system reaches the stable state with $S^\infty - 1 = b \times \tau$ where

$$b = V \times \left[-\frac{1}{H} + \gamma_a \times \frac{\partial \ln p_s}{\partial T}\right]$$

and V is an effective velocity of upward motions

$$V(z) = -\frac{K(z)}{f_{H_2SO_4}} \times \frac{\partial f_{H_2SO_4}}{\partial z};$$

H is gas scale height and γ_a — lapse rate. Substitution of characteristic values of parameters in equation (4) leads to the result that a supersaturation of $(S^\infty - 1) \sim 10^{-4}$ settles in a rising parcell in a short time $\sim 10^2$ s. The rate of condensational growth of a droplet is given by the expression

$$\frac{dv}{dt} = \frac{4 \times \pi \times r \times D \times V_m}{k \times T} \times p_s \times (S - 1).$$

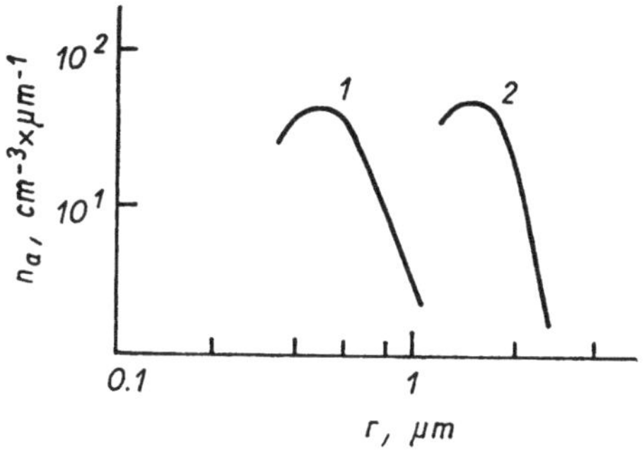

Fig. 4c. Same as Fig. 4a but for the altitude 58 km

And the one can evaluate the time that takes the droplet to grow up to the radius r

$$\Delta t = \frac{k \times T}{6 \times D \times V_m \times p_s \times (S-1)} \times r^2 = 1.7 \times 10^{14} \times \frac{r^2\,[\mathrm{cm}]}{(S-1)}.$$

At low values of supersaturation $(S-1) \sim 10^{-4}$ obtained for the middle cloud region the growth of one micron in size droplet will take $\sim 10^{10}$ s, much greater than turbulent mixing time. That's why condensational growth of particles in the middle cloud seems to be negligible. Time constant for turbulent mixing and sedimentation remains approximately the same as in the upper cloud (Fig. 5).

This analysis leads us to some conclusions about the role of various microphysical processes in the formation of observed properties of Venus' aerosol.

1. The birth of new acid particles due to the homogeneous nucleation is likely to take place in the upper cloud layer only.
2. The condensational growth of droplets is significant higher than 57 km where the photochemical processes can maintain large value of supersaturation ($S = 10-100$). Below this level the supersaturation is too small for condensation of particles.
3. The observed maximal droplet size is determined by the balance of sedimentation, condensation and mixing. These processes provide long lifetime for the droplets with sizes of 2—3 μm. Acid particles larger than 4 μm are quickly removed from the cloud by sedimentation.

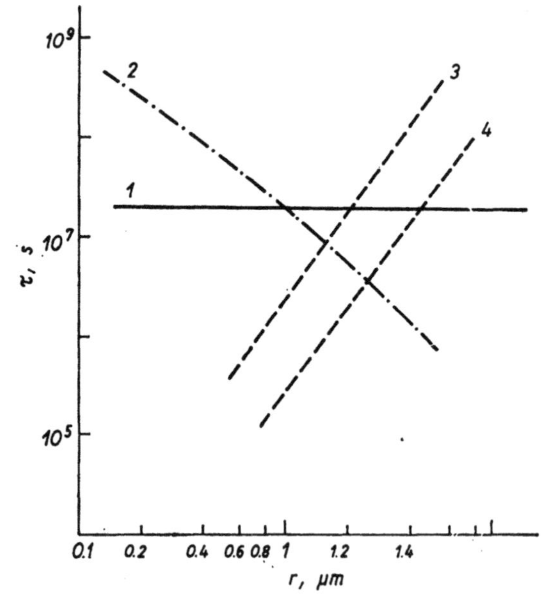

Fig. 5. The comparison of characteristic times of microphysical processes in the upper cloud layer ($\sim$62 km):

1 — turbulent mixing time τ_t; and coagulation time $\tau_{coag} \sim \tau_t$; 2 — sedimentation time τ_s; 3 and 4 — the time of condensational growth τ_{cond} for the aerosol system with number densities 100 and 10 cm^{-3} respectively

4. Coagulation monitors the particle number density. It accounts for the fact that observed number density does't exceed 10^3-10^4 cm^{-3}.
5. Unimodal size spectra in the range $0.1-1.0$ μm are formed in the upper portion of the cloud. The mixing of particles originated in different conditions may lead to the wide unimodal or bimodal size distributions resembling that observed in the experiments.

References

[1] Marov, M. Ya., Lystsev, V. E., Lebedev, V. E. et al.: The structure and microphysical properties of Venus clouds: Venera 9, 10 and 11 data. Icarus **44** (1980) 3, 608.
[2] Marov, M. Ya., Byvshev, B. V., Baranov, B. P. et al.: The investigation of Venus' cloud structure by means of nephelometers on board Venera-13 and Venera-14. Kosmich. Issled. **21** (1983) 2, 269 (Russian).
[3] Ragent, B., Blamont, J.: The structure of the clouds of Venus: results of the Pioneer-Venus nephelometer experiment. J. Geophys. Res. **85** (1980) A 13, 8089.
[4] Titov, D. V.: The microstructure of the Venus' cloud layer from the results of spectrometry on board Venera-11 lander. Kosmich. Issled. **22** (1984) 1, 120 (Russian).
[5] Zasova, L. V., Spänkuch, D., Moroz, V. I. et al.: Infrared experiment aboard the automatic interplanetary stations Venera 15 and Venera 16. 3. Some conclusions on the structure of clouds, based on the analysis of spectra. Veröffentlichungen des Forschungsbereichs Geo- und Kosmoswissenschaften, Heft 18, 1990.
[6] Knollenberg, R. G., Hunten, D. M.: The mycrophysics of the clouds of Venus: results of the Pioneer-Venus particle size spectrometer experiment. J. Geophys. Res. **85** (1980) A 13, 8039.
[7] Moshkin, B. E., Moroz, V. I., Gnedykh, V. I. et al.: Preliminary results of optical investigations of aerosol media at 30–60 km in the Venus atmosphere on board Vega-1, 2 landers. Pis'ma v Astron. Zh. **12** (1986) 1, 85 (Russian).
[8] Toon, O. B., Ragent, B., Colburn, D. et al.: Large solid particles in the clouds of Venus: do they exist? Icarus **57** (1984) 2, 143.
[9] Krasnopolskijj, V. A., Parshev, V. A.: Photochemistry of Venus atmosphere at the altitudes more than 50 km. 2. The results of calculations. Kosmich. Issled. **19** (1981) 2, 261 (Russian).
[10] Yung, Y. L., Demore, W. B.: Photochemistry of the stratosphere of Venus: implications for atmospheric evolution. Icarus **51** (1982) 2, 199.
[11] Hamill, P., Kiang, C. S., Cadle, R. D.: The nucleation of $H_2SO_4-H_2O$ solution aerosol particles in the stratosphere. J. Atmos. Sci., **34** (1977) 1, 150.
[12] Gmitro, J. I., Vermeulen, T.: Vapor-liquid equilibria for aqueous sulfuric acid. A.I.Ch.E. Journ. **10** (1964) 5, 740.
[13) Ayers, G. P., Gillet, R. W., Gras, J. L.: On the vapor pressure of sulfuric acid. Geophys. Res. Lett. **7** (1980), 433.
[14] Rossow, W. B.: Clouds microphysics: analysis of the clouds of Earth, Venus, Mars and Jupiter. Icarus **36** (1978), 1.
[15] Friedlander, S. K.: Smoke, dust and haze. Fundamentals of aerosol behavior. Wiley, New York, 1977.

18. Importance of Ionospheres for Neutral-Atmosphere Processes of Earth-like Planets

J. TAUBENHEIM

(Extended Summary)

The existence of an ionized component modifies the structure and dynamics of a (fictitious) neutral (non-ionized) atmosphere of a planet in the following three respects:

(1) **Superthermal escape flux:** As the mean kinetic energy of the ions (ion temperature) is generally greater than that of the neutrals, charge exchange processes in the exosphere between 'hot' H^+ ions and 'cold' H atoms (and/or O^+ and O, respectively) lead to an energy transfer to the neutrals, which increases the flux of neutral H (and/or 0) escaping from the atmosphere over that given by the classical Jeans' formula of escape flux for the same exospheric temperature.

(2) **Ion drag:** The presence of a magnetic field introduces an anisotropy of the mobility of the ions in the atmosphere, which acts like an internal friction ('ion drag') upon the wind component perpendicular to the magnetic field direction. The (thermally driven) wind circulation in the ionosphere of a planet having a permanent magnetic field, therefore, may essentially differ from that of a non-magnetic planet.

(3) **Ion chemistry:** Chemical reactions between ions and neutrals, as well as dissociative recombination of molecular ions are important for the production of neutral minor constituents in the thermospheres of planets. As an example, this is pointed out in more detail for the nitrogen oxides which are found in the atmospheres of all earth-like planets.

Neutral nitric oxide (NO) is important not only in forming the D region of the Earth's daytime ionosphere, ionized by direct solar radiation at the Lyman-alpha line (121.6 nm), but also for the ionization of the atmospheres on the night sides of all earthlike planets, due to solar Lyman-alpha photons backscattered from the night-side tail of the planet's hydrogen corona. In the atmospheres of Venus and Mars, the abundance of neutral NO has not yet been measured, but data on the concentration of NO^+ ions are available (see, e.g., in Krasnopolsky's monograph [1]). Estimates with well-known ion-chemical rate constants clearly show, that charge exchange processes from other ions (O^+, N_2^+, N^+) are not adequate to explain the observed concentrations of NO^+ ions quantitatively, but rather there must be a sufficient amount of neutral NO to be photoionized by solar Lyman-alpha. The source of neutral NO is the reaction of excited nitrogen atoms in the $N(^2D)$ state with molecules of the atmospheric major constituents, i.e.

$$O_2 + N(^2D) \rightarrow O + NO \quad \text{for the Earth}$$

$$CO_2 + N(^2D) \rightarrow CO + NO \quad \text{for Venus and Mars.}$$

Thus, $N(^2D)$ atoms play a key rôle in understanding the minor constituent chemistry of the upper atmospheres and ionospheres of all earth-like planets. Their origin is partly from the dissociation of neutral nitrogen molecules (N_2) by a photopredissociation mechanism

(a) $N_2 + h\nu \longrightarrow N(^2D) + N$,

by impact of fast electrons (e^*) from space, or fast photoelectrons,

(b) $N_2 + e^* \longrightarrow N(^2D) + e$

or (less important) by dissociative photoionization

(c) $N_2 + h\nu \longrightarrow N(^2D) + N^+ + e$.

Another part, however, of processes in which $N(^2D)$ atoms are generated, is immediately connected with the existence of the ionosphere, namely, by dissociative recombination of N_2^+ or NO^+ ions,

(d) $N_2^+ + e \longrightarrow N(^2D) + N$,

(e) $NO^+ + e \longrightarrow N(^2D) + O$ (78%),
$\searrow N(^4S) + O$ (22%),

and by charge exchange of N_2^+ ions with neutral O atoms

(f) $N_2^+ + O \longrightarrow N(^2D) + NO$ (90%),
$\searrow N(^4S) + NO$ (10%).

Table 1

Estimated Production Rates of $N(^2D)$, Daytime, in $cm^{-3} \times sec^{-1}$

Planet	Height (km)	Non-ionosph. Rate (a) + (b) + (c)	Ionospheric Rate (d) + (e) + (f)	Dominant Production Process
Venus	125	320	470	$NO^+ + e$
	150	80	1400	$NO^+ + e$
	175	13	82	$N_2^+ + O$
	200	2	23	$NO^+ + e$
Earth	125	370	490	$N_2^+ + O$
	150	840	2000	$N_2^+ + O$
	175	1040	2700	$N_2^+ + O$
	200	690	1900	$N_2^+ + O$
Mars	125	40	54	$NO^+ + e$
	150	52	10	$NO^+ + e$
	175	43	3	$NO^+ + e$
	200	19	2	$N_2^+ + O$

We may call (d), (e), and (f) 'ionospheric' processes of $N(^2D)$ production, while (a), (b), and (c) are 'non-ionspheric', as they would take place even if there were no ionosphere. We now compare the production rates ($cm^{-3} \times s^{-1}$) of 'ionospheric' and 'non-ionospheric' generation of (N^2D), which are listed in Table 1 for Venus, Earth, and Mars under daylight conditions. Reaction rate coefficients and concentrations of ions and neutrals in Venus' and Mars' atmospheres have been taken from the compilations in [1], for the Earth's atmosphere from [2] and [3]. This comparison shows impressively the importance of the existence of the ionosphere for the atmospheric chemistry, as the production of $N(^2D)$ by 'ionospheric' reactions is much larger than by 'non-ionospheric' processes both in the Earth's and Venus' atmospheres, and at least at lower altitudes in Mars' atmosphere, too.

For satisfactory quantitative modeling of the chemical structure of the upper atmospheres (above 100 km) of Venus and Mars, new experimental data are desirable with respect to

— neutral nitric oxide (NO) abundance,
— ion densities of N^+ and NO^+, and
— photoelectron fluxes and energies

in dependence on height, latitude, and time of day.

References

[1] V. A. Krasnopolsky: Photochemistry of the Atmospheres of Mars and Venus (Physics and Chemistry in Space, Vol. 13). Springer-Verlag, 1986.
[2] U.S. Standard Atmosphere, 1976, ed. by NOAA, NASA, and USAF. Washington, D.C., 1976.
[3] International Reference Ionosphere — IRI 79. Report UAG-82, World Data Center A. Boulder, Co., 1981.

List of Authors

Dipl.-Ing. W. Bärwald
AdW, IKF, Rudower Chaussee 5,
Berlin, DDR-1199

Dr. rer. nat. H. Becker-Ross
AdW, ZOS, Rudower Chaussee 5,
Berlin, DDR-1199

Dipl.-Ing. I. D. Cerenin
AN SSSR, IKI, Profsojuznaja 84/32,
Moscow, 117810

Dr. rer. nat. K. Dethloff
AdW, HHI, Rudower Chaussee 5,
Berlin, DDR-1199

Dipl.-Phys. A. V. D'jachkov
AN SSSR, IKI, Profsojuznaja 84/32,
Moscow, 117810

Dr. rer. nat. H. Driescher
AdW, IKF, Rudower Chaussee 5,
Berlin, DDR-1199

Dipl.-Meteor. W. Döhler
MD, MHO, Telegrafenberg, Potsdam,
DDR-1561

Dipl.-Phys. R. Dubois
AdW, HHI, Rudower Chaussee 5,
Berlin, DDR-1199

Dr. rer. nat. A. P. Ehkonomov
AN SSSR, IKI, Profsojuznaja 84/32,
Moscow, 117810

Prof. Dr. Larry W. Esposito
Laboratory for Atmospheric and Space Physics and Department of Astrophysical, Planetary and Atmospheric Sciences
University of Colorado, Boulder,
CO 80309-0392

Dipl.-Math. G. Fellberg
AdW, IKF, Rudower Chaussee 5,
Berlin, DDR-1199

Dipl.-Phys. V. I. Gnedykh
AN SSSR, IKI, Profsojuznaja 84/32,
Moscow, 117810

Dr. rer. nat. H. Goehring
AdW, HHI, Rudower Chaussee 5,
Berlin, DDR-1199

Dipl.-Phys. A. V. Grigor'ev
AN SSSR, IKI, Profsojuznaja 84/32,
Moscow, 117810

Dr. rer. nat. J. Güldner
MD, MHO, Telegrafenberg, Potsdam,
DDR-1561

Dr. rer. nat. R. Haus
AdW, HHI, Rudower Chaussee 5,
Berlin, DDR-1199

Dipl.-Ing. S. P. Ignatova
AN SSSR, IKI, Profsojuznaja 84/32,
Moscow, 117810

Prof. Dr. sc. nat. H. Jahn
AdW, IKF, Rudower Chaussee 5,
Berlin, DDR-1199

Dr.-Ing. H. Kaiser
ADW, IKF, Rudower Chaussee 5,
Berlin, DDR-1199

Dr. sc. nat. V. V. Kerzhanovich
AN SSSR, IKI, Profsojuznaja 84/32,
Moscow, 117810

Dipl.-Ing. N. G. Khavenson
AN SSSR, IKI, Profsojuznaja 84/32,
Moscow, 117810

Dipl.-Ing. L. I. KHLJUSTOVA
AN SSSR, IKI, Profsojuznaja 84/32,
Moscow, 117810

Dipl.-Ing. R. S. KREMNJOV
AN SSSR, IKI, Profsojuznaja 84/32,
Moscow, 117810

Dr. rer. nat. S. S. LIMAYE
Space Science & Engineering Center,
1225 West Dayton Street University of
Wisconsin-Madison, Wisconsin 53796, USA

Dr. sc. nat. V. M. LINKIN
AN SSSR, IKI, Profsojuznaja 84/32,
Moscow, 117810

Dipl.-Ing. A. N. LIPATOV
AN SSSR, IKI, Profsojuznaja 84/32,
Moscow, 117810

Dipl.-Phys. I. A. MACYGORIN
AN SSSR, IKI, Profsojuznaja 84/32,
Moscow, 117810

Prof. Dr. sc. nat. V. I. MOROZ
AN SSSR, IKI, Profsojuznaja 84/32,
Moscow, 117810

Dr. rer. nat. B. E. MOSHKIN
AN SSSR, IKI, Profsojuznaja 84/32,
Moscow, 117810

Dr. rer. nat. J. NOPIRAKOWSKI
AdW, IKF, Rudower Chaussee 5,
Berlin, DDR-1199

Prof. Dr. s.··· nat. D. OERTEL
ADW, IKF, Rudower Chaussee 5,
Berlin, DDR-1199

Dipl.-Phys. D. V. PACAEV
AN SSSR, IKI, Profsojuznaja 84/32,
Moscow, 117810

Prof. Dr. sc. nat. K. SCHÄFER
AdW, HHI, Rudower Chaussee 5,
Berlin, DDR-1199

Dr. rer. nat. R. SCHUSTER
AdW, IKF, Rudower Chaussee 5,
Berlin, DDR-1199

Dipl.-Ing. A. A. SHURUPOV
AN SSSR, IKI, Profsojuznaja 84/32,
Moscow, 117810

Dipl.-Ing. W. SKRBEK
AdW, IKF, Rudower Chaussee 5,
Berlin, DDR-1199

Dr. sc. nat. D. SPÄNKUCH
MD, MHO, Telegrafenberg, Potsdam,
DDR-1561

Dipl.-Ing. W. STADTHAUS
AdW, ZWG, Rudower Chaussee 6,
Berlin, DDR-1199

Dr. rer. nat. H. STUDEMUND
AdW, IKF, Rudower Chaussee 6,
Berlin, DDR-1199

Prof. Dr. rer. nat. habil. J. TAUBENHEIM
AdW, HHI, Rudower Chaussee 5,
Berlin, DDR-1199

Dr. sc. nat. F. W. TAYLOR
Department of Atmospheric, Oceanic and
Planetary Physics, Oxford University,
Oxford OXI 3 PU, England

Dr. rer. nat. D. V. TITOV
AN SSSR, IKI, Profsojuznaja 84/32,
Moscow, 117810

Dipl.-Ing. M. UHLIG
AdW, IKF, Rudower Chaussee 5,
Berlin, DDR-1199

Dr. rer. nat. E. A. USTINOV
AN SSSR, IFA, Pychevskijj per. 3,
Moscow, 109017

Dipl.-Astron. L. V. ZASOVA
AN SSSR, IKI, Profsojuznaja 84/32,
Moscow

Dipl.-Ing. I. A. ZELENOV
AN SSSR, IKI, Profsojuznaja 84/32,
Moscow, 117810

List of abbreviations

AdW: Akademie der Wissenschaften der DDR

AN SSSR: Akademija Nauk SSSR

HHI: Heinrich-Hertz-Institut für Atmosphärenforschung und Geomagnetismus

IFA: Institut Fisiki Atmosfery

IKF: Institut für Kosmosforschung

IKI: Institut Kosmichekikh Issledovanija

MD: Meteorologischer Dienst der DDR

MHO: Meteorologisches Hauptobservatorium

ZOS: Zentralinstitut für Optik und Spektroskopie

ZWG: Zentrum für wissenschaftlichen Gerätebau

Bei Fragen zur Produktsicherheit wenden Sie sich bitte an:
If you have any questions regarding product safety,
please contact:

Walter de Gruyter GmbH
Genthiner Straße 13
10785 Berlin
productsafety@degruyterbrill.com